Astrophysics and Stellar Astronomy

SPACE SCIENCE TEXT SERIES
A. J. Dessler and F. C. Michel, Editors
Rice University, Houston, Texas

Carl-Gunne Fälthammar, Associate Editor
Royal Institute of Technology, Stockholm, Sweden

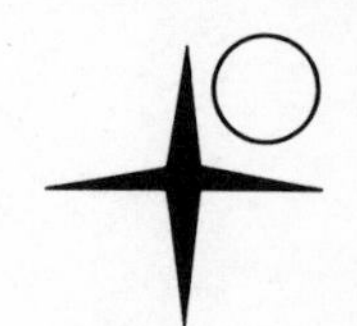

William M. Kaula
AN INTRODUCTION TO PLANETARY PHYSICS

Thomas L. Swihart
ASTROPHYSICS AND STELLAR ASTRONOMY

Frank D. Stacey
PHYSICS OF THE EARTH

R. C. Whitten and I. G. Poppoff
FUNDAMENTALS OF AERONOMY

Robert C. Haymes
INTRODUCTION TO SPACE SCIENCE

A. J. Dessler and F. C. Michel
PARTICLES AND FIELDS IN SPACE

Astrophysics and Stellar Astronomy

THOMAS L. SWIHART
University of Arizona
Tucson, Arizona

John Wiley & Sons, Inc.
New York London Sydney Toronto

Library of Congress Catalog Card Number: 68–57103
SBN 471 83990 6
Printed in the United States of America
10 9 8 7 6 5 4 3 2

To Merna, Gail, David, and Jennifer

Preface

I have long felt that many schools could profitably offer a general astronomy course designed for their science majors. Such a course ought to be more of a challenge than the descriptive course aimed at filling only the cultural needs of the nonscience students. This book is an outgrowth of that belief. It is the result of my experience in teaching the second semester of the general astronomy course offered to science majors at the University of Arizona.

Since astronomy is essentially physics and mathematics applied in an astronomical setting, the work of astronomers cannot be properly appreciated through qualitative reasoning alone. The student who has a modest understanding of physics and mathematics can find in astronomy, as in other fields, stimulating concepts that are simply not available to students not so prepared.

Mathematics is used through calculus. Even though most of the physics which is needed is developed in the book, it will be helpful for the student to have completed the sophomore-level general physics course.

The units commonly used by astronomers are often peculiar to astronomy and unheard of in other sciences. This forces a difficult choice upon anyone who wishes to communicate astronomical information to a nonastronomical audience. I do not wish to unnecessarily complicate the presentation, but I do want to prepare the reader for further study in astronomy; accordingly, I have kept rather close to standard astronomical usage.

I have divided the subject matter of the book into sections, each of which is developed in a logical order. This requires that many topics be mentioned briefly before they can be discussed in detail, but this is necessary when discussing any subject consisting of closely related parts.

I have not gone far out of the way to report the latest data or the newest speculations, as the book is intended as a teaching text rather than a report on the present state of astronomy. Little emphasis is given to factual data for their own sake. For example, the reader who is interested in the 315 (or is it 437?) different spectral peculiarities of QQ Vulpeculae-type stars will have to look elsewhere.

The Introduction develops selected topics of physics that are of special importance to astronomy and that are used considerably in later parts of the book.

Chapters I (Positions and Magnitudes of the Stars) and II (Binary and Variable Stars) cover the "service" areas of astronomy, i.e., areas whose main importance lies in the gathering of data for other areas.

Chapters III (Astrophysics) and IV (Galaxies and Cosmology) form the main part of the book, and the preceding material is only an introduction for the final two chapters of the book.

It would be a mistake for the instructor to spend so much class time on the early parts of the book that the coverage of Chapters III and IV had to be rushed. The Introduction contains background material and should not be covered with the same thoroughness as the later material. In fact, the Introduction can be used only for reference purposes and the class can start directly in Chapter I, if the students have a sufficiently good physics background.

Problems and references are given at the end of each chapter, and various kinds of basic data and complete solutions of the problems are included in the appendices. The problems generally require a good deal of thought, and it is strongly recommended that students study the given problems and solutions as a regular part of the course work.

This text was written for the intermediate-level general astronomy course, but it could also be used with other courses. For example, the school which has no astronomy department but which wishes to offer an intermediate-level astrophysics course through the physics department could use this as the textbook.

I am indebted to many persons in the preparation of this book. In particular I wish to mention the following: H. A. Abt, B. J. Bok, F. I. Boley, D. R. Brown, R. H. Hilliard, E. D. Howell, B. T. Lynds, T. E. Margrave, A. B. Meinel, F. C. Michel, A. G. Pacholczyk, B. E. Westerlund, R. J. Weymann, R. E. White, and R. E. Williams.

THOMAS L. SWIHART

November 1968
Tucson, Arizona

Contents

Introduction

This introduction develops selected topics in physics which are essential to the understanding of much of the material of this book. The topics are presented with their astronomical applications in mind, so the emphasis and point of view are somewhat different from those of most physics textbooks. The reader with a good physics background can skip this introduction without any loss of continuity, although he will probably need to refer occasionally to some of the equations.

In Section 1 the basic parameters of a radiation field are defined, and the properties of thermodynamic equilibrium radiation are described in Section 2. In Section 3 the Bohr theory is developed in some detail because it introduces concepts vital to the understanding of virtually all astrophysical research. Sections 4 and 5 describe some of the more important properties of a gas in thermodynamic equilibrium, and Section 6 gives the background needed for an understanding of nuclear reactions in stars.

1. Measuring Radiation Energy

Solid Angle. Just as a plane angle is a measure of the opening between two straight lines, a solid angle is a measure of the opening in a cone. At the left of Figure 1.1 are two straight lines meeting at the vertex V. The angle θ, measured in radians, is found by constructing a circle of radius r centered on V. If s is the arc length of this circle which is intersected by the lines, then $\theta = s/r$. At the right of the same figure is the analogous picture of a cone, not necessarily circular, with the vertex V. The solid angle of the cone ω, measured in steradians, is found by constructing a sphere of radius r about V as center. Then, if A is the area of the sphere which is intersected by the cone,

$$\omega = \frac{A}{r^2} \tag{1.1}$$

The solid angle of an object that covers all directions is seen to be 4π steradians.

The spherical coordinates (r,θ,ϕ), shown in Figure 1.2, are often used. The angles θ and ϕ define a direction. Angle θ is the angle between the given

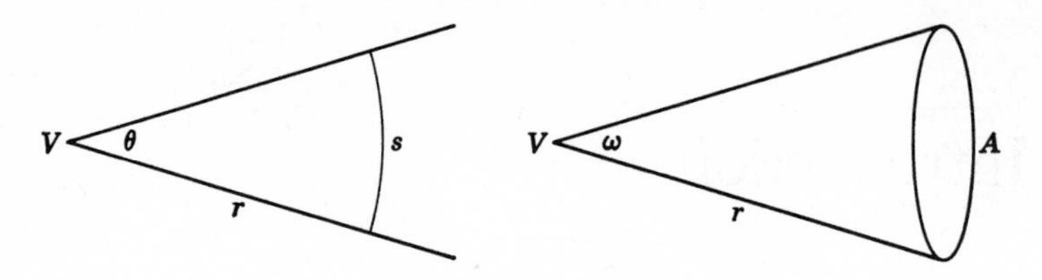

Figure 1.1 Plane and solid angles.

direction and the z axis, and ϕ is the angle between the x axis and the projection of the given direction onto the xy plane. The shaded area in the figure has its sides parallel to the directions of increasing θ and ϕ. Since this area is a distance r from the origin, its sides have lengths $r\,d\theta$ and $r \sin\theta\, d\phi$. Its area is then $r^2 \sin\theta\, d\theta\, d\phi$, and according to equation (1.1) the solid angle it subtends at the origin is

$$d\omega = \sin\theta\, d\theta\, d\phi \tag{1.2}$$

In applying equation (1.1), it must be remembered that A is the projection of the area normal to the line of sight.

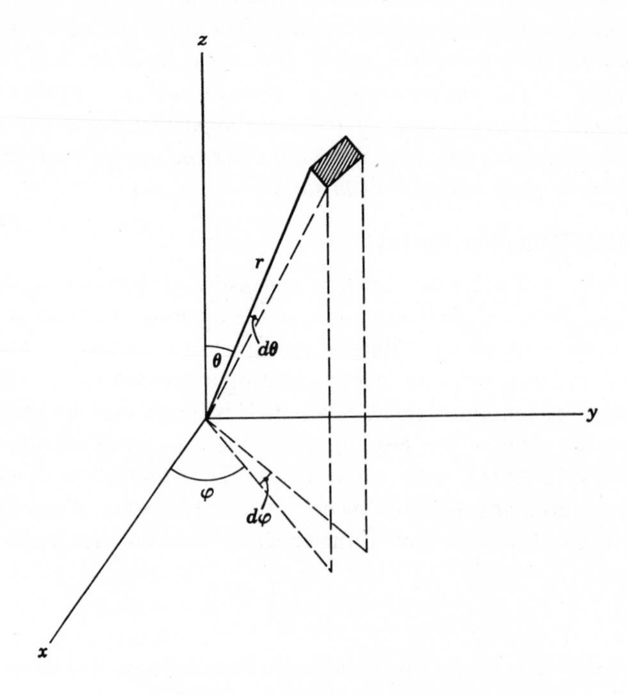

Figure 1.2 Spherical coordinates.

Intensity and Derived Quantities. The basic unit of radiation energy measurement will now be defined. In Figure 1.3 dA is an element of area with normal along N, and dA' is another element of area whose distance is large compared with the dimensions of dA. Consider dE, the radiation energy crossing the surface dA per second and directed toward dA'. This will be proportional to $dA \cos \theta$, the area projected normal to the direction of propagation, and it will also be proportional to the solid angle of dA' as seen from dA. The factor of proportionality is known as the specific intensity, or merely as the intensity, of the radiation field. The intensity I is energy per unit area, per unit time, and per unit solid angle:

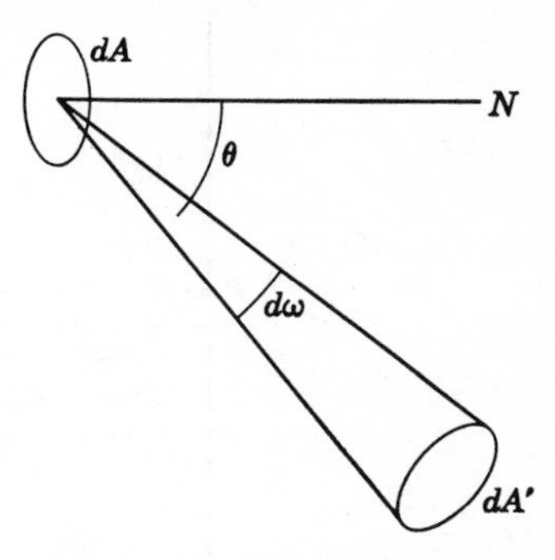

Figure 1.3 Defining specific intensity.

$$dE = I\, dA \cos \theta\, d\omega \tag{1.3}$$

Intensity I is a function of position, direction, and (possibly) time.

Another quantity of interest is the mean intensity J. This is the average of I with respect to direction:

$$J = \frac{\int I\, d\omega}{\int d\omega} = \frac{1}{4\pi} \int I\, d\omega \tag{1.4}$$

These integrals are to be carried out over all directions. If the spherical angles of equation (1.2) are used, the mean intensity is

$$J = \frac{1}{4\pi} \int_0^{2\pi} \int_0^{\pi} I \sin \theta\, d\theta\, d\phi \tag{1.5}$$

The limits are apparent from Figure 1.2. Note that if I is independent of direction, it can be taken out of the above integrals, with the result that $J = I$. A radiation field in which I does not depend on direction is said to be isotropic.

Another quantity of great interest is the flux F. The flux at a given point and in a given direction is the net amount of radiant energy crossing unit area (measured normal to that direction) per second. In Figure 1.4, P is a point on the surface AB, seen edgewise, and PN is the normal to the surface at that point. If I is the intensity of radiation at P in the direction of $d\omega$ as shown, then equation (1.3) indicates that the radiation within this solid angle contributes the amount $I \cos \theta\, d\omega$ to the energy which crosses the surface AB per unit area per unit time. The flux in the direction of PN is found by

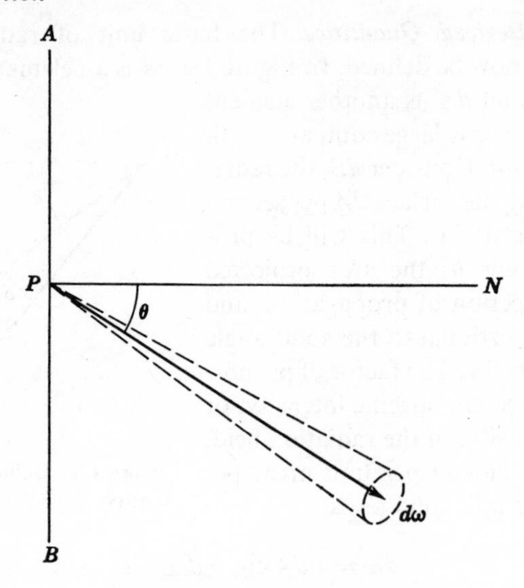

Figure 1.4 Obtaining the flux.

integrating this expression over all directions:

$$F = \int I \cos \theta \, d\omega \tag{1.6}$$

If one lets direction PN correspond to the z axis of Figure 1.2 so that the θ of equation (1.6) is the same as the θ of equation (1.2), then

$$F = \int_0^{2\pi} \int_0^{\pi} I \cos \theta \sin \theta \, d\theta \, d\phi \tag{1.7}$$

It is often convenient to change variables from θ to μ, where $\mu = \cos \theta$. If this is done, equation (1.7) becomes

$$F = \int_0^{2\pi} \int_{-1}^{+1} I\mu \, d\mu \, d\phi \tag{1.8}$$

It should be noted that the flux is the net rate of transfer of energy in the given direction, for it represents the excess of the energy in the given direction over that in the opposite one. In an isotropic radiation field, i.e., one in which I does not depend on direction, the flux is zero.

Consider the elementary surface dA, which is normal to the direction of $d\omega$. The energy which crosses this surface in time interval dt and which is confined to the solid angle $d\omega$ is $I \, dA \, d\omega \, dt$. This energy is contained in the cylinder of cross-section area dA and length $c \, dt$, where c is the speed of light. Then $I \, d\omega/c$ is the energy per unit volume contributed by the solid angle $d\omega$, and the total energy density u is obtained by integrating this over all directions:

$$u = \frac{1}{c}\int I\,d\omega = \frac{4\pi}{c}J \tag{1.9}$$

Monochromatic Quantities. The quantities considered so far have involved the total energy carried by waves of all wavelengths or frequencies. One is often interested in what is happening at a particular wavelength, however, and quantities which depend on wavelength must be introduced for this. One cannot consider directly an intensity at a particular wavelength because of the fact that it is not possible to completely isolate one wavelength from all others. One must always deal in practice with a number of waves which cover a range of wavelengths in the neighborhood of the wavelength of interest. The monochromatic intensity I_λ, therefore, is defined as the intensity per unit wavelength range. Thus $I_\lambda\,d\lambda$ is the intensity due to all waves having wavelengths between λ and $\lambda + d\lambda$. It is the product $I_\lambda\,d\lambda$, not I_λ by itself, which is the quantity of physical significance. In a similar fashion one can define the monochromatic intensity per unit frequency range, I_ν.

The relation between the two different forms of the monochromatic intensity follows from the relation between wavelength and frequency. Since $\lambda = c/\nu$, the wavelength interval $d\lambda$ corresponds to the frequency interval $d\nu$ if $d\lambda = -c\nu^{-2}\,d\nu$. The minus sign means only that wavelength increases as frequency decreases. It follows that

$$I_\nu = I_\lambda\left|\frac{d\lambda}{d\nu}\right| = \frac{c}{\nu^2}I_\lambda = \frac{\lambda^2}{c}I_\lambda \tag{1.10}$$

The intensity due to all wavelengths or frequencies, often called the integrated intensity, is given by

$$I = \int_0^\infty I_\lambda\,d\lambda = \int_0^\infty I_\nu\,d\nu \tag{1.11}$$

Identical arguments hold concerning the monochromatic and integrated mean intensities, fluxes, and energy densities.

Relations with Electromagnetic Fields. Consider a light wave traveling in the positive x direction. Because of the transverse nature of light waves, electric and magnetic fields in the y and z directions are associated with this wave. The fields of such a wave can often be represented by the simple equations

$$\begin{aligned} E_y &= E_{y0}\sin(kx - \omega t - \delta_1) \qquad & E_z &= E_{z0}\sin(kx - \omega t - \delta_2) \\ H_z &= H_{z0}\sin(kx - \omega t - \delta_1) \qquad & H_y &= H_{y0}\sin(kx - \omega t - \delta_2) \end{aligned} \tag{1.12}$$

where E_y and H_y are the y components of the electric and magnetic fields, respectively, and E_{y0} and H_{y0} are the amplitudes of these fields. There is a similar interpretation for the z components of these quantities. Also, k and ω are related to the wavelength and the frequency of the wave by $k = 2\pi/\lambda$, $\omega = 2\pi\nu$. The δ's are called phase constants, and their numerical values determine the sizes of the fields that go with any given x and t values. Note that a given component of the electric field is similar to the perpendicular component of the magnetic field in that they have the same phase constants, and the similarity does not end with this. Electromagnetic theory indicates that the amplitudes of these perpendicular components are also related to each other. If gaussian units are used and if the material medium does not have too large a density, then

$$H_{z0} = E_{y0} \qquad H_{y0} = -E_{z0} \tag{1.13}$$

Equations (1.12) are called plane waves, since the fields are constant on planes which are perpendicular to the direction of propagation. If the amplitudes and the phase constants are strictly constants, or if they fluctuate in such a manner that

$$\frac{E_{y0}}{E_{z0}} = \text{constant} \qquad \delta_1 - \delta_2 = \text{constant} \tag{1.14}$$

then the waves are said to be polarized. If the fluctuations in E_{y0} are completely independent of those in E_{z0}, and likewise for δ_1 and δ_2, then the wave is unpolarized.

The radiation energy per unit volume due to an electric or a magnetic field is equal to the square of the field divided by 8π. Then the energy densities of the fields of equations (1.12), in view of equation (1.13), are

$$u_E = u_H = \frac{E^2}{8\pi} = \frac{1}{8\pi}(E_y^2 + E_z^2) \tag{1.15}$$

The average of the total energy density is obtained by taking the time average of the $\sin^2$ terms over a period. This latter is $\frac{1}{2}$, so

$$\bar{u} = \bar{u}_E + \bar{u}_H = \frac{E_0^2}{8\pi} \tag{1.16}$$

where

$$E_0^2 = E_{y0}^2 + E_{z0}^2 \tag{1.17}$$

If the amplitudes E_{y0} and E_{z0} fluctuate with time, the averages of the squares should appear in (1.17).

The energy contained in a rectangular volume of cross-sectional area A and length s, due to the above wave, is $(AsE_0^2/8\pi)$. If s is along the direction of propagation (the x axis), this same amount of energy is carried across the area A in the time interval s/c. The flux of this wave in the direction of its propagation is, therefore,

$$F = \frac{c}{8\pi} E_0^2 \tag{1.18}$$

The plane wave considered above moves only in one discrete direction, while intensity is defined only for radiation which covers a range of directions; therefore the intensity of this wave is not defined. Also, only a single wavelength was considered. In practice, a range of wavelengths will always be involved, and one must define the amplitudes per unit wavelength interval. Thus $E_0(\lambda)\, d\lambda$ should replace E_0 in equations (1.16) and (1.18), and the left sides of these equations should be replaced by the proper monochromatic quantity multiplied by $d\lambda$.

2. Black-Body Radiation

Thermodynamic Equilibrium. It is well known that all material objects give off some electromagnetic radiation: the hotter they are, the more radiation they emit. When light shines on an object, it helps to heat up the object, causing it to give off more radiation of its own. The radiation an object emits does not include that which it simply scatters or reflects from some other direction. A building, for example, is visible because of the sunlight it reflects, not because of its own radiation.

If one shines radiation on an object from all directions, if this radiation is of the same nature (same I_λ for all λ) as that which the object emits, and if the object is given enough time to become accustomed to this environment, the object achieves a very special state known as thermodynamic equilibrium, or TE. No object is ever in perfect TE, but it is possible to come very close. An object kept in an adiabatic enclosure, i.e., one that does not allow any energy to pass through its walls, will soon come to TE, and nearly adiabatic enclosures do exist. The properties of matter and radiation simplify considerably when TE conditions exist, and so a good first step toward understanding these properties is the study of TE conditions. Some of the properties of matter in TE are discussed in Sections 4 and 5. The concern here is with radiation in TE.

The intensity of the radiation which an object in TE emits depends only on the temperature of the object and the wavelength of the radiation. It does not depend on the size, shape, or composition of the object. This is one form of Kirchhoff's law, which is usually stated in terms of the absorption and emission properties of matter. It has been found that an object which absorbs (rather than reflects or transmits) most of the radiation incident on it also emits very much as if it were in TE. Since such an object appears black if its temperature is low enough that it emits very little visible light, the radiation of an object in TE is usually called black-body radiation.

The Planck and Related Functions. The intensity of black-body radiation is known as the Planck function, after Max Planck, who first derived theoretically the form of TE radiation. It is usually designated by $B(T)$, the T indicating that it is a function of temperature. The monochromatic intensities emitted by black bodies are given by the following expressions:

$$B_\nu(T) = \frac{2h\nu^3/c^2}{e^{h\nu/kT} - 1} \qquad B_\lambda(T) = \frac{2hc^2/\lambda^5}{e^{hc/\lambda kT} - 1} \tag{2.1}$$

In the above, h is Planck's constant (6.63×10^{-27} erg sec), c is the vacuum speed of light (3.00×10^{10} cm sec^{-1}), and k is the Boltzmann constant (1.38×10^{-16} erg °K^{-1}). When the numerical values of the physical constants are inserted (see Appendix A for more accurate values), equations (2.1) become

$$B_\nu(T) = \frac{1.474 \times 10^{-47}\nu^3}{e^{(4.799\times 10^{-11}\nu)/T} - 1} \qquad B_\lambda(T) = \frac{1.191 \times 10^{-5}\lambda^{-5}}{e^{(1.439/\lambda T)} - 1} \tag{2.2}$$

In equations (2.2), ν is measured in sec^{-1}, λ in centimeters, and T in °K. The units of $B_\lambda(T)$ are erg cm^{-2} sec^{-1} ster^{-1} per wavelength interval of one cm, which is the same as erg cm^{-3} sec^{-1} ster^{-1}. The units of $B_\nu(T)$ are the same except that it is per frequency interval of one sec^{-1}, that is, erg cm^{-2} ster^{-1}. The steradian has no physical dimensions, so intensity, mean intensity, and flux all have the same physical dimensions; however, it is best that one keep in mind the per unit steradian that goes with intensity.

One of the important properties of TE is that the radiation field is exactly the same at all points and in all directions. This means that the mean intensity J also equals the Planck function, and the flux in any direction is zero, since there is no net transfer of energy. It is often useful to consider the partial flux into one hemisphere, ignoring the equal amount directed in the opposite way. If F^+ is defined as this partial flux, and if θ is measured from the outward direction, then equation (1.7) indicates that

$$F_\lambda^+ = \int_0^{2\pi} d\phi \int_0^{\pi/2} d\theta I_\lambda \cos\theta \sin\theta \tag{2.3}$$

In TE the intensity equals the Planck function, which is independent of direction; therefore

$$F_\lambda^+ \text{ (black-body)} = \pi B_\lambda(T) \tag{2.4}$$

This is the rate at which energy per unit wavelength interval escapes from unit area of a black body. Of course, in TE energy is incident on the body at the same rate.

Equations (1.9) and (2.1) indicate that the monochromatic energy densities of black-body radiation are given by

$$u_\nu = \frac{8\pi h\nu^3/c^3}{e^{h\nu/kT} - 1} \qquad u_\lambda = \frac{8\pi hc\lambda^{-5}}{e^{hc/\lambda kT} - 1} \tag{2.5}$$

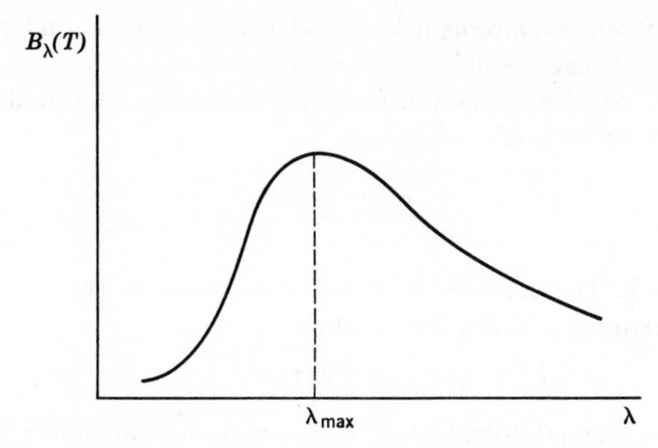

Figure 2.1 The Planck Function.

The units of u_ν are erg cm^{-3} sec, and those of u_λ are erg cm^{-4}. Physicists seem to use energy density more often than astronomers, and one sometimes sees the quantities in equations (2.5), instead of the intensities in equations (2.1), referred to in physics textbooks as Planck functions.

If one plots $B_\lambda(T)$ against λ, as in Figure 2.1, one finds a curve that rises to a maximum and falls off for very large λ. The curve goes to zero for both very large and very small wavelengths. The plot of $B_\nu(T)$ versus ν looks very much the same.

The height of the curve and the wavelength or the frequency of the maximum are functions of the temperature. The maximum of the Planck function can be found by setting its derivative equal to zero. Thus if one evaluates $dB_\lambda/d\lambda$ from equation (2.1) and sets it to zero, one finds

$$\frac{xe^x}{e^x - 1} = 5$$

where $x = hc/\lambda kT$. The solution is $x = 4.965$, and this corresponds to the wavelength λ_{max} for which B_λ is maximum. It follows that

$$\lambda_{max} T = \frac{hc}{4.965k} = 0.290 \text{ cm °K} \tag{2.6}$$

At room temperature T is about 290°K and λ_{max} is around 10^{-3} cm $= 10^5$ Å, in the infrared. If our eyes were sensitive to this wavelength region, summer nights would not appear very dark. At higher temperatures, λ_{max} moves to shorter values, and eventually, an appreciable amount of visible light is given off. As T goes up, the apparent color changes from black to deep red, to orange, to yellow, to white, and finally to bluish white. Equation (2.6) is known as the Wien displacement law, and it helps one to understand the apparent colors of black bodies.

If one works with frequencies instead of wavelengths, one finds that the condition for B_ν to be maximum is that

$$\frac{xe^x}{e^x - 1} = 3$$

where $x = h\nu/kT$. The solution to this is $x = 2.821$, so the frequency for which B_ν is maximum satisfies the equation

$$h\nu_{\max} = 2.821kT \tag{2.7}$$

Note that the wavelength for which B_ν is maximum is quite different from the wavelength for which B_λ is maximum. It is not B_ν or B_λ that is important, but rather $B_\nu\,d\nu$ or $B_\lambda\,d\lambda$, or the integral of these over some finite range.

There are two approximations to the Planck function which are often useful. If $x = h\nu/kT = hc/\lambda kT$ is very large, then e^x is very much larger than unity, and the minus one in the denominators of equations (2.1) can be ignored. This leads to the following:

$$\left.\begin{aligned} B_\nu(T) &= \frac{2h\nu^3}{c^2}\,e^{-h\nu/kT} \\ B_\lambda(T) &= \frac{2hc^2}{\lambda^5}\,e^{-hc/\lambda kT} \end{aligned}\right\} \quad \frac{h\nu}{kT} = \frac{hc}{\lambda kT} \gg 1 \tag{2.8}$$

This is known as the Wien distribution, and it is valid for very short wavelengths (high frequencies) if the temperature is not too high.

At the other extreme, if x is very small, then $e^x = 1 + x$ to a good approximation. If this approximation is substituted into the denominators of equations (2.1), the results are:

$$\left.\begin{aligned} B_\nu(T) &= \frac{2\nu^2 kT}{c^2} \\ B_\lambda(T) &= \frac{2ckT}{\lambda^4} \end{aligned}\right\} \quad \frac{h\nu}{kT} = \frac{hc}{\lambda kT} \ll 1 \tag{2.9}$$

This is the Rayleigh-Jeans distribution and is valid for very long wavelengths (low frequencies), if the temperature is not too low.

The intensity radiated in all wavelengths or frequencies by a black body is found by integrating B_λ over all λ (or B_ν over all ν). With the substitution $x = h\nu/kT$, this gives

$$B(T) = \int_0^\infty B_\nu(T)\,d\nu = \frac{2k^4T^4}{h^3c^2}\int_0^\infty \frac{x^3\,dx}{e^x - 1}$$

The last integral above has the value $\pi^4/15$, so

$$B(T) = \frac{\sigma}{\pi} T^4 \tag{2.10}$$

where

$$\sigma = \frac{2\pi^5 k^4}{15h^3c^2} = 5.67 \times 10^{-5} \text{ erg cm}^{-2} \text{ sec}^{-1} \,{}^\circ\text{K}^{-4} \tag{2.11}$$

is the Stefan-Boltzmann radiation constant. A different radiation constant, $a = 4\sigma/c$, is often used instead of σ. The integrated Planck function $B(T)$ is seen to be quite sensitive to the temperature. $B(T)$ is simply the area under the curve of Figure 2.1.

3. Atomic Spectra

Kirchhoff's Laws. The spectrum of a light source, i.e., the detailed wavelength distribution of the radiation, is an important property of the source. In this section some of the physical principles governing the characteristics of spectra are discussed.

A qualitative description of spectra was given by Gustav Kirchhoff over a century ago. This was given in the form of three laws which can be expressed as follows:

1. An incandescent solid, liquid, or extremely high-pressure gas has a continuous spectrum; i.e., it gives off radiation of all wavelengths.
2. An incandescent gas, if the pressure is not too high, has a bright line or emission spectrum consisting of radiation at only certain discrete wavelengths. These wavelengths are characteristics of the material in the gas.
3. If the light from a source of continuous radiation is passed through a gas, the lines of the gas will be superposed on the continuous background. If the gas is hotter than the source, the lines will be in emission (more intense than the neighboring continuum); if the gas is cooler than the source, the lines will be in absorption (less intense than the neighboring continuum). The latter is called a dark line, an absorption, or a Fraunhofer spectrum.

The continuous spectrum of the first law will be black-body radiation if the source is in thermodynamic equilibrium; otherwise, it will deviate to some extent from the radiation discussed in the last section. The fact that the lines mentioned in the second and third laws occur at wavelengths which are characteristic of the material in the gas is the basis for identifying the chemical makeup of astronomical objects.

The nature of black-body radiation could be explained by Planck only by assuming that the then-accepted "classical" picture of physics was imperfect. Specifically, he had to assume that light is composed of particles, called photons, and that the energy of a photon is related to its classical frequency and wavelength by the relations

$$E = h\nu = \frac{hc}{\lambda} \tag{3.1}$$

where h is Planck's constant. On this picture a photon of blue light has nearly twice the energy of one of red light, since its frequency is nearly twice as great. This assumption about the nature of light was also needed to explain the observed photoelectric effect, as Einstein demonstrated some years later.

The main features of the continuous spectra were explained by Planck's analysis of black-body radiation, but the bright- and dark-line spectra proved to be more of a puzzle. It could not be understood why the radiation of atoms ever had to be limited to certain discrete wavelengths. It was not until 1913 that Bohr was able to explain the relatively simple spectrum of hydrogen by using Planck's assumption of equation (3.1), plus some further assumptions of his own which were also in disagreement with classical physics.

The Bohr Atom. In 1911 Rutherford performed some scattering experiments which suggested that atoms consist of a heavy, positively charged central part called the nucleus, and that this is surrounded by much lighter, negatively charged electrons. An atom will normally have just enough electrons to exactly balance the positive electric charge of the nucleus, so the atom is under normal circumstances electrically neutral. But electric charges exert forces on each other. If e_1 and e_2 are two electric charges measured in esu (electrostatic units), the force that each exerts on the other in vacuum is given by Coulomb's law:

$$F = \frac{e_1 e_2}{r^2} \tag{3.2}$$

The force F is measured in dynes, and r is the distance between the charges in centimeters. The force is attractive if the charges are unlike, repulsive if they are of the same kind.

The coulomb force would cause the electrons to fall into the nucleus if they did not move, so the electrons must be moving around the nucleus. Equation (3.2) shows the same dependence on distance as the gravitational force, and so, if classical physics were correct, the electrons would move in orbits similar to those of gravitating masses. If the forces between the different electrons could be neglected, then the electron orbits would obey Kepler's laws of motion (see Section 12). In order to avoid the complications of the forces between electrons, the atom will be assumed to have only one electron. This is the normal state of the hydrogen atom, but other types of atoms can sometimes have only one electron.

As the electron moves in its orbit, it is constantly being accelerated toward the nucleus. But classical physics says that an accelerated electric charge should radiate energy, so the orbiting electron should constantly lose energy and eventually spiral into the nucleus. This collapse into the nucleus should take only a very small fraction of a second, and so there is the problem of how atoms can exist at all. Why do atoms not radiate energy all of the time?

It was apparent that a break had to be made with classical physics in order to explain line spectra and related atomic phenomena. Bohr used classical physics as far as he could, then he made whatever additional assumptions he found necessary without regard for the reasons for them. In this fashion he came up with an empirical model of the hydrogen atom, and it turned out to be quite basic in the later development of quantum mechanics.

Bohr assumed that the electron moves in a circular orbit. (The generalization to elliptical orbits was made some years later by A. Sommerfeld.) Let an electron of charge $(-e)$ move in a circular orbit of radius r about the nucleus of charge $(+Ze)$. Z is the atomic number of the nucleus; $Z = 1$ for hydrogen, $Z = 2$ for helium, and so on. Any mass moving in a circle is accelerated toward the center by the amount v^2/r, where v is the orbital speed. Then, using (3.2), one has

$$F = -\frac{Ze^2}{r^2} = -\frac{mv^2}{r} \tag{3.3}$$

Equation (3.3) then yields

$$mv^2r = Ze^2 \tag{3.4}$$

as the relation between the size of the orbit and the orbital speed, where m is the mass of the electron.

The energy of the atom is composed of two parts: the kinetic energies of the electron and the nucleus, and the potential energy due to the force of attraction between electron and nucleus. The motion of the atom as a whole is not important in the present discussion, so the concern is only with the kinetic energy of the orbital motion. The light electron has a much greater orbital motion than the heavy nucleus, so virtually all of the kinetic energy of the atom is due to the electron, and $KE = mv^2/2$. (In practice, the correction for the motion of the nucleus is small but not negligible.)

The potential energy (PE) of the atom is the energy needed to move the electron from a very distant point up to its present position:

$$PE = -\int_{\infty}^{r} F\,dr \tag{3.5}$$

The negative sign indicates that the work is done against the prevailing force F of the system. With the coulomb force then,

$$PE = -\frac{Ze^2}{r} \tag{3.6}$$

The total energy of the atom is the sum of the kinetic and the potential energies:

$$E = \tfrac{1}{2}mv^2 - \frac{Ze^2}{r} = -\frac{Ze^2}{2r} \tag{3.7}$$

The latter relation follows from equation (3.4).

The potential energy was defined to be zero when the charges are so far apart that they have essentially no influence on each other, and a result of this is that the lower limit of the integral in equation (3.5) is infinity. This is a very convenient definition in the present problem, but it is by no means essential. The only important property of *PE* is its difference at two given points, not its actual value at any point. It is not significant that the *PE* in the present case is negative; it is significant that the *PE* increases as the electron moves away from the nucleus. The attractive coulomb force tends to draw the electron towards the nucleus, toward smaller values of the *PE*. If the two particles had like charges, both positive or both negative, then the *PE* would be opposite in sign to the present case. The repulsive force would then tend to move the particles apart, again toward smaller values of the *PE*.

Equations (3.3), (3.4), and (3.7) describe the assumed circular motion of the electron about the nucleus. Only classical physics has been used so far, but this must be changed immediately. As explained above, it must be assumed that the electron does not necessarily radiate even though it is being constantly accelerated by the nucleus. At this point Bohr makes another major assumption: the electron cannot exist in an orbit of arbitrary size, which means that it cannot have arbitrary values of the energy. It can exist only in those orbits for which the angular momentum is given by some positive integer times $h/2\pi$, where h is again Planck's constant. This looks like a highly artificial assumption, and it is; Bohr found it necessary in order to get the right answer. Since the angular momentum in a circular orbit is mvr, this means

$$mvr = \frac{nh}{2\pi} \qquad n = 1, 2, \ldots \tag{3.8}$$

Later, when quantum mechanics became an established part of physics, equation (3.8) ceased to be an arbitrary assumption and became a logical part of much more basic theory.

The integer n is what is known as a quantum number, and the complete state of the atom can be determined from its value. For example, from equations (3.4) and (3.8) one obtains

$$mv^2r = \frac{1}{mr}(mvr)^2 = \frac{1}{mr}\left(\frac{nh}{2\pi}\right)^2 = Ze^2$$

or

$$r_n = \frac{n^2h^2}{4\pi^2mZe^2} = 0.529 \times 10^{-8}n^2 \quad \text{cm} \tag{3.9}$$

The radius of the nth Bohr orbit is proportional to n^2. The first Bohr orbit has a radius of about 0.5 Å, and it is a basic unit of length in atomic physics. The velocity in the nth orbit follows from equations (3.8) and (3.9):

$$v_n = \frac{2\pi Z e^2}{nh} \tag{3.10}$$

and the energy E_n comes from (3.7) and (3.9):

$$E_n = -\frac{2\pi^2 m Z^2 e^4}{n^2 h^2} = -13.6\frac{Z^2}{n^2} \quad \text{eV} \tag{3.11}$$

The eV (electron volt) is the unit of energy commonly used in atomic physics. One eV equals about 1.6×10^{-12} erg.

The above equations give the characteristics of the allowed orbits in terms of the quantum number n. Classically, any of a continuous set of values of radius, energy, angular momentum, or the like should be possible; Bohr's postulate (3.8) says that only a certain discrete set of these parameters is permitted. These quantities have been quantized. It is important to note that these special rules apply only when the electron is bound to the nucleus, i.e., when the total energy is negative. When the electron has enough kinetic energy to be free and travel in a hyperbolic orbit, there is no quantization, and any of a continuous set of positive energy values is permitted.

The orbit for $n = 1$ has the lowest possible energy, and this is called the ground state. For hydrogen ($Z = 1$) the ground state has an energy of -13.6 eV on the scale in which zero energy corresponds to escape velocity of the electron. A hydrogen atom in the ground state must, therefore, somehow receive at least 13.6 eV in order for the electron to become free. This process of causing the electron to become free is called ionization, and the energy needed to do this from the ground state is known as the ionization potential of the atom.

Figure 3.1 is an energy level diagram for hydrogen. The permitted orbits are shown on a scale that is linear with energy. It is seen that the negative energy levels crowd closer together as n gets larger, and they reach the limit $E = 0$

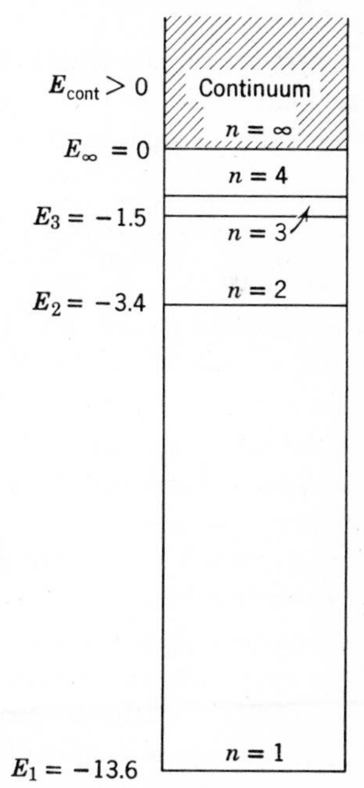

Figure 3.1 Energy-level diagram for hydrogen.

as n becomes infinite. For larger energies yet, there is no quanti zation and the continuous set of allowed energies is indicated.

As stated above, Bohr assumed that the atom will not necessarily radiate energy when it is in one of its allowed orbits. He then postulated further that, under certain circumstances, the electron can change from one permitted orbit to another, either of which can be either bound or free. When the electron jumps from one orbit to a lower one, the atom loses energy; when it changes to a higher orbit, the atom gains energy. These energy gains and loses may be supplied by neighboring atoms, but they can also be in the form of radiation. In other words, it is possible for the electron to drop to a lower orbit as light is emitted by the atom, and it is also possible for light passing near the atom to be absorbed as the electron jumps to a higher orbit. In both of these cases, the energy of the light absorbed or emitted is exactly the same as the difference in the energy levels of the electron jump or transition. Bohr agreed with Planck that light is composed of particles or photons and that equation (3.1) expresses the relation between the energy of a photon and its frequency. Whenever there is a transition from a level of energy E to one of energy E' with the consequent absorption or emission of a photon, then this photon will have a frequency given by

$$h\nu = |E - E'| \tag{3.12}$$

There are three possible types of transition, depending on whether the initial and final levels are bound or free. In a bound-bound transition, both energy levels satisfy equation (3.11). If n and n' are the two quantum numbers, then

$$\nu = \frac{2\pi^2 m Z^2 e^4}{h^3}\left|\frac{1}{n^2} - \frac{1}{n'^2}\right| = 3.290 \times 10^{15} Z^2 \left|\frac{1}{n^2} - \frac{1}{n'^2}\right| \tag{3.13}$$

It is seen that equation (3.13) corresponds to a discrete set of frequencies as n and n' take on all possible integral values. The bound-bound transitions, therefore, correspond to the line emission and absorption of Kirchhoff's second and third laws. The other two types of transition are bound-free and free-free. In the latter cases at least one of the levels is in the continuum, and so they both give rise to a continuous set of possible frequencies. It is seen that Kirchhoff's laws are not sufficiently general, for they do not mention this continuum which a low pressure gas can absorb and emit.

The crude Bohr picture of the hydrogen atom gave an amazingly accurate agreement with the observed hydrogen spectrum. It not only predicted the general form of the spectrum, it also predicted the precise wavelengths or frequencies. It was apparent that there must be a good deal of important physics behind the postulates that Bohr had to make.

Consider equation (3.13) as n is held constant and n' takes all possible values greater than n. For $n' = n + 1$, the frequency corresponds to the energy difference between the two neighboring levels. As n' becomes larger, the energy difference becomes greater, but by a smaller and smaller amount. In the limit as $n' \to \infty$, the energy reaches the amount needed to ionize the atom from the nth level, and the frequency reaches its corresponding limit. This series of possible transitions from fixed n to all greater values of n' represents an infinite number of lines which crowd closer and closer together as the series limit is approached. The series with $n = 1$ is called the Lyman series, and successive lines are designated by letters of the Greek alphabet. Thus the lines arising from the transitions from $n = 1$ to $n' = 2, 3, 4$, and so on are commonly designated as $L\alpha$, $L\beta$, $L\gamma$, and so on. Line $L\alpha$ is the strongest line in the hydrogen spectrum, but it falls in the far ultraviolet at about 1215 Å. The higher Lyman lines crowd together toward the Lyman limit at 912 Å.

The hydrogen series of lines having $n = 2$ and $n' \geqslant 3$ is known as the Balmer series. Since they occur in the easily observed visible region of the spectrum, and since they are a very strong series of lines under many astronomical conditions, the Balmer lines are among the most important to astronomers. Individual Balmer lines are given the letter H plus a Greek letter, $H\alpha$ at 6563 Å being the strongest. The other Balmer lines converge to the Balmer limit at 3647 Å. The Paschen series consists of transitions in which the lower level is $n = 3$, and its series limit is at 8206 Å. It is seen that there are an infinite number of such series, each higher one moving to longer wavelengths and smaller frequencies.

The Bohr atom is unsatisfactory from several points of view. As mentioned above, it makes some radical assumptions which are not consistent with the classical physics which it otherwise uses, and the only justification is that they seem to work. Also, the predictions are accurate only for one-electron system. For normal helium or more complicated atoms, the predictions fail. Still, the Bohr atom is one of the major concepts in the historical development of quantum mechanics, a development which was realized in the following decade.

In quantum mechanics all ideas of a classical picture of an atom are abandoned, and the Bohr assumptions appear as logical parts of a much larger, self-consistent theory. The Bohr theory is still a useful conceptual aid in the understanding of many atomic phenomena, although errors can result if one takes this picture too literally. The fact remains that spectral lines exist because the bound energy states are quantized.

4. Excitation and Ionization

An atom may be pictured as composed of a positively charged nucleus surrounded by negatively charged electrons. There are normally enough electrons to make the atom as a whole electrically neutral. It can happen, however, that one or more of the electrons will somehow receive enough energy to escape from the nucleus, leaving one or more free electrons plus a positive ion. This process is known as ionization, and an atom can obviously be ionized a number of times equal to the number of electrons it normally possesses. The electrons which remain bound to the nucleus must exist in one of the allowed energy states which were suggested by the Bohr theory. The lowest bound energy state of an atom or ion is known as the ground state, and all of the others are called excited states. A complete statement of the energy situation of an atom must include both the number of electrons which have been stripped off and the energy levels which the remaining electrons are in; i.e., it must include both the ionization and the excitation conditions of the atom.

The excitation levels of an atom are what determine the line spectra of the atom, and these depend on the ionization state the atom is in. As an example, neutral helium or He I has two bound electrons, and this results in a rather complicated spectrum. Once-ionized helium or He II has only one bound electron, like neutral hydrogen, and so its excited levels and resulting spectrum have exactly the same form as those of hydrogen. The He II electron is held with a nuclear charge twice as large ($Z = 2$ in the equations of Section 3) as in the case of hydrogen, and this causes a distortion of the levels to energy values different from those of hydrogen, but they still have the same relative separation. But He III, which has lost both of its electrons, has none left and has no atomic spectrum. Ions which have the same number of bound electrons belong to what is known as an isoelectronic series, and all such members have similar appearing spectra. Thus H I, He II, Li III, and so on belong to the isoelectronic series having one bound electron, and they all have hydrogen-like spectra. Similarly, He I, Li II, Be III, and so on are members of the two-electron series, and their spectra are similar to each other but quite different from those of the one-electron series. This property of isoelectronic series allows one to determine the appearance of spectra which cannot be observed in the laboratory. For example, Fe XV has not been produced in the laboratory, yet its spectrum can be predicted from the observed spectrum of Mg I.

If an atom in some excited state could be isolated from all outside effects, it would very quickly drop to the ground state, possibly by way of other excited states, with the emission of the appropriate photon or photons. This is known as spontaneous emission, and if atoms are not disturbed, they will emit photons spontaneously until they are in as low an energy state as possible. If the atom is placed in a radiation field, the radiation can cause the atom to make transitions to other energy levels in two separate ways. If a photon has the proper frequency, it can be absorbed by the atom, causing the atom to jump to a higher excited or ionized level. Also, if the photon has the proper frequency, it can induce the atom to emit an identical photon by dropping to a lower excited or ionized level. This latter process is known as induced or stimulated emission. There are then three kinds of radiative transitions which an atom can make: absorption, spontaneous emission, and induced emission. The rate at which these transitions take place depends on how many atoms are in the relevant energy levels, how likely the given transitions are, and—in the cases of absorption and induced emission—the intensity of the radiation field. The likelihood of a given type of transition between two given levels is measured by the so-called transition probability. Transition probabilities need to be calculated from quantum mechanics or measured in experiments, and there are some important cases in which they are not very well known.

The atom can also make transitions without radiation being involved. A neighboring atom may collide with it, knocking it into a higher energy level, or the collision may take up some of the excitation or ionization energy, causing the atom to drop to a lower level. In these collisional transitions, the change in excitation or ionization energy is taken from or added to the kinetic energy of the collision. Since kinetic energy is not conserved during these collisions, they are known as inelastic collisions. One can have collisional excitations or de-excitation, ionization or de-ionization. Spontaneous emissions tend to drop an atom quickly to its lowest energy state, but if it is disturbed often enough by photons and other atoms, it may spend an appreciable part of the time in higher excited or ionized levels.

The Excitation Equation. In any given physical situation, some of the atoms are undergoing radiative and collisional transitions between different levels, and the number of atoms in a given level at any time depends on how violent the conditions are. It has long been known that if a system is in thermodynamic equilibrium, classical physics predicts that the probability that a particle in the system has energy E is proportional to $e^{-E/KT}$, where k is Boltzmann's constant and T is the temperature. This again illustrates the tendency to favor small energies. As the temperature increases, the density of high-energy photons gets larger as does the frequency of high-energy collisions, and so the higher energy levels become more frequently populated.

The same exponential dependence on energy is valid for quantum mechanical systems, at least if the density is not too large, with an important modification. In order to understand this modification, one must understand what is meant by the quantum mechanical state of an atom.

According to the Bohr theory, there is only one independent quantity describing the excitation state of an atom. The independent quantity can be identified as the quantum number, and in terms of that number the energy of the atom, the orbital angular momentum, the orbital speed, and anything else can be determined; however, the Bohr theory is not complete. For example, the Bohr theory is restricted to circular orbits, but elliptical orbits could also be considered. It can easily be shown that the orbital energy depends only on the size of the orbit, but the angular momentum depends on both the size and the shape of the orbit. If elliptical orbits are allowed, therefore, two independent quantities are needed to completely describe the excitation state of the atom, and this leads to the requirement of two quantum numbers instead of one. Quantum mechanics claims that some information about the orientation of the orbit and the spins of the particles is also important, and these data lead to yet other quantum numbers. A complete quantum mechanical description of the excitation state of an atom is given when the values of all of these quantum numbers are given. Each possible set of values of the quantum numbers makes up what is known as a quantum mechanical state of the atom.

Often one is interested only in the energy of an atom, not in its complete quantum mechanical state. Since there may be a number of states which have the same energy, one may wish to lump together all states which belong to each of the allowed energy levels. The point of this is that, other things being equal, an energy level that corresponds to many quantum mechanical states is more likely to be occupied than one which belongs to only a few of them. The number of states which have a common energy level is known as the statistical weight of that level. The classical expression must then be changed so that the probability that an atom has energy E is proportional to $g(E)e^{-E/kt}$, where $g(E)$ is the statistical weight of energy level E. If E is not one of the allowed energy levels, then $g(E) = 0$, and no particles can have that particular energy.

Let N_j and N_k be the number of atoms per unit volume which have energies E_j and E_k, respectively. Then the relative populations of the two levels are given by

$$\frac{N_j}{N_k} = \frac{g_j}{g_k} e^{-(E_j - E_k)/kT} \tag{4.1}$$

where the g's are the statistical weights. Actually, equation (4.1) is correct when the E's are the total energies of all kinds which the atoms possess; however, it can also be applied to each separate kind of energy the atoms have, and in the present section E will represent only the excitation and ionization energies.

Let N_{ij} represent the number of atoms per unit volume in the ith ionization stage and in the jth excitation level. Then i will run from 1 (for the neutral atom) up to $Z + 1$ (for the completely stripped nucleus), where Z is the number of electrons in the neutral atom. The ground state of any ionization stage is $j = 1$, and the higher excited levels have integral j values through

$j = \infty$. In Section 3 it was convenient to let the free electron define the point of zero energy, but now this will be changed in order to stay with the usual conventions. The ground state of the ionization stage of interest is now defined to have zero energy, so the higher excited levels have positive energy. [Note that equation (4.1) depends only on the difference between the energy levels, so it is not changed if a constant is added to all energy levels.] It follows that equation (4.1) becomes

$$\frac{N_{ij}}{N_{i1}} = \frac{g_{ij}}{g_{i1}} e^{-E_{ij}/kT} \tag{4.2}$$

As defined above, E_{ij} is the excess energy of level j over that of the ground state, and it is known as the excitation potential of level j.

If N_i is the total number of atoms in ionization stage i per unit volume, then

$$\begin{aligned} N_i &= \sum_{j=1}^{\infty} N_{ij} \\ &= \frac{N_{i1}}{g_{i1}} \sum_{j=1}^{\infty} g_{ij} e^{-E_{ij}/kT} \end{aligned}$$

The above sum is known as the partition function of the given ionization stage, and it is usually denoted by B_i:

$$B_i = \sum_{j=1}^{\infty} g_{ij} e^{-E_{ij}/kT} \tag{4.3}$$

In terms of the partition function, one has

$$\frac{N_{ij}}{N_i} = \frac{g_{ij}}{B_i} e^{-E_{ij}/kT} \tag{4.4}$$

In logarithmic form this becomes

$$\log \frac{N_{ij}}{N_i} = \log \frac{g_{ij}}{B_i} - \frac{5040}{T} E_{ij} \tag{4.5}$$

In equation (4.5), T must be measured in degrees Kelvin and E_{ij} in electron volts.

Equations (4.2), (4.4), and (4.5) are different forms of what is known as the excitation equation or the Boltzmann equation. The atomic structure must be known before the equations can be used; i.e., the bound levels must be identified and their energies and statistical weights must be known. This information is fairly well known for most ions of interest. For hydrogen-like spectra the index j can be identified with the quantum number n of the last section. For other spectra there is no such simple interpretation, and j should be thought of as just an index used to keep track of the bound levels. In all cases there is an infinite number of these discrete levels.

Equation (4.3) shows that the partition function is the sum over all bound levels of the statistical weight times the Boltzmann factor $e^{-E/kT}$. In many cases conditions are so mild that very few atoms or ions get excited to higher levels, so that most of them are in the ground state. When this holds, it is seen that the first term in (4.3) is the dominant one, and the partition function is about equal to the statistical weight of the ground level. This is usually a good approximation for cases like hydrogen and helium in which the first excited level is far above the ground state. Otherwise, the higher terms in (4.3) must be included.

It may be somewhat discomforting to note that the sum in equation (4.3) diverges. The excitation potentials tend toward a finite limit equal to the ionization potential, and the statistical weights are all positive integers. The answer to this paradox is that the higher excited levels do not actually exist. Perturbations of neighboring atoms and ions prevent the electrons from existing in orbits which are too large, so each term in the series should be multiplied by the probability that the given level does exist. The probability must go to zero for large enough values of the index j, and this has the effect of cutting off the series after a finite number of terms. In most instances of importance the partition function is insensitive to precisely how this cutoff is chosen, and the values are accurately known; however, there are some exceptions to this.

As stated above, for hydrogen the index j can be taken as the quantum number n. The statistical weight of the nth level is $2n^2$ (it is $4n^2$ if the nuclear spin is included, but this does not affect the ratio of the statistical weights), and the excitation potentials can be found from equation (3.11); however, in the present energy scale, 13.6 eV must be added to the values given in (3.11):

$$E_n = 13.6 \frac{n^2 - 1}{n^2} \qquad \text{eV} \tag{4.6}$$

With the assumption that nearly all of the neutral hydrogen is in the ground state, the partition function $B_1 = 2$, and equation (4.5) becomes

$$\log \frac{N_{1n}}{N_1} = 2 \log n - \frac{68{,}500}{T} \frac{n^2 - 1}{n^2} \tag{4.7}$$

Table 4.1 shows some numerical values calculated from equation (4.7). The increased population of the higher levels with increased temperature is apparent. It is also seen that, for the lower temperatures, the approximation that all neutral hydrogen is in the ground state is a very good one; however, it breaks down as T gets very large. It should be emphasized that the numbers in Table 4.1 are relative to the abundance of neutral hydrogen only; if

TABLE 4.1. N_{1n}/N_1 FOR HYDROGEN

	$T = 5040°K$	$T = 10{,}080°K$	$T = 20{,}160°K$
$n = 2$	2.5×10^{-10}	3.2×10^{-5}	1.1×10^{-2}
$n = 3$	6.9×10^{-12}	8.1×10^{-6}	8.3×10^{-3}
$n = 4$	2.8×10^{-12}	6.8×10^{-6}	1.0×10^{-2}
$n = 5$	2.2×10^{-12}	7.6×10^{-6}	1.3×10^{-2}
$n = 6$	2.1×10^{-12}	8.9×10^{-6}	1.8×10^{-2}

absolute abundances are desired, then the ionization conditions of hydrogen must also be known.

The Ionization Equation. If equation (4.1) is applied to levels in two successive stages of ionization, with the proper definitions of statistical weights of the continuum levels, and if the result is summed over all of the excited levels of both ionization stages, then one obtains

$$\frac{N_{i+1}N_e}{N_i} = \left(\frac{2\pi mkT}{h^2}\right)^{3/2} \frac{2B_{i+1}}{B_i} e^{-I_i/kT} \tag{4.8}$$

where N_e is the number of free electrons per unit volume, and I_i is the ionization potential of the ith ionization stage. This is the ionization equation, or Saha equation, and it relates the number of atoms in two successive stages of ionization to the quantities that are relevant.

This relation can also be understood from a different point of view. In thermodynamic equilibrium, which must hold if the excitation and ionization equations are to be valid, the number of atoms in a given level must not change with time. Thus the rate at which atoms in the ith stage are being ionized to the $(i + 1)$st stage must equal the rate at which ions in the $(i + 1)$st stage are combining with free electrons to form ions in the ith stage. The former is obviously proportional to N_i, and it also depends on temperature; the latter is proportional to the product $N_{i+1}N_e$, and it too depends on the temperature. Equation (4.8) simply expresses the fact that these ionization and recombination processes take place at the same rate.

If one substitutes the numerical values of the constants and takes logarithms, equation (4.8) becomes

$$\log\left(\frac{N_{i+1}N_e}{N_i}\right) = 15.38 + \log\left(\frac{2B_{i+1}}{B_i}\right) + 1.5 \log T - \frac{5040}{T} I_i \tag{4.9}$$

The numerical factors in equation (4.9) are such that, as before, temperature must be measured in degrees Kelvin and the ionization potential in electron volts, while N_e is in particles per cubic centimeter. An alternative form of the ionization equation is obtained by introducing the electron pressure. Each separate type of particle in a gas makes its own contribution to the total pressure in the gas. As is explained in Section 5, in most astronomical cases the contribution is equal to the number of particles per unit volume times kT. The free electrons in the gas, therefore, produce a pressure given by $P_e = N_e kT$. If this relation is used to eliminate N_e from equation (4.9), the result is

$$\log \frac{N_{i+1}P_e}{N_i} = -0.48 + \log \frac{2B_{i+1}}{B_i} + 2.5 \log T - \frac{5040}{T} I_i \qquad (4.10)$$

The hydrogen partition functions are $B_1 = 2$, $B_2 = 1$, and equation (4.10) leads to the following:

T	5040°K	10,080°K	20,160°K
$\frac{N_{\mathrm{H\,II}}}{N_{\mathrm{H\,I}}} P_e$	1.5×10^{-5}	5.4×10^{2}	7.6×10^{6}

For electron pressures in the range 1–10 dyn cm^{-2} hydrogen changes from almost completely neutral at 5040°K to almost completely ionized at 10,080°K. These numbers can be combined with those of Table 4.1 to yield the populations of the excited levels in terms of all hydrogen, neutral plus ionized. The results are given in Table 4.2 for $P_e = 1$. Note that while the

TABLE 4.2. $N_{1n}/(N_1 + N_2) = N_{1n}/N_H$ FOR HYDROGEN WITH $P_e = 1$

	T = 5040°K	T = 10,080°K	T = 20,160°K
$n = 2$	2.5×10^{-10}	6.0×10^{-8}	1.4×10^{-9}
$n = 3$	6.9×10^{-12}	1.5×10^{-8}	1.2×10^{-9}
$n = 4$	2.8×10^{-12}	1.3×10^{-8}	1.3×10^{-9}
$n = 5$	2.2×10^{-12}	1.4×10^{-8}	1.7×10^{-9}
$n = 6$	2.1×10^{-12}	1.6×10^{-8}	2.4×10^{-9}

abundances in the excited levels relative to the neutrals increases tremendously with increasing temperature, the abundances relative to all hydrogen change much less, and they fall off at the higher temperatures.

Another instructive example involves helium. Here $Z = 2$, and so there are three stages of ionization. The relevant data are $B_1 = B_3 = 1$, $B_2 = 2$ for the partition functions, and $I_1 = 24.58$ eV, $I_2 = 54.41$ eV for the ionization potentials. Equation (4.10) can then be written twice as follows:

$$\log \left(\frac{N_2 P_e}{N_1}\right) = 0.12 + 2.5 \log T - \frac{1.239 \times 10^5}{T}$$

$$\log \left(\frac{N_3 P_e}{N_2}\right) = -0.48 + 2.5 \log T - \frac{2.742 \times 10^5}{T}$$

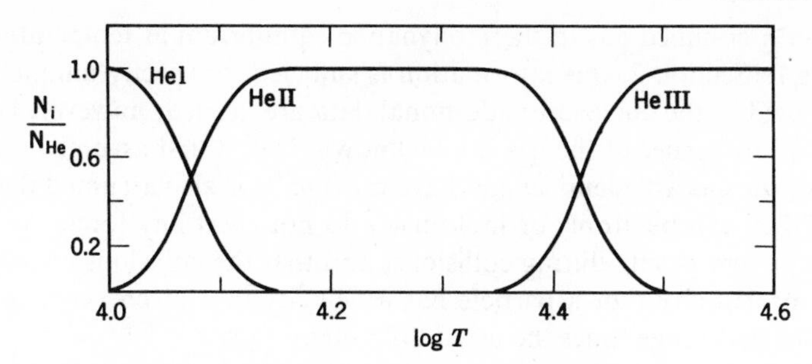

Figure 4.1 Ionization of helium.

The data from these equations are illustrated in Figure 4.1 for $P_e = 1$. The figure shows the percentage of all He atoms in the various stages of ionization plotted against log T. For temperatures below 10,000°K, essentially all the He is neutral. Between about 10,000 and about 14,000°K, it changes from nearly all He I to nearly all He II, and it remains once-ionized to about 22,000°K. Above 22,000°K an appreciable amount becomes ionized a second time, and above about 30,000°K it is almost all completely ionized to form He III. Note that the He I is gone long before He III becomes important, so one usually does not have to be concerned with more than two stages of ionization at one time. The transition regions in which two successive stages are both important are rather narrow, so usually one stage dominates all others. In any physical situation, the electron pressure will also vary, and this can cause the relative widths of the different regions to vary some; however, the general form of Figure 4.1 will hold for any successive ionization stages of any element.

The Bohr theory, or more accurately quantum mechanics, can be used to determine what lines an atom can absorb and emit. The excitation and ionization equations can be used to determine what percentage of the atoms can do this absorbing and emitting as a function of the physical conditions in the source. This helps make it possible to determine both the chemical composition and the physical conditions which exist in astronomical sources where given lines are observed. Some ionization potentials and partition functions of astronomical interest are given in Appendix D.

5. Gas Laws

The Maxwell-Boltzmann Distribution. Since most of the matter in the visible part of the universe is in a gaseous state, it is not surprising that astronomers are quite interested in the properties of a gas. Many properties of interest can be determined by appealing to thermodynamic equilibrium, much as was done in Section 2 for radiation.

Consider a confined gas in thermodynamic equilibrium at temperature T. According to Section 2, this information is sufficient to specify uniquely the radiation field in the enclosure; additional data are needed, however, before the relevant properties of the gas can be known. It is at first assumed that all particles of the gas are identical and have mass m. It is also assumed that the particles (they can be atoms or molecules) do not exert any forces on each other except very briefly during collisions, and that the collisions are elastic. Then the probability that a particle has a velocity in a given range is proportional to that range times the usual Boltzmann factor $e^{-E/kT} = e^{-mv^2/2kT}$. If v_x, v_y, and v_z are the velocity components along the rectangular axes, then the velocity range is given by

$$dv_x\, dv_y\, dv_z = v^2\, dv\, d\omega$$

where v is the speed and $d\omega$ is the element of solid angle of the velocity range. The solid angle is related to the spherical angles (θ,ϕ) by equation (1.2). Then the probability that a particle has a speed between v and $v + dv$ and is directed within $d\omega$ is given by

$$p(v,\theta,\phi)\, dv\, d\omega = \left(\frac{m}{2\pi kT}\right)^{3/2} e^{-mv^2/2kT} v^2\, dv\, d\omega \tag{5.1}$$

The constant $(m/2\pi kT)^{3/2}$ is the factor needed to make the probability function equal to one when integrated over all speeds and directions.

Equation (5.1) is known as the Maxwell-Boltzmann velocity distribution, and it is closely related to the distribution of excitation and ionization energies, equation (4.1), and to the distribution of photon energies, equation (2.1). Figure 5.1 shows a schematic plot of this function against speed for a fixed direction. Very small speeds are not likely because of the v^2 factor, whereas the probability falls off for very large speeds because of the exponential. The result looks very much like the Planck function of Figure 2.1.

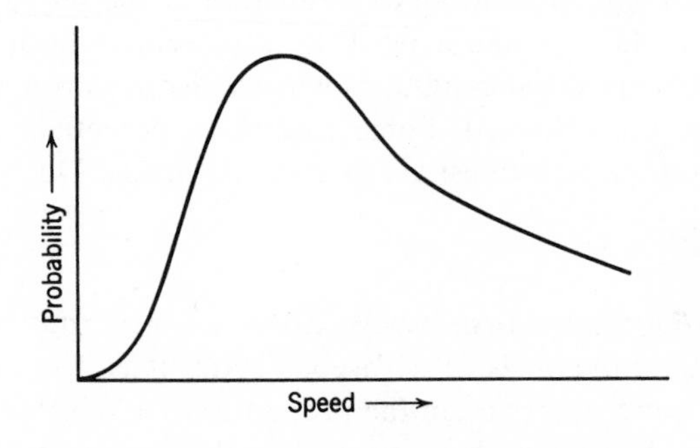

Figure 5.1 The Maxwell-Boltzmann distribution.

The most probable speed v_p can be found from the fact that the function (5.1) has zero slope at $v = v_p$. It is easily seen that

$$v_p = \left(\frac{2kT}{m}\right)^{1/2} \tag{5.2}$$

Heavy particles are seen to move more slowly than light ones at a given temperature, and all tend to move faster with increased temperature.

Equation (5.1) gives the probability that a given particle has a certain velocity, but it can also give the number of particles which have this velocity. Let N be the total number of particles of the gas per unit volume, and let $N(v,\theta,\phi)\,dv\,d\omega$ be the number per unit volume with speeds in the range dv and directions within $d\omega$. Then it is apparent that

$$N(v,\theta,\phi)\,dv\,d\omega = Np(v,\theta,\phi)\,dv\,d\omega \tag{5.3}$$

Thus Figure 5.1 can represent both what a given particle will do at different times and what the whole gas is doing at any one time. Every time a particle has a collision, its speed and direction are likely to change. Most of the time its speed is near v_p; only rarely will it be much greater or much less than this. Likewise, at any time most of the particles have speeds near v_p; only a small percentage of them have speeds much greater or much less than this. Collisions are constantly changing the speeds of all of the particles, but the relative number having a given speed does not change with time.

The Maxwell-Boltzmann distribution can be written in terms of other quantities besides velocity. For example, the kinetic energy of a particle is $E = mv^2/2$, and one can determine from (5.1) the probability that the kinetic energy lies in any range dE. Since energy does not depend on direction, equation (5.1) is first integrated over all solid angles. Then, using

$$v^2\,dv = \frac{1}{2}\left(\frac{2}{m}\right)^{3/2} E^{1/2}\,dE$$

one finds

$$p(E)\,dE = 2\pi\left(\frac{1}{\pi kT}\right)^{3/2} e^{-E/kT} E^{1/2}\,dE \tag{5.4}$$

One can find directly from equation (5.2) or by differentiating equation (5.4) that the most probable energy is given by

$$E_p = kT \tag{5.5}$$

Equation (5.5) is analogous to the Wien displacement law or, more specifically, equation (2.7), which is valid for photons. The main difference between the Planckian and Maxwell-Boltzmann distributions is that the density of photons in thermodynamic equilibrium is fixed by the temperature, but the density of particles is not. Thermodynamic equilibrium at a given temperature can occur for any material density.

A very important quantity is the average energy per particle. Since the average of any group of quantities is their sum weighted by their probability, it follows that

$$\bar{E} = \int_0^\infty Ep(E)\, dE$$

$$= 2\pi\left(\frac{1}{\pi kT}\right)^{3/2} \int_0^\infty e^{-E/kT} E^{3/2}\, dE$$

The result of the integration is

$$\bar{E} = \tfrac{3}{2}kT \tag{5.6}$$

The average energy is 50% larger than the most probable energy. Equation (5.6) can be compared with the average energy of a photon, which can be shown from the Planck function to be approximately $2.701kT$. (See problem 2 at the end of this chapter.)

The Perfect-Gas Equation of State. One of the more important properties of a gas is its pressure. The pressure at a point is the force which would be exerted on unit area of a material surface placed at that point, or equivalently, it is the rate with which momentum is passed through a unit geometric surface at that point. If the Maxwell-Boltzmann velocity distribution is used to calculate the pressure, the result is

$$P = NkT \tag{5.7}$$

This important relation gives the pressure of a simple Maxwell-Boltzmann gas in terms of its temperature and the number of gas particles per unit volume. It is known as the equation of state of a perfect or ideal gas, and it can be expressed in several other equivalent forms. For example, if ρ is the average mass per unit volume, then $\rho = Nm$, and one has

$$P = \frac{k}{m}\rho T \tag{5.8}$$

The molecular weight μ of the gas is the dimensionless ratio m/m_0, where m_0 is the mass of unit atomic weight. Formerly, m_0 was defined as $\frac{1}{16}$ of the mass of the normal oxygen nucleus, but physicists have since decided that they prefer it to be $\frac{1}{12}$ of the mass of the normal carbon nucleus. In either case, it is very close to the mass of the hydrogen atom. The gas constant R is defined as the ratio k/m_0 ($R = 8.31 \times 10^7$ erg g^{-1} °K^{-1}), so

$$P = \frac{k}{\mu m_0}\rho T = \frac{R}{\mu}\rho T \tag{5.9}$$

If there are n gram molecules or moles of the gas, $n = \mathcal{M}/\mu = \rho V/\mu$, where $\mathcal{M}$ and V are the total mass and volume of the gas; therefore,

$$PV = nRT \tag{5.10}$$

Equations (5.7)–(5.10) are all equivalent, and which form is to be used is simply a matter of convenience. It should be pointed out that physicists generally give molecular weight the dimensions of grams per mole, so the gas constant comes out as erg per (mole-degree), but this is only another way of stating the same thing as above. Numerical values are not affected by this.

It has been assumed that there is only one kind of particle in the gas. If the gas is a mixture of different kinds of particles, equation (5.1) can be applied separately to each kind as long as the assumptions about no forces, elastic collisions, and so on are maintained. Since this relation depends on the particle mass, each kind will have its own velocity distribution; equation (5.4) does not depend on mass, however, so all particles will have the same energy distribution. This is not surprising, since a gas in thermodynamic equilibrium should be in energy equilibrium with its radiation field, and it is mentioned in Section 2 that the latter does not depend on the kind of material it is in equilibrium with. As (5.6) shows, all particles move—on the average—with the same kinetic energy, so the light ones have greater speeds than the heavy ones. If N_i is the number of particles of type i per unit volume, then they will produce a partial pressure given by

$$P_i = N_i kT \tag{5.11}$$

and the total pressure is simply the sum of all of the partial pressures:

$$P = \sum_i P_i = \sum_i N_i kT = NkT \tag{5.12}$$

The other equations are still valid if m is defined as the average mass per particle and if μ is the average molecular weight per free particle.

Consider the example of pure helium gas. At low temperatures the gas is neutral, and it has a molecular weight of about 4. If the temperature is raised somewhat, some of the helium atoms will become once-ionized, and each will free an electron. This produces a mixture of He atoms, He^+ ions, and free electrons. There are now more particles per original atom, but there is no more mass than before; therefore, the mean molecular weight is somewhat less than 4. At extremely high temperatures, the helium is almost all twice-ionized, and there are now twice as many free electrons as there are helium nuclei. The molecular weight has decreased to about $\frac{4}{3}$, and $\frac{2}{3}$ of the total pressure is supplied by the electrons, only $\frac{1}{3}$ by the nuclei. Collisions between the different kinds of particles cause each kind to have the same average kinetic energy, so the electrons have a much larger mean speed than the helium nuclei. It is seen that, in general, the mean molecular weight of a gas cannot be considered a constant.

Another important property of a gas is its thermal or internal energy. This consists of the kinetic energies of all the particles plus any potential energy involved in binding or holding together an individual particle. This latter includes excitation and ionization energies and, if the particles are molecules, the binding energies of the atoms in the molecules. The nuclear binding energy is usually considered as separate, but this is a matter of personal preference. It is noted in Section 3 that potential-energy differences are the quantities of importance, not the energy values themselves. This means that in any region in which the various potential energies are all essentially constant, they can be conveniently set to zero and ignored, and the thermal energy is just the kinetic energy of the particle motions. If U is the thermal energy per gram of material, then equations (5.6) and (5.9) show that

$$U = \frac{3}{2}\frac{kT}{m} = \frac{3}{2}\frac{RT}{\mu} = \frac{3}{2}\frac{P}{\rho} \tag{5.13}$$

If the potential energies are not all constant, however, then their effects must be included. For example, if an abundant substance such as hydrogen or helium is partially ionized, then the ionization energy must also be included. Normally a change in thermal energy will cause a certain change in temperature, as indicated by (5.13), but if ionization is also varying, this will absorb a good part of the energy change. As a result, the temperature is much less sensitive to the thermal energy than (5.13) indicates, and the specific heats are much larger than normal. Such ionization regions are quite important in the structure of stars. It should be noted that this effect is in addition to the effect of ionization causing the molecular weight μ to be variable.

Interparticle Forces. The properties of a Maxwell-Boltzmann gas are determined under the assumptions that particles do not exert forces upon each other except for the very short intervals during collisions and that the gas is in thermodynamic equilibrium. Since these assumptions are never rigorously correct, it is important to estimate the effects of the deviations from them which are found in different astronomical situations. The effects of interparticle forces are examined first.

If two particles pass close to each other, the forces that they exert on each other will modify somewhat their velocities. If this effect is strong enough, it will produce deviations from the Maxwell-Boltzmann distribution. The importance of these interactions can be estimated by comparing typical values of the interaction energy and kinetic energy. If the absolute value of the potential energy of interaction is very much less than the mean kinetic energy of the particles, then the motions will not be appreciably affected. This potential energy, denoted by PE, depends on the separation r of the particles.

If $\bar{r}$ is the mean separation, then $PE(\bar{r})$ is a typical value to be compared with $(\frac{3}{2})kT$. It appears that large deviations from the Maxwell-Boltzmann distribution will not occur if

$$\tfrac{3}{2}kT \gg |PE(\bar{r})| \tag{5.14}$$

For any given situation these energies can be calculated, and one can see whether the above inequality is satisfied. The magnitude of the potential energy increases as the particles get closer together, so high densities tend to cause a violation of (5.14). On the other hand, high temperatures tend to favor (5.14) because they correspond to large kinetic energies.

The potential energy can be easily calculated for a pure hydrogen gas. The situation for neutral hydrogen is quite different from that for ionized hydrogen. For ionized hydrogen, the particles are protons and electrons, each having a charge of magnitude $e = 4.80 \times 10^{-10}$ esu. The coulomb interaction gives $|PE| = e^2/r$, as indicated by equation (3.6). A neutral hydrogen atom has a proton and an electron separated by a distance d, and two such atoms will affect each other slightly because of the different distances of the different charges. In this case the potential energy is about d^2e^2/r^3 (see problem 8 at the end of this chapter), the exact value depending on the relative orientations. The mean separation $\bar{r}$ is related to the mean particle density N by

$$\tfrac{4}{3}\pi\bar{r}^3N = 1 \tag{5.15}$$

so if this is used to eliminate $\bar{r}$, equation (5.14) becomes

$$\tfrac{3}{2}kT \gg \begin{cases} e^2(\frac{4}{3}\pi N)^{1/3} & \text{(ionized)} \\ d^2e^2(\frac{4}{3}\pi N) & \text{(neutral)} \end{cases} \tag{5.16}$$

If d is identified with the first Bohr orbit ($d = 5.29 \times 10^{-9}$ cm), then numerical values in (5.16) give

$$T \gg \begin{cases} 2 \times 10^{-3}N^{1/3} & \text{(ionized)} \\ 10^{-19}N & \text{(neutral)} \end{cases} \tag{5.17}$$

As long as inequality (5.17) is strongly satisfied, one can be confident that interparticle forces do not cause very appreciable deviations from the Maxwell-Boltzmann distribution. In interstellar space the particle densities are very low, N rarely being as large as 10^6 particles per cm^3. Temperatures are generally about 10^4°K in regions where hydrogen is ionized and at least a few degrees elsewhere; consequently, (5.17) is strongly satisfied in interstellar space. In the atmosphere of the Sun, typical values are $T = 6000$°K, $N = 10^{17}$ cm^{-3}, with the hydrogen neutral; again, (5.17) is strongly satisfied. According to R. L. Sears (1964), the center of the Sun has the approximate values $N = 9.5 \times 10^{25}$, $T = 1.6 \times 10^7$, and the hydrogen is ionized. These values indicate that the left side of (5.17) is somewhat less than 20 times the right side. This is getting rather marginal in importance, and deviations from the Maxwell-Boltzmann distribution of a few per cent can be expected from this effect in the central regions of stars. The van der Waals equation of state is an example of a relation that is sometimes used by physicists when the perfect gas laws are not sufficiently accurate; however, this is used very little if at all by astronomers because most deviations from perfect gas laws of astronomical importance are due to the effect to be considered next.

The Fermi-Dirac Distribution. There is another type of particle interaction which can cause deviations from the Maxwell-Boltzmann equations. This is a quantum mechanical effect known as degeneracy. According to quantum mechanics a material particle is a sort of localized wave; i.e., it consists of a wave which is appreciably nonzero over some finite region. If this region of the wave is extremely small compared with other distances of interest, then the wave nature can be ignored; otherwise, the wave nature is very important in determining the properties of the particle, and quantum mechanics must be used for calculating these properties. The extent of this region of the wave represents a basic uncertainty in the position of the particle.

The uncertainty principle of W. Heisenberg states that the uncertainty in the position of a particle is related to that of its momentum. If the x component of the momentum of a particle is uncertain by the amount Δp_x, then the x coordinate itself must be uncertain by an amount Δx which satisfies

$$\Delta x \gtrsim \frac{h}{2\pi\Delta p_x} \tag{5.18}$$

Similar relations hold for the y and z components.

For a classical particle it is apparent that

$$\overline{p_x^2} = \tfrac{1}{3}\overline{p^2} = \tfrac{2}{3}m\bar{E} = mkT \tag{5.19}$$

Any particle will have a momentum at any time that is approximately equal to its average value, usually within a factor of two or three, and so the uncertainty in momentum is of the order of its average value. Thus $\Delta p_x \approx (mkT)^{1/2}$, and (5.18) gives

$$\Delta x \gtrsim \frac{h}{2\pi(mkT)^{1/2}} \tag{5.20}$$

In order for quantum mechanical effects to be negligible, Δx must be very small compared with other distances of interest, in particular, the mean distance between particles $\bar{r}$. Equations (5.15) and (5.20) indicate that quantum mechanical effects can be ignored if

$$T \gg \frac{h^2}{k(36\pi^4)^{1/3}} \frac{N^{2/3}}{m} \tag{5.21}$$

The numerical factor on the right side of (5.21) is 2.1×10^{-38} in cgs units, but a more accurate calculation shows that these quantum effects are negligible if the "much greater than" is replaced by "greater than about" and if the numerical factor is multiplied by about 10:

$$T \gtrsim 2.5 \times 10^{-37} \frac{N^{2/3}}{m} \tag{5.22}$$

It is interesting that the above relation depends on the particle mass. It is possible for conditions to be such that the relatively heavy protons satisfy (5.22), while the much lighter electrons do not. In fact the conditions given for the center of the Sun are such that (5.22) is mildly violated for electrons. Under these conditions the electrons form what is called a partially degenerate gas. In a certain type of star known as a white dwarf, (5.22) is strongly violated and the electrons form a highly degenerate gas.

When (5.22) is violated, the gas particles cease to follow the Maxwell-Boltzmann distribution (5.1), and instead they obey what is known as the Fermi-Dirac distribution:

$$p(v, \theta, \phi)\, dv\, d\omega = \frac{2m^3}{Nh^3} \frac{v^2\, dv\, d\omega}{\exp\left(\alpha + \dfrac{mv^2}{2kT}\right) + 1} \tag{5.23}$$

In this equation, α is a quantity that depends upon N and T and is such that the integral of (5.23) over all speeds and directions is unity. It can be seen that this reduces to (5.1) if $\alpha \gg 1$.

A gas that obeys (5.23) has different properties from a Maxwell-Boltzmann gas, and in particular, it does not satisfy the perfect gas equation of state. Equation (5.23) in fact leads to the equation of state

$$P = NkT \times \frac{2}{3} \frac{F_{3/2}(\alpha)}{F_{1/2}(\alpha)} \tag{5.24}$$

in which

$$F_n(\alpha) = \int_0^\infty \frac{x^n\, dx}{e^{\alpha+x} + 1} \tag{5.25}$$

The properties of a degenerate or Fermi-Dirac gas are best understood in terms of the energy levels allowed by quantum mechanics. It is noted in Section 3 that an electron which is bound to an atom cannot have arbitrary values of the energy, but must be in one of the permitted quantum mechanical states. The same is true of free particles. No two electrons in the same atom can be in the same state, and likewise no two identical free particles per unit volume can be in the same quantum mechanical state. This statement is known as the Pauli exclusion principle.

At low densities for a given temperature, the number of particles per unit volume is quite small compared to the number of low energy states available to them; as a result the quantum mechanical restrictions do not appreciably affect the gas, and it follows the classical Maxwell-Boltzmann distribution. At higher densities for the same temperature, however, the lower energy levels become nearly filled up. As more particles are squeezed into the same volume, they must take on energies that are much higher than average, since the Pauli principle will not allow them to share the lower levels with the particles already in them. A degenerate gas has a greater energy per particle than that indicated by equation (5.6). When the density is so great that all levels below a certain high energy level are completely filled, then the properties of the gas become dependent only on the density and the positions of the energy levels. The material is completely degenerate, and its characteristics no longer depend upon the temperature. By raising the temperature enough, however, the degeneracy can be removed. Because of the large energies of degenerate gases, the velocities often become quite large, and corrections for special relativity must be made. Thus equations (5.23) and (5.24) are only valid for a limited density range, and if the density is too large, new equations valid for relativistic degeneracy must be used.

There are actually two different quantum mechanical distributions which gases can obey: the Fermi-Dirac distribution described above and the Bose-Einstein distribution. Black-body radiation described in Section 2 follows a special form of the Bose-Einstein distribution. Both Fermi-Dirac and Bose-Einstein distributions tend to the classical Maxwell-Boltzmann distribution in the limit as the quantum effects become ignorable.

Deviations from Thermodynamic Equilibrium. All of the equations of this section have been considered under the assumption of thermodynamic equilibrium; however, one is often interested in situations which are very far from TE, and the validity of these relations would then be in some doubt. The velocity distribution of gas particles is fixed in a complicated way by the various radiative and collisional processes which occur. The elastic collisions tend to set up the same kind of distribution as in TE, but the inelastic ones, in which there is an energy exchange between kinetic and excitation, ionization, or some other form, do not. The reason is that inelastic collisions usually tend to take, preferentially, energy from particles whose energy is above some threshold. Also, radiative ionizations and recombinations tend to upset the TE distribution; however, under most conditions of astronomical interest elastic collisions are far more numerous than both inelastic ones and radiative transitions, and the result is that the velocity distribution and anything that depends on it are usually the same as in TE at some particular temperature. This temperature is known as the kinetic temperature (often it is called the electron temperature). The point is that when TE does not prevail, temperature is not a uniquely defined quantity. The temperature of black-body radiation which has the same relative energy distribution as occurs in a region is known as the color or radiation temperature of that region. The TE temperature for which two given excitation or ionization levels have the same relative populations as in an actual region is known as the excitation or ionization temperature of that region. In TE all of these different definitions of temperature are equivalent, but in a non-TE situation, these temperatures can be very different from each other.

To summarize the effects of deviations from TE, it can and often does happen that the characteristics of the radiation field and the populations of the excitation and ionization levels are very different from the same quantities in TE at any temperature; however, the strong effects of the elastic collisions usually keep the velocity distribution of free particles very close to TE values at some temperature, that temperature being defined as the kinetic temperature.

6. Properties of the Nucleus

The purpose of this section is to provide a brief background for the understanding of nuclear reactions in stars. No attempt is made to cover the complete field of nuclear physics.

Potential energy is discussed many times in this text, and it is also quite basic in the present subject. The nucleus of an atom may consist of a number of particles which exert forces on each other, and this gives rise to a potential energy. Under certain circumstances this potential energy can be changed, and this results in the absorption or emission of nuclear energy.

Nuclei are composed of protons and neutrons. The proton has the same charge as the electron, except that it is opposite in sign, while the neutron is electrically neutral. Both the proton and the neutron have a mass of approximately 1 amu (atomic mass unit) each, which means that each is about $\frac{1}{12}$ as massive as the carbon nucleus which has 6 protons and 6 neutrons. The number of protons in a nucleus is indicated by Z and is called the atomic number. The total number of protons plus neutrons is the mass number A. A chemical element is defined by its atomic number. Thus hydrogen has $Z = 1$, helium has $Z = 2$, lithium has $Z = 3$, and so on. There are usually approximately the same number of protons and neutrons in a nucleus, but there is some variation on this. Nuclei having the same number of protons but different numbers of neutrons are known as isotopes. Natural hydrogen has two isotopes of mass numbers $A = 1$ and 2, and an unstable hydrogen isotope of mass number 3 has been produced artificially. Isotopes are usually indicated by their chemical symbol with the mass number as a superscript. Thus H^1 is the most common form of hydrogen, and H^2 (deuterium) and H^3 (tritium) are the other hydrogen isotopes.

The forces which hold nuclei together are obviously not electrostatic, for the neutron is neutral and the protons should repel each other because of their like charges. Experiments show that the nuclear forces are extremely strong when the particles are no more than a few times 10^{-13} cm apart, but they fall off very rapidly for larger distances. The electrostatic forces cause two nuclei to repel each other if they are far enough apart; however, if they can be forced close enough together so that the nuclear forces between them become important, i.e., if they can overcome their coulomb barrier, then they will attract each other.

Since the potential energy gets smaller as two particles approach each other under an attractive force, nuclei of mass number $A \geqslant 2$ have negative potential energy compared with the free protons and neutrons. In other words, if it is possible to put Z protons and $(A - Z)$ neutrons together to form a nucleus, there must be some energy released in the process. This is a necessary but not sufficient condition for the nucleus to be stable. When nuclei come together to form one or more new nuclei, the process is known as a nuclear reaction.

Nuclear Reactions. Consider two nuclei having atomic and mass numbers (Z_1,A_1), and (Z_2,A_2), respectively. If they collide with enough relative kinetic energy to overcome the coulomb barrier, then there is a chance that they will combine to produce a new nucleus having $(Z_1 + Z_2, A_1 + A_2)$. (This can happen by means of the so-called tunnel effect even if the coulomb barrier is not completely overcome, but the collision still must be energetic enough for the particles to get very close together.) The lost potential energy plus any excess kinetic energy must somehow be released. It may go to the relative kinetic energy of the same two types of particles that had the collision in the first place, in which case the net effect is simply an elastic scattering of the two particles. Otherwise, part of this energy may be emitted in the form of high-energy photons known as γ rays, and the new nucleus may remain intact for at least a little while. Also, part of this energy may be given to some other particle (Z_3,A_3) which is then ejected from the nucleus, leaving behind the nucleus $(Z_1 + Z_2 - Z_3, A_1 + A_2 - A_3)$. The probabilities of these different types of reaction depend upon the Z and A values that are involved. Any new nucleus formed in such a nuclear reaction may or may not be stable; if not, it is said to be radioactive. A radioactive nucleus will decay to another nucleus by emitting some particle or particles after a time which is approximately equal to what is known as its half-life. The half-lives of radioactive nuclei vary from extremely small fractions of a second to an extremely large number of years. A stable nucleus has an infinite half-life. As stated above, the stable nuclei tend to have about equal numbers of protons and neutrons, but for heavy nuclei the large electrostatic charge causes the stable isotopes to have an excess of neutrons over protons.

Consider the example of two colliding protons. If they can overcome their coulomb barrier they should form the nucleus $Z = 2$, $A = 2$, which is the helium isotope He^2. This isotope is strongly unstable, however, so usually the two protons simply scatter off with no net nuclear reaction. Very rarely such a collision will result in the emission of two different types of particles, a positron and a neutrino, leaving an H^2 nucleus behind. The positron ($Z = 1$, $A = 0$) is identical to the electron except it has positive charge, and the neutrino ($Z = A = 0$) has neither charge nor mass. If the positron is indicated by the symbol β^+ (electrons and positrons are often called β particles), then this nuclear reaction can be represented by the relation

$$H^1 + H^1 \rightarrow H^2 + \beta^+ + \text{neutrino} \tag{6.1}$$

This reaction illustrates that the elementary particles are not necessarily permanent. The proton is an elementary particle in the sense that it cannot be considered as being made out of yet more elementary particles; however, there are circumstances in which it can change into other types of particles, as in (6.1), in which one of the protons changes into a neutron plus a positron.

The reaction (6.1) represents conservation of charge, since there is a net charge of $Z = 2$ on each side. It also has, at least approximately, conservation of mass, since the total mass number $A = 2$ on each side. It is not explicitly indicated in (6.1), but one might expect any such reaction to conserve energy, momentum, and angular momentum (or spin) also. It is because of this expectation that the existence of the neutrino was discovered long before it was directly detected experimentally. Reactions involving the emission of β particles did not experimentally conserve energy or spin, so the neutrino was hypothesized in order to maintain the assumption that these quantities are conserved. The neutrino has very little interaction with matter, so it is very difficult to detect.

Mass-Energy Equivalence. Consider the reaction

$$C^{13} + H^1 \rightarrow N^{14} + \gamma \tag{6.2}$$

where γ represents a γ ray photon. It is apparent that charge and mass number are conserved. According to previous arguments there must be a decrease of potential energy as the C^{13} and H^1 combine to make N^{14}, and this plus the excess kinetic energy should equal the energy of the γ ray. This is similar to a free-bound recombination of an ion and electron which results in a neutral atom plus a photon. The energy of the γ ray in (6.2) is the excess binding energy the N^{14} nucleus has over the C^{13} nucleus. One could check this, at least in principle, by hitting N^{14} nuclei with particles which cause the nuclei to break up into C^{13} plus H^1. The minimum bombardment energy for which this is possible is this excess binding energy. In this way the binding energies of nuclei can be found even though the precise nature of the nuclear forces is not known.

It is experimentally verified that reactions similar to (6.2) do conserve momentum, spin, and charge; however, they do not conserve mass or energy. Experiments show that the N^{14} nucleus has slightly less mass than the sum of the masses of C^{13} and H^1, and the difference is found to be equal to the energy of the γ ray divided by the square of the speed of light. In general, if m_p and m_n are the masses of proton and neutron, and if $m(Z,A)$ and $E(Z,A)$ are the mass and binding energy of the nucleus with atomic number Z and mass number A, then

$$m(Z,A) = Zm_p + (A - Z)m_n + E(Z,A)/c^2 \tag{6.3}$$

It should be noted that $E(Z,A)$ is a negative quantity, so the mass of any nucleus is always less than the sum of the masses of the appropriate number of nucleons (protons and neutrons). This is an example of the mass-energy equivalence predicted by Einstein in his special relativity.

The above suggests that whenever potential energy is stored in a system, the mass of the system increases by the amount

$$\text{mass} = \frac{\text{energy}}{c^2} \tag{6.4}$$

This is not limited to nuclear phenomena. The hydrogen atom in the ground state should be less massive than one in an excited state, and both should be less massive than a free proton plus electron. A mouse trap weighs more when set than when sprung, and a person is more massive at the top of a flight of stairs than at the bottom. The difference is that nuclear forces are strong enough to cause very large potential energy changes, large enough to change the masses by a measurable fraction of the masses involved. There is a tendency to interpret (6.2) as an example of converting mass to energy, while the recombination of a proton and an electron is generally regarded as an example of the release of potential energy. Either interpretation is equally valid for both examples.

There is another reaction which may appear to be more directly the conversion of mass into energy. Positrons do not last very long, since they soon find electrons and the particles annihilate each other with the emission of γ rays. Electrons and positrons have measurable masses, and the energy of the γ rays satisfies equation (6.4). This again suggests that mass can be considered as a form of potential energy and vice versa.

As is noted in Section 21, reactions (6.1) and (6.2) are both important in stars. The amount of energy they release can easily be found from the atomic masses. (See Appendix C.) The H^2 atom (including one electron) is 2.0141 amu, while 2 protons plus two electrons total 2.0156 amu. The difference is

0.0015 amu, and this must be the mass equivalent carried off by the positron, neutrino, and the extra electron in reaction (6.1). This extra electron and the positron then combine to annihilate each other as described above, and the net effect is the production of a neutrino plus γ rays. Since the amu is about 1.66×10^{-24} g, the energy released is $0.0015 \times 1.66 \times 10^{-24} \times (3 \times 10^{10})^2 = 2.2 \times 10^{-6}$ erg. In (6.2) the C^{13} plus H^1 atoms exceed the N^{14} atom by 0.0082 amu, so the energy released in this reaction is 1.2×10^{-5} erg. Since 1.6×10^{-6} erg $= 1$ MeV (million electron volts), the energy released in these examples is a few MeV. These energies are more or less typical.

The temperatures and densities in the centers of stars are quite high, so numerous high-energy collisions between the particles take place. Some of these result in nuclear reactions similar to those discussed here, and nuclear energy is thereby released. This energy heats the material and provides the balance for the energy radiated away into space. This balance is discussed in some detail in Section 21.

Reaction Rates in the Sun. It is common to grossly overestimate the rate at which nuclear reactions take place in the center of a star like the Sun. Compared with modern high-energy particle accelerators, conditions in the center of the Sun are extremely mild, and nuclear reactions take place very slowly. One way to illustrate this is to consider the reaction rate needed to produce the solar luminosity. The Sun emits about 4×10^{33} erg/sec, and since typical nuclear reactions in the Sun release a few MeV = a few $\times 10^{-6}$ erg, the number of nuclear reactions taking place per second must be of the order of 10^{39}. While this is a very large number, there are some 10^{57} nuclei in the Sun, of which some 10 per cent or so are in the center where the reactions take place. Thus only about one nucleus in 10^{17} in the center of the Sun is expected to undergo a nuclear reaction each second. In other words, an average nucleus can expect to wait for about 10^{17} sec $= 3 \times 10^9$ yr before having a reaction. Although this number is not very accurately established, it is one way of estimating the maximum age of the Sun.

It is instructive to compare the energy requirements of a nuclear reaction with typical particle energies. It is stated above that nuclear forces are important only if the particles are closer together than a few times 10^{-13} cm. In other words, two nuclei must come closer together than about 10^{-12} cm in order for a reaction to have any chance of occurring. Two nuclei of charges Z_1 and Z_2 have potential energy due to their electric charges of $Z_1 Z_2 e^2/r$ if r is their separation. For $r = 10^{-12}$ cm, PE $= 2.3 \times 10^{-8} Z_1 Z_2$ erg. If the particles are to get this close together, they must have a relative kinetic energy this great. Since all particles have the same kinetic energy, on the average, it is seen that there is a discrimination against reactions involving particles with large charges. If the center of the Sun has a temperature of 1.6×10^7 °K, then the mean kinetic energy per particle is $3kT/2 = 3.3 \times 10^{-9}$ erg. Even for two protons with $Z_1 = Z_2 = 1$, the energy needed to overcome the coulomb barrier is nearly 10 times the mean particle energy. Only an extremely small percentage of the particles have energies this much greater than average, which is another way of saying that the reaction rate is very slow.

PROBLEMS

1. A spherical star of radius R emits radiation of uniform intensity I in all directions. Determine the mean intensity and the flux of the radiation at a distance r from the star.

2. Determine the average energy of photons in thermodynamic equilibrium at temperature T.

3. Find the fraction of the total energy of a black body which is radiated at wavelengths longer than a given wavelength. Assume that the temperature is low enough that the Rayleigh-Jeans approximation is valid at the given wavelength.

4. The Pickering series of once-ionized helium is a series of spectral lines corresponding to transitions between the level $n = 4$ and higher levels. Show that every other line of this series has the same wavelength as a line of the Balmer series of hydrogen.

5. The lines of the Balmer series crowd closer together as higher series members are considered. If each line were exactly one Å wide, how many Balmer lines would be individually visible without overlapping other lines?

6. The temperature of a gas is changed by a small amount. Derive an expression for the amount by which the electron pressure must be changed in order that the relative abundances of the first two ionization stages of a given element be unchanged.

7. A gas is composed of 50% hydrogen and 50% helium by mass. If it is in thermodynamic equilibrium at 10^4 °K and if the density is 10^{-7} g cm^{-3}, determine the pressure, the electron pressure, and the mean molecular weight of the mixture.

8. An electric dipole consists of two opposite electric charges, $+e$ and $-e$, separated by the distance d. Determine the potential energy change needed to bring two such dipoles within a distance r of each other. Assume that all four charges are in a straight line, and assume $r \gg d$.

9. A 200-lb man climbs a mountain 10^4 ft high. By how much does his mass change as a result of his increase in potential energy?

REFERENCES

Most of the material in this chapter can be found in any of a large number of elementary and intermediate-level physics text books. The physical conditions at the center of the Sun were taken from the following:

1. Sears, R. L. *Astrophys. J.*, **140,** 477, 1964.

I

Positions and Magnitudes of the Stars

The measurements of stellar distances, brightnesses, and motions are described in this chapter. There is nothing difficult about these subjects, but the units and quantities defined by astronomers may take some getting used to. The most fundamental method of measuring stellar distances is the subject of Section 7. Section 8 relates the flux received at the Earth from a star to the relevant intrinsic properties of the star, and Section 9 describes the magnitude systems used by astronomers. The effects of the Earth's atmosphere on apparent brightnesses of stars are discussed in Section 10. Section 11 defines the basic parameters of stellar motions and describes some of the ways in which these motions can provide further useful information.

7. Trigonometric Parallaxes

About the only direct way of obtaining the distance of a star is the trigonometric parallax method. This is illustrated in Figure 7.1, where S is the position of a star, and O represents the Sun. As the Earth moves around the Sun, there are two times about six months apart when the line joining Earth and Sun is perpendicular to the line joining Sun and star. In the figure, E and E' are such positions. Note that the assumption is not made that the star is in the plane of the Earth's orbit. If the star is not too far away, angle ESE' can be found by measuring the position of S with respect to the much more distant background stars, and this leads directly to the distance of the star at S.

As the Earth moves around the Sun, the apparent direction to the star changes. The magnitude of this change, reduced to what it would be if the observer moved from Sun to Earth, defines the parallactic angle or parallax of the star. The parallax is indicated by p in Figure 7.1. In practice, a large number of photographs of the star are taken over a period of years. For maximum accuracy, the photographs should be taken when the Earth is as near as possible to positions E or E'. On each photograph the position of the

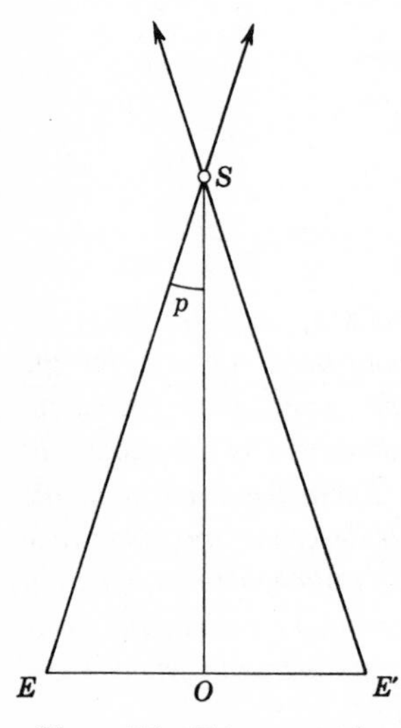

Figure 7.1 Trigonometric parallax.

star is measured with respect to the positions of many other stars in the same picture, and all of these measures are combined to produce the best value of the parallax which will represent the data. In this way measuring errors are reduced considerably over what they would be if only a few pictures were taken. A small correction for the fact that the reference stars are not at an infinite distance can be made statistically.

In Figure 7.1, *OE* is about one astronomical unit (AU), the mean distance between the Earth and the Sun, and *OS* is the distance r of the star. It follows that $\tan p = 1/r$, if r is measured in AU. In practice p is such a small angle that if it is measured in radians, it is equal to its tangent. It is more common to measure p in seconds of arc ($''$), so

$$r\ (\text{AU}) = 2.063 \times 10^5/p'' \qquad (7.1)$$

The parsec (pc) is defined as the distance of a star whose parallax is $1''$, so

$$r\ (\text{pc}) = 1/p'' \qquad (7.2)$$

A parsec is obviously 2.063×10^5 AU, and one can easily convert to other distance units:

$$1\ \text{pc} = 2.063 \times 10^5\ \text{AU} = 3.086 \times 10^{18}\ \text{cm} = 3.262\ \text{ly} \qquad (7.3)$$

A light year (ly), the distance light travels in a vacuum in one year, is sometimes used for expressing star distances, but the parsec is much more frequently used by astronomers.

The nearest stars to the Sun are those in the triple system of Alpha Centauri. The main star in this system is, by coincidence, almost exactly like the Sun. It has a parallax of $0\rlap{.}''76$, and so its distance is 1.3 pc. Three-fourths of a second of arc is the apparent size of a penny seen $3\frac{1}{4}$ miles away, and all other stars have parallaxes which are smaller than this!

Using the trigonometric method, astronomers can measure stellar parallaxes with an error of about $0\rlap{.}''005$ (an amazing accuracy!), but very few stars are close enough for this to provide high accuracy in measuring their distance. For example, a star must be within 4 pc in order for the trig parallax to be known with better than a 2% accuracy. Only 33 stars satisfy this condition. There are only about 700 stars within 20 pc, the maximum distance such that the trig parallax is good to at least 10%. Most stars are so far away that there is no hope of obtaining a meaningful trig parallax. It should be kept in mind, therefore, that when one says that the distance of a star is known, one usually means that the accuracy is no better than 10–20% at best. Since most other methods of measuring stellar distances depend on trig parallaxes for their calibration, the importance of the long, tedious, and uninspiring work of measuring trig parallaxes cannot be overestimated.

8. Stellar Radiation

The radiation parameters of a star are defined in terms of a spherical star of radius R. The luminosity L is the total energy radiated per unit time in all directions and in all wavelengths, and L_λ or L_ν is the corresponding monochromatic quantity. The quantity $F(R)$ is the surface flux of the star and equals the luminosity divided by the surface area:

$$F(R) = \frac{L}{4\pi R^2} \tag{8.1}$$

As defined in Section 1, the flux is the integral over the outward hemisphere of the surface intensity times the cosine of the angle with the outward direction. In Section 2 it was noted that the integrated intensity of a black body is proportional to T^4. From equations (2.4) and (2.10), one has

$$F(R) = \sigma T^4 \tag{8.2}$$

if the star is a black body. The luminosity then follows from the above two equations:

$$L = 4\pi\sigma R^2 T^4 \tag{8.3}$$

The temperature of a black body is seen to be a measure of the energy it radiates per unit area per second. Now, a star is not a black body, but one can still define a quantity which is a similar measure of the energy radiated per area per second. This is the effective temperature of a star T_e. The effective temperature is defined so that equations (8.2) and (8.3) hold for the actual star:

$$F(R) = \sigma T_e^4 \tag{8.4}$$

$$L = 4\pi\sigma R^2 T_e^4 \tag{8.5}$$

The effective temperature of a star is the same as the temperature of a black body that emits energy at the same rate per unit area.

All stars except the Sun are so far away that it is not possible to separate the radiation which comes from different parts of the surface. The radiation which one can measure is an integration over the visible surface of the star of the light it gives off. In Figure 8.1, let A be the area of a telescope lens at Earth which is collecting and measuring the starlight. Since the telescope is

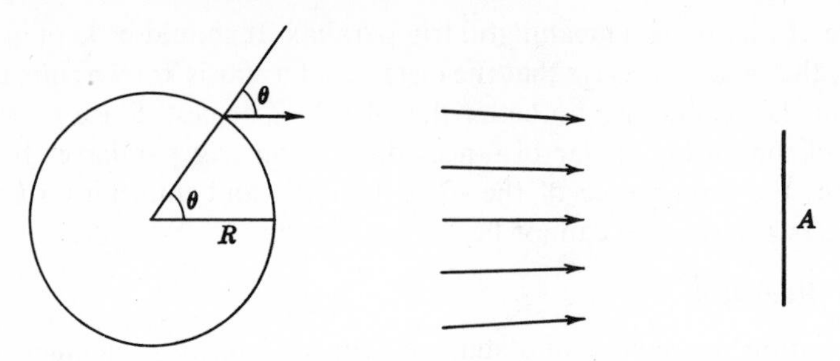

Figure 8.1 Measuring starlight.

pointing toward the star, the light falls on the lens essentially on a perpendicular. If E is the energy striking the lens per second, then

$$E = A \times \text{flux at Earth} = A \int I \cos \alpha \, d\omega$$

where I is the intensity of starlight at the Earth, and α is the angle between the direction of the light and the normal to the lens. As α is a very small angle, $\cos \alpha$ can be set equal to 1. The above integration is over the solid angle of the star, and the ratio E/A is the flux at the Earth $F(r)$. From the definition of solid angle given in Section 1, one can write

$$d\omega = \frac{d\Sigma \cos \theta}{r^2}$$

where $d\Sigma$ is an element of surface area of the star, and θ is the angle between the normal to $d\Sigma$ and the direction to Earth. In other words, $(d\Sigma \cos \theta)$ is the element of surface area of the star projected normal to the direction toward Earth. With ϕ defined as the azimuthal angle around the direction toward Earth,

$$d\Sigma = R^2 \sin \theta \, d\theta \, d\phi$$

and the expression for $F(r)$ becomes

$$F(r) = \frac{R^2}{r^2} \int_0^{2\pi} d\phi \int_0^{\pi/2} d\theta \, I \cos \theta \sin \theta \tag{8.6}$$

While this looks like the integrals involved in obtaining the surface flux, the appearances are misleading; the surface flux is the integral over outward directions, evaluated at a given point on the star, and the above is the integral over the star's surface. It is only when the star has spherical symmetry, so that $I(\theta)$ is the same function for all points on the star's surface, that the integrals in (8.6) are the same as the surface flux $F(R)$. If the star does have spherical symmetry, then (8.6) becomes

$$F(r) = \frac{R^2}{r^2} F(R) = \frac{R^2}{r^2} \sigma T_e^4 = \frac{L}{4\pi r^2} \tag{8.7}$$

What is actually measured is the flux at the Earth's distance, $F(r)$, in erg cm^{-2} sec^{-1}. If the distance of the star is known, the luminosity follows. If the effective temperature is also known, then the radius R can be found. There are many complications in the application of this procedure, as later sections indicate.

If a star does not have spherical symmetry, the above relations are not completely correct. A rapidly rotating star is flattened out of a spherical shape, and it does not radiate equally in all directions. In such cases the relations of this section are still correct on the average, but one can modify them in the rare instances in which there is enough information to do so.

9. Magnitude Systems

Apparent Magnitude. The apparent brightness of a star depends on its flux at the Earth's position; however, astronomers usually do not measure the apparent brightness of stars in flux units, but rather in magnitudes. There are many different kinds of magnitudes, and each kind or system is defined by the scale, the wavelength sensitivity of the measuring equipment, and the zero point.

The scale is the same for all magnitude systems. If m_1 and m_2 are the magnitudes of two stars in any system, and if F_1 and F_2 are the fluxes of the same stars at the Earth, then

$$m_2 - m_1 = 2.5 \log \frac{F_1}{F_2} \tag{9.1}$$

Magnitudes are logarithmic, and they are defined so that the brighter star has the smaller magnitude. A factor of 100 in the observed flux corresponds to a difference of five magnitudes.

Over 2000 years ago the Greek astronomer Hipparchus divided the naked-eye stars into six classes or magnitudes according to their apparent brightness. Stars belonging to the brightest group were known as first magnitude stars, and the faintest were of magnitude 6. When instruments became available for measuring apparent brightness more accurately, it was found that the system of Hipparchus is approximately logarithmic in the flux. In the 1850s, N. R. Pogson suggested that the system whose scale is defined by equation (9.1) be adopted, since a mathematically precise system was needed, and since (9.1) is in good accord with the Hipparchus system.

For any one star,

$$m = \text{constant} - 2.5 \log F \tag{9.2}$$

The constant appearing in equation (9.2) is the zero point, and it fixes the numerical value of the magnitude for a star of given apparent flux.

Both the constant and the type of flux being considered define the magnitude system. If the flux is monochromatic at some wavelength, then m is a monochromatic magnitude at that wavelength. If F is the integrated flux including the energy in all wavelengths, then m is known as a bolometric magnitude. It is not generally possible to measure directly either a monochromatic or a bolometric magnitude, although astronomers would certainly like to be able to do this. Instead they measure magnitudes which depend on the wavelength sensitivity of the equipment, including telescope, filter, photographic plate or photocell, and anything else in the optical system. Let $\varphi(\lambda)$ be the efficiency of the equipment for responding to radiation of wavelength λ, so it has a value between 0 and 1 for all λ. Then the flux actually measured is related to the flux incident on the telescope by

$$F\,(\text{measured}) = \int_0^\infty \varphi(\lambda) F_\lambda(r)\, d\lambda \tag{9.3}$$

The response function $\varphi(\lambda)$ does not include the effects of the Earth's atmosphere. This must be accounted for separately, as discussed in Section 10.

The response function and the constant of equation (9.2) define the magnitude system being used. A bolometric magnitude has $\varphi = 1$ for all wavelengths. If φ is zero everywhere except in a very narrow wavelength range, then the corresponding magnitude is nearly monochromatic. It is quite common to use the three kinds of magnitudes defined by H. L. Johnson and known collectively as the (U,B,V) system. The letters stand for ultraviolet, blue, and visual, and they have greatest sensitivity at wavelengths of about 3600, 4500, and 5500 Å, respectively. The half-widths are all a few hundred Angstroms. Thus $\varphi_V(\lambda)$ is a certain known function that has been tabulated by Johnson, and

$$F_V = \int_0^\infty \varphi_V(\lambda) F_\lambda(r)\, d\lambda \tag{9.4}$$

$$m_V = V = \text{constant}\,(V) - 2.5 \log F_V \tag{9.5}$$

Equations (9.4) and (9.5) are for the visual magnitude, and similar relations hold for the blue and ultraviolet. It is usual to denote these magnitudes by the letters U, B, and V, although m_U, m_B, and m_V are also used.

The equipment of an observer is such that response functions cannot be exactly duplicated. If one wishes to measure visual magnitudes, one should obtain a filter such that the equipment has a response function as close as is possible to that of the visual system. The measurements are then made in a system that is very close to, but not identical with, the visual system. The observer must also measure a number of stars whose magnitudes in the visual system are already known, and this allows a calibration to be made between the two systems. Thus any object which is measured can have its magnitude transferred to the visual system; this standardization is necessary in order for the results to be meaningful to other observers.

The above considerations point out the importance of establishing a large number of standard stars in all parts of the sky, stars whose magnitudes in the various standard magnitude systems are known. The observations are made relative to these standard stars. This also points out the method of fixing the zero point for a given system. Some star or group of stars is simply defined as having a certain magnitude on the system, and all other stars are then measured relative to it. The zero points are such that magnitudes do not differ very much from those set up by Hipparchus some 2000 years ago; i.e., the faintest stars visible to the naked eye are of about magnitude 6, and the brightest stars are somewhere around magnitude 1. Of course, each system has its own zero point set up in some unique way.

Using equations (8.7) and (9.2), one finds

$$\begin{aligned} m &= \text{constant} - 2.5 \log \left(\frac{L}{4\pi r^2}\right) \\ &= \text{constant} + 5 \log r - 2.5 \log L \end{aligned} \tag{9.6}$$

In the second relation of (9.6) the factor involving 4π in the logarithm has been absorbed into the constant. The magnitudes which have been discussed so far depend on the amount of flux reaching the Earth, and so are called apparent magnitudes. The flux reaching Earth depends both on the distance of the star and on the intrinsic brightness or luminosity of the star, and equation (9.6) shows how an apparent magnitude depends on them. This relation is valid in any magnitude system as long as L is defined in terms of the appropriate response function.

Absolute Magnitude. Astronomers often express the intrinsic brightness as well as the apparent brightness of stars in terms of magnitudes. Such a quantity is called an absolute magnitude, and it can be defined in any system. The absolute magnitude of a star in any system is numerically the same as what the apparent magnitude in the same system would be if the star were at a distance of 10 pc. Using M for absolute magnitude, one has, from equation (9.6),

$$M = \text{constant} + 5 - 2.5 \log L \tag{9.7}$$

where the constants in (9.7) and the second part of (9.6) are the same. Taking the difference between these, one finds

$$m - M = 5 \log r - 5 = 5 \log \left(\frac{r}{10}\right) \tag{9.8}$$

The distance r, as usual, is measured in parsecs. The quantity $(m - M)$ is a function of distance alone and is called the distance modulus. It is obvious that the distance modulus of a star is the same in all magnitude systems.

It should again be emphasized that it has been assumed in this section that there is no absorbing material between the star and the telescope. In practice this assumption is not valid, and suitable corrections must be made. The corrections made necessary by the Earth's atmosphere are considered in Section 10, while the effects produced by the tenuous material between the stars are examined in Section 23.

10. Atmospheric Extinction

It was pointed out in Section 9 that observations of stellar magnitudes which are made from the ground have to be corrected for the effects of the Earth's atmosphere. Even in the best of observing conditions, this correction is quite important. The observed magnitudes must be adjusted to the values they would have if the measurements were made from above the atmosphere.

In Figure 10.1, O is the position of an observer on the ground, S is in the direction toward a star, and Z is in the direction toward the observer's zenith. The angle between these directions is z, the zenith distance of the star. As the starlight passes through the Earth's atmosphere, some of it is absorbed and some scattered into other directions. These processes are described by the absorption coefficient σ, measured in cm^{-1}. The relative loss of flux of the starlight on traveling the distance ds is $\sigma\, ds$:

$$dF = -F\sigma\, ds \tag{10.1}$$

The minus sign indicates that flux is lost as the distance traversed increases, so as written above, ds is in the direction in which the light is traveling, from

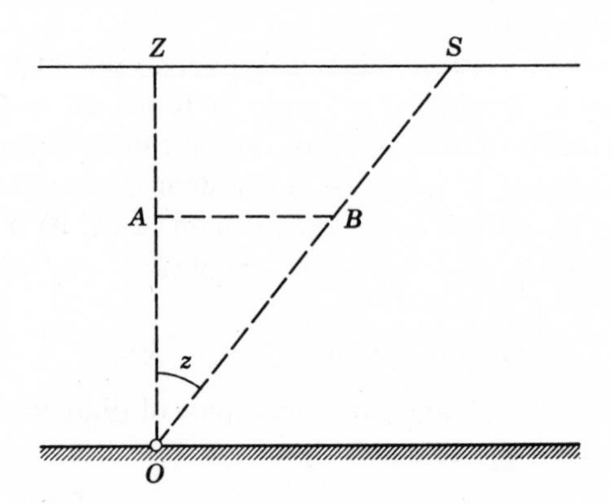

Figure 10.1 Atmospheric extinction.

S toward O. The absorption coefficient will, of course, depend on position. The lower atmosphere is more dense than that higher up, and so it will absorb and scatter more radiation per unit path length. If the point S is taken high enough that essentially all of the absorption takes place below it, then (10.1) can be integrated along the line SO:

$$F_o = F \exp\left(-\int_S^O \sigma \, ds\right) \tag{10.2}$$

where F_o is the flux actually observed on the ground, and F is that incident on the top of the atmosphere at S. The integration in equation (10.2) is along the path traveled by the light, and atmospheric refraction causes this path to be slightly curved rather than exactly straight.

The way in which σ depends on position must be known before the integral in (10.2) can be evaluated; however, this integral can be expressed in terms of a quantity τ_S, called the optical thickness of the atmosphere along the light path from S to O:

$$\tau_S = \int_S^O \sigma \, ds \tag{10.3}$$

With this definition, equation (10.2) becomes

$$F_o = Fe^{-\tau} \tag{10.4}$$

The significance of optical thickness can be seen from equation (10.4). When it is a very small number, there is very little absorption or scattering of the radiation, and the region over which it is defined is nearly transparent. When $\tau_S = 1$, the radiation getting through is $(1/e)$ of that which is incident, so the losses due to absorption and scattering are becoming appreciable. When the optical thickness is very large, the corresponding region is nearly opaque to the radiation. Optical thickness is a very important quantity, and it is referred to many times in future sections.

The observations are usually made in magnitudes, so equation (10.4) can be combined with equation (9.1) to yield

$$m_o = m + 1.086\tau_S \tag{10.5}$$

In the above, m_o and m are the magnitudes as seen from the ground and from above the atmosphere, respectively.

The observations give m_o, but m is the value that is desired. The difference $(m_o - m)$ is the atmospheric extinction in magnitudes, and this must somehow be determined if the observations are to yield m. Suppose that there are some standard stars for which the true magnitudes m are known. Then observations of m_o for these standard stars will immediately give the extinction $(m_o - m)$. Now the zenith distance z of any star is constantly changing, so the extinction for that star is also changing with time. If the standard stars are observed throughout the night, then the extinction can be determined as a function of zenith distance and then applied to the star under investigation. For greatest accuracy the program star should be quite close in the sky to at least one standard star, which means that a large number of good standard stars are needed.

How are the true magnitudes of the standard stars obtained in the first place? Suppose that the atmosphere were plane-parallel. This means that its curvature could be neglected and that its properties would depend on the distance above the ground but not on horizontal position. Thus the absorption coefficient σ would be the same for the two points A and B in Figure 10.1. If ds is again an element of distance along the light path between S and O, and if dx is the corresponding change in height above the ground, then it is apparent from the figure that $ds = \sec z\, dx$. With the plane-parallel assumption, equation (10.3) gives

$$\tau_S = \sec z \int_Z^O \sigma\, dx = \tau_o \sec z \tag{10.6}$$

In equation (10.6), the quantity τ_o is the optical thickness of the atmosphere in the direction of the zenith. Equations (10.5) and (10.6) then give

$$m_o = m + 1.086\tau_o \sec z \tag{10.7}$$

The plane-parallel approximation should actually be a very good one for small zenith distances, and this can be easily checked. A given star can be observed over several hours, giving m_o as a function of zenith distance z. If equation (10.7) is valid and if m is constant, then a plot of m_o versus $\sec z$ will be a straight line. In practice such a plot is a good straight line for small z. The slope of this line gives τ_o, and the intercept of it with the m_o axis (found by extrapolation) determines m. For large zenith distances the plane-parallel assumption overestimates the amount of atmosphere in the line of sight, so equation (10.7) will give too large a correction. Once m is known for a star, that star can be used as a standard, as explained above.

For accurate photometry, the Earth's atmosphere is sufficiently variable that the extinction should be measured independently each night, and there are often large variations within a single night. The amount of the extinction depends on the magnitude system, and so the standard stars must be observed in each magnitude system that is being used. Quite a few stars have variable m, and great care must be taken to insure that such stars are not used as standards. It should be emphasized that nothing has been assumed about the height of the atmosphere or about the vertical variation of σ. The absorption coefficient can vary with height in any manner, and no changes in this section will be necessary.

11. Stellar Motions

The Doppler Effect. Wavelength and frequency are not absolute properties of a photon, but depend on the motion of the observer. In other words, two observers having a relative motion will see different wavelengths and frequencies for the same photon. The effect of the state of motion on the observed wavelength or frequency is known as the doppler effect. The doppler effect is common to all waves, and a quantitative derivation for classical waves is given in any elementary physics textbook. The case of light is much more complicated, as an application of special relativity is required.

Let a source of radiation emit a photon whose wavelength is λ and whose frequency is ν for any observer at rest with respect to the source. An observer who is moving with velocity V with respect to the source will detect a wavelength and frequency of λ' and ν' given by

$$\frac{\lambda'}{\lambda} = \frac{\nu}{\nu'} = \frac{1 + V_r/c}{(1 - V^2/c^2)^{1/2}} \tag{11.1}$$

where V_r is the component of V along the line connecting source and observer. The quantity V_r is known as the radial velocity between the two, and it is taken as positive for recession and negative for approach. In most astronomical cases the velocities are far less than the speed of light, and the denominator of equation (11.1) is essentially unity. Then one has approximately

$$\frac{\lambda' - \lambda}{\lambda} = \frac{\nu - \nu'}{\nu} = \frac{V_r}{c} \tag{11.2}$$

and the wavelength and frequency shifts depend only on the radial velocity, not on the total velocity V.

Emission and absorption lines which have been identified in the spectra of sources make it possible to measure the radial velocities of the sources. For if a spectral line is identified as corresponding to a known transition, then its wavelength λ relative to the source is known. The measurement of the apparent wavelength λ' then determines the radial velocity according to equation (11.2).

Stellar Velocities. Stars move with respect to each other, and the measurement and analysis of these motions or an important part of stellar astronomy. It is usual to divide the motion of a star into a part along the line of sight and a part perpendicular to the line of sight. The latter is in the plane of the sky, and Figure 11.1 shows this division. In the figure, O is the position of the observer, S is the position of the star at some time, and S' is the position of the same star unit time later. The velocity of the star with respect to the observer is represented by the length SS'. The components of the velocity

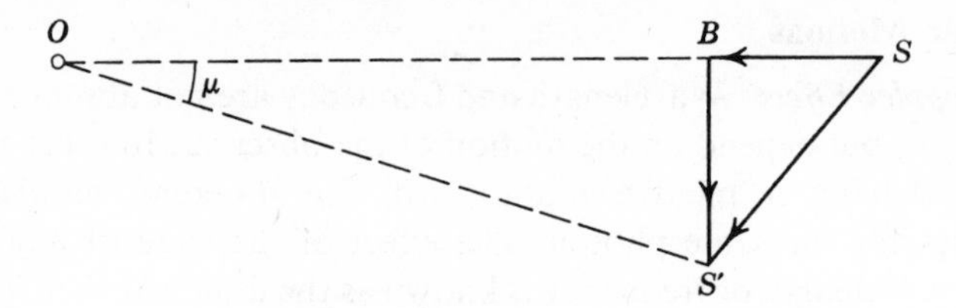

Figure 11.1 Stellar velocity.

along and normal to the line of sight OS are SB and BS'; SB is the radial velocity of the star, and it can be measured by means of the doppler effect. The BS' component is called the tangential velocity of the star, and it is not directly measurable; however, the angle BOS', designated by μ, is the angular velocity of the star and can be directly measured. The angular velocity is usually measured in seconds of arc per year and is known as the proper motion of the star.

The proper motion as observed from the Earth is a sinusoidal curve, being the superposition of the parallactic motion (Section 7) plus the angular velocity with respect to the Sun. Likewise, the observed radial velocity includes the reflected orbital motion of the Earth around the Sun. Published values of proper motion and radial velocity, however, are always reduced to what they would be if the measurements were made from the Sun. Unless stated otherwise, it will be assumed that these reductions have been made.

The velocity of a star with respect to the Sun is known as its space velocity. Let V, V_t, and V_r be the space, tangential, and radial velocities of a star. By convention, V and V_t are always positive, while V_r is positive for recession, negative for approach, as mentioned above. Obviously,

$$V^2 = V_t^2 + V_r^2 \tag{11.3}$$

If r is the distance of the star, it appears from Figure 11.1 that $V_t = \mu r$ if the appropriate units are used. With these quantities measured in their usual units, that is, with V_t in km/sec, μ in ″/yr, and r in pc, one has

$$\begin{aligned} V_t &= 4.74\mu r \qquad \text{km/sec} \\ \mu &= 0.211 V_t p \qquad ''/\text{yr} \end{aligned} \tag{11.4}$$

In the second relation above, p is the parallax in seconds of arc. Both V_r and μ can be directly measured, although it must not be assumed that these measures are trivial; however, the distance of the star must be known if the tangential velocity or the space velocity is to be found.

Solar Motion. The measured quantities described above are due to the relative motion of the Sun and a star. Is there some way the effects of the Sun can be eliminated so that the absolute motions of the stars can be determined?

The direct answer is no, since there is no such thing as absolute motion; however, the Sun certainly does have a special place in these measurements, and something can be done about this.

The stars in the vicinity of the Sun, within 100 pc or so, define what is called the solar neighborhood. The choice of this size is due to the characteristics of the Galaxy. The effects of differential galactic rotation become important for distances much greater than this, and it is desirable to exclude these effects at present. The average velocity of all the stars in the solar neighborhood defines what is known as the Local Standard of Rest, or the LSR. It would appear that the LSR is a more fundamental reference for velocities than is the Sun.

The velocity of any star in the solar neighborhood with respect to the LSR is called the peculiar velocity of the star. The Sun's peculiar velocity is usually referred to as the solar motion. The space velocity of a star is, therefore, the vector difference between its peculiar velocity and the solar motion.

The LSR could be found by measuring the average space velocity of all stars in the solar neighborhood. This average would be the motion of the LSR with respect to the Sun, and its negative would be the solar motion. Unfortunately, very few stars have well-known space velocities; however, if the values of the peculiar velocities are not correlated with the positions within the solar neighborhood, then the observed radial velocities by themselves are sufficient to determine the solar motion.

Let V_0 and V_p be the magnitudes of the solar motion and the peculiar velocity of a star, respectively. These motions make the angles ψ and χ with the direction between Sun and star, as illustrated in Figure 11.2. The direction of the solar motion is known as the solar apex. Then the observed radial velocity V_r of the star is the resultant of the radial components of V_0 and V_p:

$$V_r = V_p \cos \chi - V_0 \cos \psi \tag{11.5}$$

This relation holds for all stars, but the only observed quantity is V_r.

Consider a group of stars all of which are seen in nearly the same direction. These stars need not be physically connected with each other, since they may be at widely different distances. This is where the big "if" mentioned above comes in: if the peculiar velocities are distributed in a fashion that is independent of position, then the small group in question will have the same

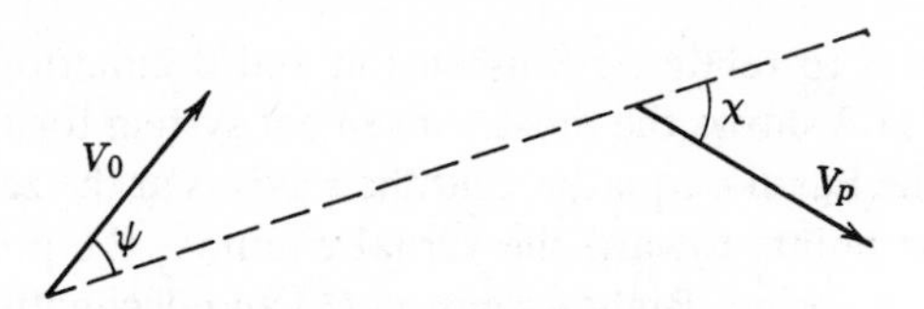

Figure 11.2 Solar and stellar motions.

distribution of peculiar velocities as the solar neighborhood as a whole. But the average V_p for the whole solar neighborhood is zero by definition. If the above assumption is correct, then V_p must also vanish when averaged over any group of stars large enough to have statistical significance. For the group chosen above, when equation (11.5) is averaged over the group, the V_p term is zero:

$$\bar{V}_r = -V_0 \cos \psi \tag{11.6}$$

The group was chosen so that it covers only a small area on the sky, so the angle ψ is essentially a constant. Now a direction is conveniently specified by its direction cosines, which are the cosines of the angles it makes with the arbitrarily chosen x, y, and z axes. Let (l_0, m_0, n_0) be the direction cosines of the solar apex, and (l, m, n) be those of the group of stars in question. Then analytic geometry shows that

$$\cos \psi = ll_0 + mm_0 + nn_0$$

and equation (11.6) becomes

$$\bar{V}_r = -V_0(ll_0 + mm_0 + nn_0) \tag{11.7}$$

One simply observes the mean radial velocity $\bar{V}_r$ for each of many such groups (at least three are needed), each having a different (l, m, n). Equation (11.7) can be written once for each group, and the resulting series of equations can then be solved for V_0 and (l_0, m_0, n_0). In this fashion the solar motion and apex may be found.

It is instructive to analyze the data given by Russell, Dugan, and Stewart (Reference 1, p. 660). Using data from Campbell's radial velocity catalogue, these authors determined the mean radial velocities of six groups of stars. Each group has about 30 stars, and the approximate right ascensions declinations, and mean velocities are as follows:

Group	I	II	III	IV	V	VI
α	0°	180°	90°	270°	...	...
δ	0°	0°	0°	0°	90°	−90°
$\bar{V}_r$	4.4	0.8	23.7	−15.1	−8.2	9.3

Not all of the stars of a given group have exactly the same positions, but they are close.

The next step is to relate right ascension and declination to the direction cosines. Figure 11.3 shows the most convenient system for this. The xy plane is the plane of the Earth's equator, and the z axis is at the zenith of the North Pole. The x axis points toward the vernal equinox, the position of the Sun on the first day of spring. Right ascension (α) and declination (δ) are defined as indicated in the figure. Since the direction cosines are the cosines of the

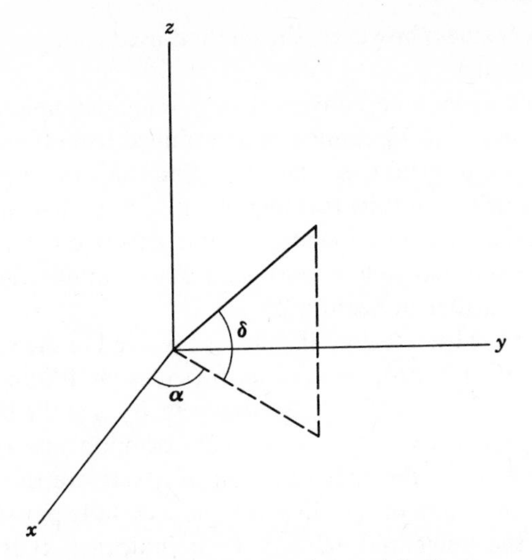

Figure 11.3 Right ascension and declination.

angles with respect to the coordinate axes, it is seen from the figure that

$$\left.\begin{aligned} l &= \cos\delta\cos\alpha \\ m &= \cos\delta\sin\alpha \\ n &= \sin\delta \end{aligned}\right\} \qquad (11.8)$$

It is apparent that the six groups are just in the directions of the coordinate axes (plus and minus), and they have direction cosines of $(\pm 1,0,0)$, $(0,\pm 1,0)$, and $(0,0,\pm 1)$. Equation (11.7) can then be written six times, once for each group:

$$\begin{aligned} -V_0 l_0 &= 4.4 \qquad & -V_0 m_0 &= 23.7 \qquad & -V_0 n_0 &= -8.2 \\ V_0 l_0 &= 0.8 \qquad & V_0 m_0 &= -15.1 \qquad & V_0 n_0 &= 9.3 \end{aligned}$$

These equations are not completely consistent, but by averaging those for opposite directions, one obtains

$$V_0 l_0 = -1.8 \qquad V_0 m_0 = -19.4 \qquad V_0 n_0 = 8.8$$

The sum of the squares of the direction cosines is unity, so the solution of the above is easily found. The result is $V_0 = 21.4$ km/sec, and $(l_0, m_0, n_0) = (-0.084, -0.907, 0.412)$. If these values are substituted into equations (11.8), the right ascension and declination of the solar apex are found to be $\alpha_0 = 265°$, $\delta_0 = 24°$. This is very close to the more accurately determined values of

$$V_0 = 20 \text{ km/sec} \qquad \alpha_0 = 270° \qquad \delta_0 = 30° \qquad (11.9)$$

This example could have been solved by a simpler approach that does not involve direction cosines; however, the method used is of general application and is worth knowing.

Proper motions cannot be converted into velocities unless the distances of the stars are known, so V_0 cannot be determined from them alone. Proper motions can be used to yield the solar apex, and they are in good agreement with the radial velocities in this. It is important to note that the direction and amount of the solar motion depends to some extent on the group of stars chosen to define them, so stellar motions are not completely random. This point is discussed further in Section 25.

Mean Parallaxes. The fact that distance is involved in the relation between tangential velocity and proper motion provides additional methods for determining stellar distances. The tangential velocity is in the plane of the sky, so it is the resultant of two separate velocity components. In other words, the three components of the space motion of a star are the radial velocity plus two more in the plane of the sky. In Figure 11.4 the paper is in the plane of the sky, and the tangential velocity V_t is indicated. It is convenient to define the two components in the plane of the sky with respect to the direction toward the solar apex. The component of V_t along the direction toward the apex is known as the υ component (υ = upsilon), and the one normal to this is the τ component. It is apparent that the τ component is not affected by the solar motion, and so it is due to the peculiar velocity of the star alone. The υ component, on the other hand, is the sum of a peculiar part and a part due to reflected solar motion. If ψ is the angle between the apex and the star, then $-V_0 \cos \psi$ is the reflected solar motion along the radial direction [equation (11.5)] and $-V_0 \sin \psi$ is that along the plane of the sky. The following relations are then apparent:

$$V_t^2 = V_\tau^2 + V_\upsilon^2 \tag{11.10}$$

$$V_\tau = V_\tau \text{ (peculiar)} \tag{11.11}$$

$$V_\upsilon = V_\upsilon \text{ (peculiar)} - V_0 \sin \psi \tag{11.12}$$

Let υ and τ be the proper motion components along and perpendicular to

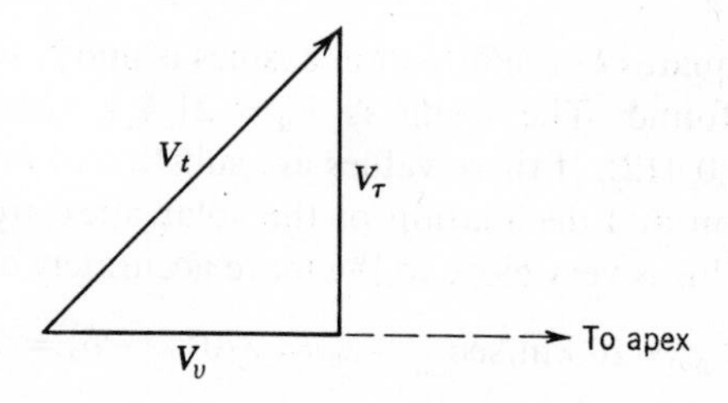

Figure 11.4 Components of transverse velocity.

the direction of the solar motion, respectively. Equations (11.4) still relate the linear and angular velocities of a star, so one has

$$V_\tau = 4.74 \frac{\tau}{p} \tag{11.13}$$

$$V_\upsilon = 4.74 \frac{\upsilon}{p} = 4.74 \frac{\upsilon\,(\text{peculiar})}{p} - V_0 \sin \psi \tag{11.14}$$

As before, p is the parallax of the star in seconds of arc. From the last two parts of (11.14), one finds

$$p = 4.74 \frac{\upsilon\,(\text{peculiar}) - \upsilon}{V_0 \sin \psi} \tag{11.15}$$

If the solar motion and apex can be assumed known, then υ can be found from the observed proper motion; however, υ (peculiar) is not known for any individual star, so equation (11.15) cannot be used directly to find its parallax.

If a group of stars has sufficiently random peculiar motions, then υ (peculiar) is as likely to be positive as negative for any member. If equation (11.15) were summed over all stars of this group, the term containing υ (peculiar) should average out to be zero. This allows the mean parallax of the group as a whole to be found. The best value of the mean parallax is found by what is known as the method of least squares. This method will not be discussed here, but the result is given by the following:

$$\bar{p} = -\frac{4.74}{V_0} \frac{\overline{\upsilon \sin \psi}}{\overline{\sin^2 \psi}} \tag{11.16}$$

The bar over a quantity means that the quantity is to be averaged over the entire group of stars.

Mean parallaxes found in this way are known as secular parallaxes. The idea is that the Sun is moving with respect to a group of stars, and this motion can be used to provide a long base line for determining the distances of the stars in the group. While trigonometric parallaxes have a base line of only 2 AU, the secular parallaxes have a base line equal to the solar motion times the time interval used.

The assumption that the values of υ (peculiar) for the group average to zero is essential. It has been stated that the solar motion depends on the kind of stars used to define it, and so the LSR should not really be used for the definitions of V_0 and the apex to be used in equation (11.16); instead, the group whose mean parallax is sought should define its own values, since the group may have a systematic motion of its own with respect to the stars which define the LSR. This is not always possible, and so the use of the LSR values can introduce some uncertainty into the secular parallaxes. This also points up the fact that the group of stars cannot be chosen on the basis of their observed proper motions. For example, if one should decide to apply this procedure to all stars of large proper motion, one would be choosing stars whose tangential velocities tend to be directed away from the apex, so that the reflected solar motion would add to and not cancel the peculiar velocities. In such a case, the average v (peculiar) would not vanish.

Another way to obtain the mean parallax of a group of stars makes use of the τ component of the proper motion. The τ component is chosen so that it will not be affected by the solar motion. If the velocities of the members of a group are random, then the average component in one direction should, on the average, equal that in any other direction. In particular, $\bar{V}_t$ should be the same as $\bar{V}_r$ (peculiar), where the latter is obtained without regard to sign. Then equation (11.13) leads to the following:

$$\bar{p} = \frac{4.74\bar{\tau}}{\bar{V}_r\ (\text{peculiar})} \tag{11.17}$$

Mean parallaxes found from equation (11.17) are known as statistical parallaxes. Statistical parallaxes should also be applied with some care, since there are a number of ways for systematic errors to enter. One can never be certain that the peculiar velocities are completely random.

It may seem that the mean parallax of a group of stars is not a very useful quantity. If the group is chosen in a random way, this is partially true; however, for groups properly chosen, the mean parallax is an extremely important and useful quantity. Suppose that a group of stars is chosen so that all of the members are of the same general type, i.e., one has reason to believe that all of the stars are of nearly the same absolute magnitude. If N is the number of stars in the group, then

$$N\bar{p} = \sum_{i=1}^{N} p_i \tag{11.18}$$

where p_i is the actual parallax of the ith star in the group. From equation (9.8) and the definition of parsec, the parallax of the ith star is related to its apparent and absolute magnitudes by the equation

$$p_i = 10^A \qquad A = 0.2(M - m_i - 5) \tag{11.19}$$

If this expression for p_i is substituted into (11.18), the following results:

$$M = 5 + 5\log(N\bar{p}) - 5\log\left(\sum_{i=1}^{N} 10^{-0.2m_i}\right) \tag{11.20}$$

Observations of the mean parallax $\bar{p}$ plus the individual apparent magnitudes m_i then allow one to find M, the absolute magnitude. Using this value of M in equation (11.19) then gives the individual parallaxes p_i. While the accuracy

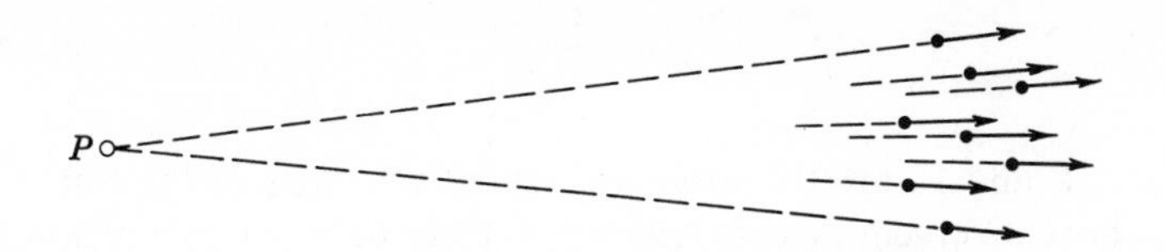

Figure 11.5 Proper motions in a moving cluster.

of this method is limited by the accuracy of the mean parallaxes and by the spread of absolute magnitude among the members of the group, the method is extremely useful. There are stars for which no other measure of distance or absolute magnitude is available.

Moving Clusters. One more method of obtaining distances which is based on stellar motions is worth describing, and that is the moving-cluster method. A star cluster consists of a large number of stars which are bound together gravitationally, and the members of the cluster are moving in nearly parallel paths as viewed from the Sun. If the cluster is not too far away, the proper motions of the individual stars will appear to diverge slightly (as illustrated in Figure 11.5) from a distant point or converge slightly toward some point. If the position of this divergent (or convergent) point P can be measured very accurately, then the distance of each member of the cluster can be found.

Since P is the point from which the cluster seems to be coming, a line directed from the observer toward P is parallel to the motion of the cluster. It is the direction of the cluster at a very long time in the past. In Figure 11.6 it is seen that α, the angle between an individual star and P, is the angle moweeen the radial direction and the direction of motion of this star. The bettions in this case are with respect to the Sun and are not to be corrected for the solar motion. Then one apparently has

$$V = \frac{V_r}{\cos \alpha} \tag{11.21}$$

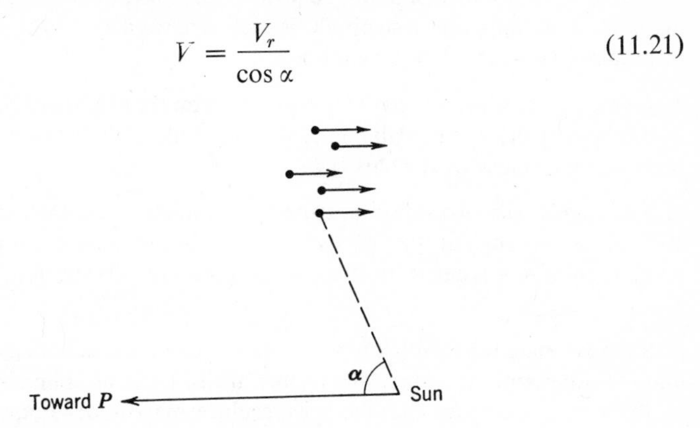

Figure 11.6 Space motions in a moving cluster.

and
$$p = \frac{4.74\mu}{V \sin \alpha} \tag{11.22}$$

As before, V and V_r are the space and radial velocities of the star. For any cluster whose divergent or convergent point can be accurately measured, the angle α can be found for each member star. A measurement of the proper motion and radial velocity for any member star then yields the distance of that star.

The peculiar motions of the stars within the cluster, i.e., the scatter in the real space velocities, will produce some uncertainty in the results, but parallaxes of extreme importance to astronomers have been found by the moving cluster method. Only a few clusters are near enough for this method to work well; among them are the Hyades cluster in Taurus and the Ursa Major group. In addition to the fact that it is of special importance to know the distance of star clusters, the moving-cluster method is valuable in that it has appreciably increased the number of stars whose distances are reasonably well known.

PROBLEMS

1. Suppose that a flat circular disk of radius R radiates like a black body at temterature T. Determine the luminosity of the disk and the flux observed at large distances.

2. A star has $T_e = 8700°\text{K}$, $M_{\text{bol}} = +1.6$, $m_{\text{bol}} = +7.2$. Calculate its distance luminosity, and radius (bol = bolometric).

3. A star has a distance of about 20 pc as determined by the trigonometric parallax method. If its apparent magnitude is well determined, what is the approximate uncertainty in its absolute magnitude?

4. A star cluster is composed of 10^5 stars, each on the average of the same luminosity as the Sun. If the apparent bolometric magnitude of the cluster as a whole is 5.0, what is the distance of the cluster?

5. The angular size of a star is measured with an interferometer, and it is found to be 10^{-5} times the angular size of the Sun. If the star has an apparent bolometric magnitude of 4.0, what is its effective temperature? (Note: $m_{\text{bol}} = -26.7$ for the Sun.)

6. Suppose there were only 2 types of stars, one of absolute magnitude $M = +5.0$ and the other with $M = 0.0$. The former are 10 times as abundant per unit volume as the latter. Compare the average absolute magnitude of the stars in a given volume with that of stars brighter than a given apparent magnitude.

7. The electric field of an electromagnetic wave traveling in the positive x direction has the form

$$E = E_o \sin (kx - \omega t)$$

for x less than x_o, and it has the form

$$E = E_o e^{-a(x-x_o)} \sin (kx - \omega t)$$

for x greater than x_o. Determine the absorption coefficient of the material which exists in the region $x \geqslant x_o$.

8. Three stars have apparent visual magnitudes of +2.0, +2.5, and +3.0. What is the apparent magnitude of the combined light of the three stars?

9. If the three stars in the above problem have a mean parallax of 0."010, and if all 3 have the same absolute visual magnitude, find M_V and the individual parallaxes.

10. In analogy with absolute magnitude, a quantity H can be defined as

$$H = m + 5 + 5 \log \mu$$

where μ is the proper motion. What can one say about the distance of a group of stars all of which have unusually large values of H?

11. The first four lines of the Balmer series of hydrogen have the following wavelengths: Hα, 6563 Å; Hβ, 4861 Å; Hγ, 4340 Å; and Hδ, 4102 Å. A star has four absorption lines whose measured wavelengths are 6530 Å, 4835 Å, 4330 Å, and 4080 Å. If these are the doppler-shifted Balmer lines, what is the approximate radial velocity of the star? Which measurement is most likely in error?

REFERENCES

An excellent reference for the material in this chapter is the following:

1. Russell, H. N., R. S. Dugan, and J. Q. Stewart. *Astronomy*, Vol. II, Ginn, Boston, 1938.

Other references which contain material relevant to this chapter include:

2. Hiltner, W. A. (Ed.). *Astronomical Techniques*, Univ. of Chicago Press, Chicago, 1962.
3. Smart, W. M. *Spherical Astronomy*, Fifth Edition, University Press, Cambridge, England, 1962.
4. Strand, K. Aa. (Ed.). *Basic Astronomical Data*, Univ. of Chicago Press, Chicago, 1963.

Atmospheric extinction is covered in the article by Robert H. Hardie on p. 178 of Reference 2 above.

II

Binary and Variable Stars

Binary and variable stars are particularly important because they offer many unique ways of obtaining useful information about stars. The understanding of how this information is obtained is the main purpose of this chapter. Section 12 contains a brief summary of the two-body problem, Sections 13–15 discuss the three main types of binaries, and variable stars are covered in Section 16.

12. Summary of the Two-Body Problem

Kepler's Laws. This section summarizes the two-body problem with emphasis on those points relevant to binary stars. The main points of the two-body problem are contained in Kepler's three laws, modified by Newton:

1. The relative orbit of the two bodies is a conic section with one of the objects at a focus. The present concern is only with elliptical orbits.
2. The line connecting the two bodies sweeps out a constant amount of area per unit time. This is equivalent to saying that the system conserves angular momentum.
3. The product of the square of the period and the total mass of the system is proportional to the cube of the mean separation of the bodies.

The definition of an ellipse is often given as the curve traced out by all points such that the sum of their distances from two given points (the focus points) is a constant. In Figure 12.1, F and F' are the foci, and for any point P on the ellipse $FP + F'P =$ constant. This shows an easy way to draw an ellipse: fasten the two ends of a string at F and F'; then, holding the string tight with a pencil, trace out the ellipse. Because of the symmetric positions of the two foci, it is apparent from the figure that the length of the string is equal to the long axis of the ellipse. Half of this, the semi-major axis OA, is usually denoted by a. Then

$$r + r' = 2a \tag{12.1}$$

From Figure 12.1 one finds, by dropping a perpendicular from P onto $F'A$, that

$$r'^2 = r^2 \sin^2 \theta + (F'F + r \cos \theta)^2 \tag{12.2}$$

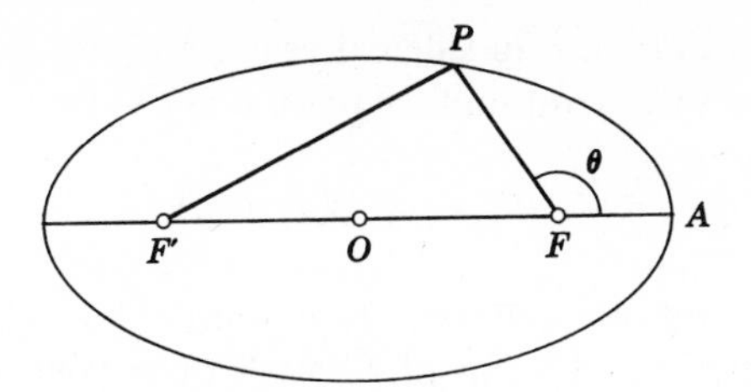

Figure 12.1 The ellipse.

The flatness or shape of an ellipse of given a is described either by the semi-minor axis b or by the eccentricity e. The latter is defined as the ratio $OF/OA = OF/a$ in the figure. The circle (for which the two foci coincide at the center) is the limiting ellipse of eccentricity zero, while the straight line connecting two foci is the other limiting ellipse with $e = 1$. The latter is the orbit of zero angular momentum. It is apparent that $F'F = 2ae$, and so equation (12.2) becomes

$$r'^2 = r^2 + 4a^2e^2 + 4rae \cos \theta \tag{12.3}$$

If r' is eliminated between equations (12.1) and (12.3), the result is the polar equation of the ellipse:

$$r = \frac{a(1 - e^2)}{1 + e \cos \theta} \tag{12.4}$$

In Figure 12.2, (r,θ) and $(r + dr, \theta + d\theta)$ are the coordinates of one body with respect to the other at times t and $t + dt$. Then $r\,d\theta$ is the element of circular arc through (r,θ), and it is also the altitude of the triangle with base $(r + dr)$. The area of this triangle is then $\frac{1}{2}r^2\,d\theta$, since the term with a product of two differentials is negligibly small in comparison, and the law of constant areal velocity requires $r^2(d\theta/dt) =$ constant. But the total area of the ellipse, $\pi ab = \pi a^2(1 - e^2)^{1/2}$, is covered by the line between the bodies in one period P; therefore,

$$r^2 \frac{d\theta}{dt} = \frac{2\pi a^2(1 - e^2)^{1/2}}{P} \tag{12.5}$$

The third law is simply

$$P^2(\mathcal{M}_1 + \mathcal{M}_2) = a^3 \tag{12.6}$$

if the masses are measured in terms of that of the Sun (more accurately, that

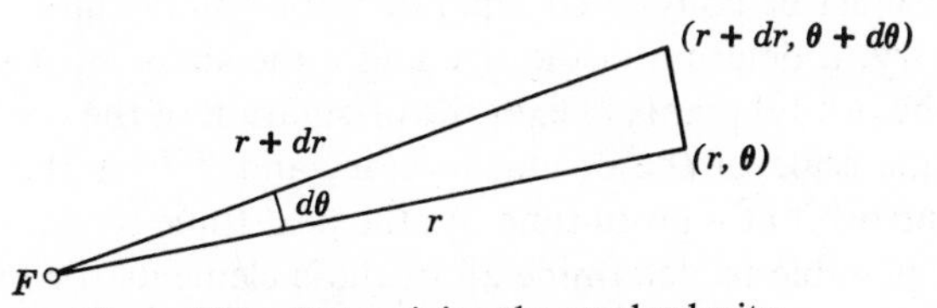

Figure 12.2 Determining the areal velocity.

of the Sun plus the Earth), P in sidereal years, and the semi-major axis a in astronomical units. In general units, equation (12.6) is

$$\frac{G}{4\pi^2} P^2(\mathcal{M}_1 + \mathcal{M}_2) = a^3 \tag{12.7}$$

where G is the constant of gravitation. Equations (12.4), (12.5), and (12.6) or (12.7) are the mathematical forms of Kepler's three laws of motion.

Orbital Elements. The basic quantities needed to uniquely specify the orbit of a pair of objects are known as the orbital elements. There are generally considered to be seven of these elements, as follows:

- a semi-major axis
- e eccentricity
- P period of revolution
- i inclination of orbital plane to some fundamental plane
- Ω position angle of node
- ω longitude of periastron
- T time of periastron passage

Figure 12.3 illustrates these elements. Let the xy plane be the fundamental plane of reference. In the planetary case this is the ecliptic, the plane of the Earth's orbit; for binary stars, this is the plane of the sky. The center of the coordinate system is located at the focus of the orbit, NN' is the line of nodes where the orbital plane intersects the xy plane, and A is the periastron point (the point on the orbit nearest to the focus). In the figure the part of the orbit below the xy plane is a dashed line. Then Ω is measured from the x axis, in the xy plane, to the node N, so $\Omega = \angle xFN$. The longitude of periastron is measured in the orbital plane from the node N to the periastron point A, and it is measured in the direction of orbital motion. Thus $\omega = \angle NFA$ or $(360° - \angle NFA)$, depending on the direction of motion. The inclination i is the angle between the orbital plane and the xy plane. It should be pointed out that the orbital elements are sometimes defined differently from this, but they are always exactly equivalent to the above (though not necessarily as reasonable!). The period is not usually considered to be a separate element for solar-system objects. The reason is that the Sun's mass dominates all others, and so according to equation (12.6), P is essentially equivalent to a. However, P must be considered separately for binary stars.

In summary, a determines the size and e the shape of the orbit, i and Ω determine the orbital plane, ω fixes the orientation of the orbit in its plane, P gives the time scale of the orbital motion, and T fixes the position of the objects in the orbit at a given time. As the next three sections point out, it is usually not possible to determine all of these elements for binary stars. The amount of information that can be obtained depends on the type of binary.

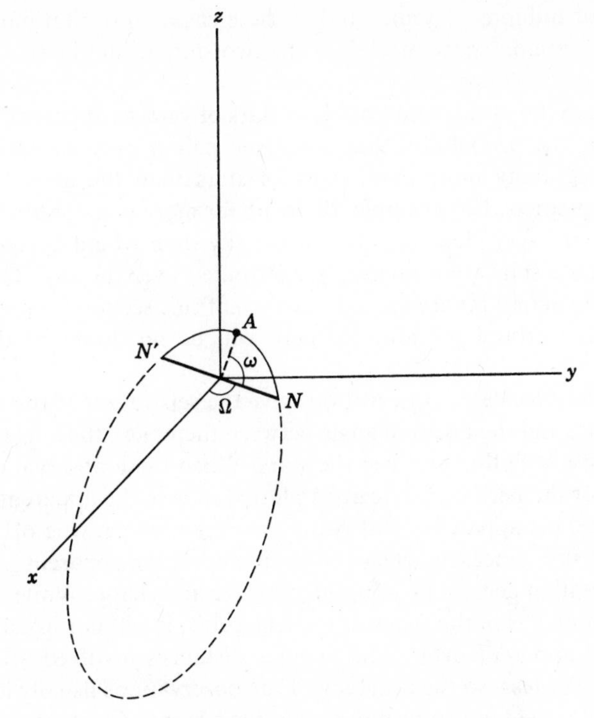

Figure 12.3 Orbital elements.

Binary stars are of interest for many reasons, and one of the most important reasons is that they provide a means of determining stellar masses. The mass of a star is the single most important characteristic it has, and it can be measured only from the effects of its gravitational field. In this regard it should be noted that the orbital elements mentioned above suffice only to fix the total mass of the two stars. Additional information is needed in order to obtain the individual masses, and this additional information involves the measuring of the center of mass of the system. Thus the position of the center of mass can be considered as the eighth orbital element of a binary system.

13. Visual Binaries

A visual binary is a system whose components can be seen separately, at least with the interferometer. A question that arises is whether the members are physically connected (binary system) or just seen by chance in the same line of sight (optical pair). If orbital motion can be detected, of course, the system is a binary; however, many such systems exist with periods of thousands or millions of years, and in these cases no orbital motion can be noticed. A common space motion of the two stars would be strong evidence that the stars are connected.

One can use the observed numbers of stars of various apparent magnitudes to determine the probability that any given pair is only an optical double. One observes many more close pairs of stars than the above probability calculation predicts. For example, there are far more pairs of fifth-magnitude stars which are very close together in the sky than would be the case if all fifth-magnitude stars were randomly distributed over the sky. Thus most of these must be actual binary stars. In any event this section is concerned with pairs showing orbital motion, so there will be no doubt of their binary nature.

The Orbital Elements. Observations give the separation of the two stars in seconds of arc and the position angle between them, i.e., the angle from north through east to the line between the stars. When these observations cover a major part of the period, they can be plotted to give the apparent orbit as in Figure 13.1. This apparent orbit is the projection of the true orbit onto the plane of the sky. Since the true orbit is an ellipse, the apparent orbit is also, although it will in general be of a different size and shape. While the brighter or primary star S is in the focus of the true orbit, it will not usually be in the focus of the apparent orbit. The angular distances involved are often one second of arc or less, so the accuracy of the observations usually leaves much to be desired. One important point is that the law of constant areal velocity holds in projection, and so Kepler's second law also holds for the apparent orbit. This increases considerably the accuracy with which the apparent orbit can be drawn.

It is not appropriate to derive here all of the equations from which the orbital elements can be obtained, but an indication of how the elements can be found from the observations is instructive. The period P, of course, follows in an obvious way if the observations extend over a long enough interval, preferably over many periods. In Figure 13.1, O is the center of the apparent orbit, and it must be the projection of the center of the true orbit. But S is the

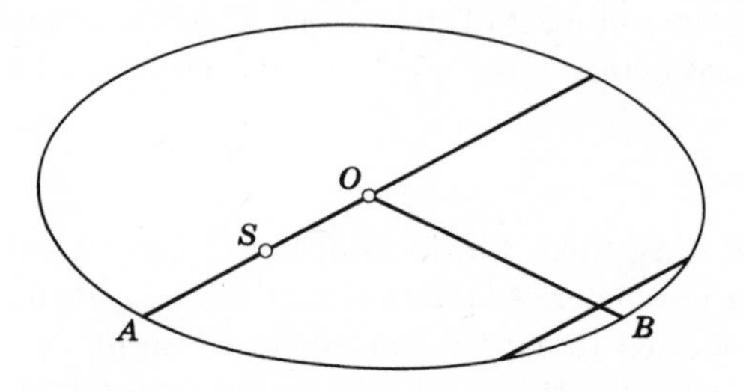

Figure 13.1 The apparent orbit of a visual binary.

projection onto the plane of the sky of the focus, so OSA is the projected semi-major axis. Referring to Figure 12.3, one can see that the projection factor involves both ω and i, so the measured size of OSA provides one relation between a'' (the value of the semi-major axis in seconds of arc), ω, and i. Also one can find OB, the projection of the semi-minor axis. This is not necessarily perpendicular to OA, but it does bisect all lines parallel to OA extended. The measured value of OB then gives a relation between a'', e, ω, and i. A third relation comes from the orbital areas. The area of the apparent orbit can be measured, and it is simply cos i times the area of the true orbit, or $\pi(a'')^2(1 - e^2)^{1/2} \cos i$. Finally, one can find the eccentricity of the true orbit directly from the apparent orbit. In Figure 13.1 OA is a a'' times its projection factor, and OS is ea'' times the same projection factor. Then the ratio $OS/OA = e$, and the three relations mentioned above suffice to determine a'', ω, and i. The position angle of the node Ω can then be found from ω and i, and the time of periastron passage T is observed directly. This discussion is intended only to show that accurate observations uniquely determine the elements; the procedure used in actual cases may be rather different.

All of the seven orbital elements mentioned in the last section are then determined for a visual binary, with one major exception: the semi-major axis can be found in seconds of arc (i.e., its angular size as seen from Earth), but not directly in linear size. To avoid confusion, in the remainder of this section A will denote the semi-major axis in AU and a'' its angular size in seconds of arc. Then

$$A = \frac{a''}{p} \tag{13.1}$$

so the parallax p must be known before the observations can yield A. In the present notation, equation (12.6) can be written

$$A^3 = \frac{a''^3}{p^3} = P^2(\mathcal{M}_1 + \mathcal{M}_2) \tag{13.2}$$

The parallax p is needed in order to obtain the mass of the system.

Even if the parallax were known, equation (13.2) would yield only the total mass of the two stars. If the individual masses are required, then additional information is needed. This additional information for visual binary stars consists of measuring the absolute orbit of each component with respect to the background stars instead of only the relative orbit of one component with respect to the other. If these absolute semi-major axes are denoted by a_1'' and a_2'', then $a_1'' + a_2'' = a''$, and the ratio a_1''/a_2'' is inversely proportional to the ratio of the stellar masses. In this way the individual masses are found.

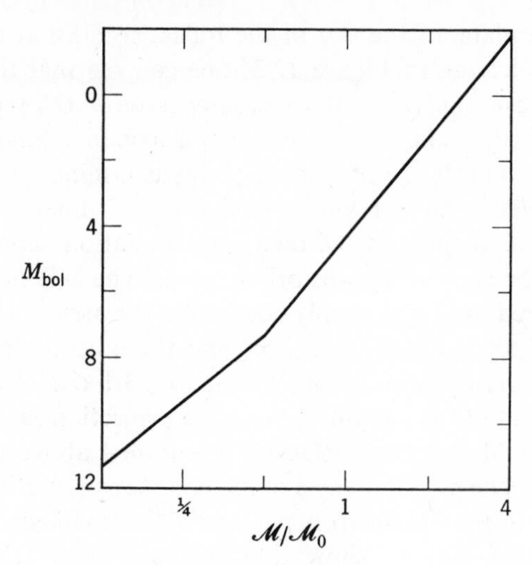

Figure 13.2 The mass-luminosity relation.

The Mass-Luminosity Relation. The number of stars for which accurate absolute orbits and accurate distances are both known is not very large. It is large enough, however, to provide some extremely important information. If certain peculiar types of stars are excluded and if one plots the absolute bolometric magnitude of the stars against the masses, one finds a fairly smooth curve, as shown in Figure 13.2. This is known as the mass-luminosity relation, and it shows that, if a star is not too peculiar, then its mass can be found if its luminosity or absolute magnitude is known, and vice versa. As is explained in Chapter III, allowances can often be made for the unusual properties of the peculiar stars which do not satisfy the normal mass-luminosity relation, so that useful information is often available for them also.

According to D. L. Harris, K. Aa. Strand, and C. E. Worley in their article in Reference 4 of Chapter I, the normal mass-luminosity relation is well represented by

$$\begin{aligned} M_{\text{bol}} &= 4.6 - 10.0 \log \mathcal{M} \qquad 0 < M_{\text{bol}} < 7.5 \\ M_{\text{bol}} &= 5.2 - 6.9 \log \mathcal{M} \qquad 7.5 < M_{\text{bol}} < 11 \end{aligned} \tag{13.3}$$

The range in mass among stars is rather small, very few having more than $10\mathcal{M}_o$ or less than $0.1\mathcal{M}_o$, where $\mathcal{M}_o$ is the solar mass.

Dynamical Parallaxes. An important application can be made of the relatively small range observed in stellar masses. Solve equation (13.2) for the parallax:

$$p = \frac{a''}{P^{2/3}(\mathcal{M}_1 + \mathcal{M}_2)^{1/3}} \tag{13.4}$$

Suppose that a visual binary has a known period but unknown parallax. If the stars are both assumed to have one solar mass, then this mass estimate will almost certainly not be in error by more than a factor of about 5 to 10, and the cube root of the total mass should be accurate to within a factor of 2. Then equation (13.4) indicates that a parallax accurate to a factor of 2 can be obtained.

It is possible to do much better than this. As is pointed out in Chapter III, astronomers can tell much about a star by studying its spectrum, and in this fashion the mass can usually be estimated with an accuracy of better than a factor of 2. Thus parallaxes with about a 20% accuracy can be obtained.

Parallaxes which are based on equation (13.4) are called dynamical parallaxes because this equation is derived from the principles of dynamics. One common method of dynamical parallaxes is based on the assumption that the stars obey the mass-luminosity relation. This assumption is sufficient for a unique solution for the parallax. Consider the hypothetical example of a binary system with $a'' = 1.0$, $P = 100$ yr, and the stars having apparent bolometric magnitudes of $m_1 = 6.0$, $m_2 = 6.5$. Then, from equation (13.4),

$$p = \frac{0''.0464}{(\mathcal{M}_1 + \mathcal{M}_2)^{1/3}}$$

If it is estimated that the stars have one solar mass each, then the parallax is $p = 0''.037$. Equation (9.8) for the absolute magnitude can be written

$$M = m + 5 + 5 \log p$$

The observed apparent magnitudes and the above parallax give absolute bolometric magnitudes of $M_1 = 3.8$, $M_2 = 4.3$. If the stars do satisfy the mass-luminosity relation, the masses can be determined from equation (13.3). The masses are then $\mathcal{M}_1 = 1.2$, $\mathcal{M}_2 = 1.1$. The original guess is seen to have been a good one, and the process can be continued until it converges. Thus one can try a new guess of $(\mathcal{M}_1 + \mathcal{M}_2) = 2.3$, and the revised value of the parallax is $p = 0''.035$. The new absolute magnitudes are $M_1 = 3.7$, $M_2 = 4.2$, and the mass-luminosity relation still gives 1.2 and 1.1 for the masses. The dynamical parallax method has thus converged on the above values.

As stated above, astronomers can often tell whether or not a star is to be expected to obey the mass-luminosity relation, and account can often be made for those which do not satisfy it. In this way accurate dynamical parallaxes can be obtained. The method does have a limitation, however: dynamical parallaxes offer accurate new information only if the period P and semi-major axis a'' are well determined. This is true only for the nearer binaries, and for many of these, trigonometric parallaxes are already available. Thus there is only a limited number of binaries for which dynamical parallaxes offer new and accurate information.

Dynamical parallaxes can also be obtained if the orbital radial velocity can be measured, but there are few cases in which the orbital velocity of a visual binary is great enough for this to be accurately applied. Equation (13.4) can also be applied in a statistical fashion to binaries whose periods are too long to be accurately known. An example of this type of method applied to pairs of galaxies is given in Section 26.

14. Spectroscopic Binaries

The spectroscopic binary is one in which the orbital motion is detected by the variable radial velocity. A majority of the spectroscopic binaries are far too close together for the components to be seen separately as a visual binary. The periodic shifting back and forth of the spectral lines gives away the binary character of the system. Usually only the spectrum of one of the stars is visible, the invisible companion showing itself only through its gravitational effects. In this case one must show some care to distinguish this one-lined binary from certain types of pulsating stars which also show a periodic doppler shifting of the spectral lines.

If the two components of a spectroscopic binary are not too different in brightness, a difference of about one magnitude being borderline, both spectra will be visible. Then the lines will be double (at least those lines which the two stars have in common) most of the time, the splitting being a periodic function of the time. These double-lined binaries are the source of considerably more information than the single-lined ones. It frequently happens that there is a considerable difference in temperature between the components, so that the lines of both stars may show up if a wide wavelength interval is examined. Thus the hot star may dominate the shorter-wavelength regions, but lines of the cooler star may appear or even dominate at much longer wavelengths.

The Velocity Curve. The plot of radial velocity vs. time for a spectroscopic binary is called the velocity curve, and Figure 14.1 shows an example. A major difference from visual binaries is that most spectroscopic binaries have relatively short periods, an obvious selection effect. As a result the observations usually extend over many periods, and the velocity curve is often well determined. The observed radial velocity, corrected to the Sun, is composed of the radial velocity of the center of mass V_m plus the orbital radial velocity V_o:

$$V_r = V_m + V_o \tag{14.1}$$

V_o is obviously the velocity of the star in its absolute orbit, not in its relative orbit. The star moves in an elliptical orbit about the center of mass, and this orbit is closed; therefore, the distance it moves in any direction must equal the distance it moves in the opposite direction during a period. Since the

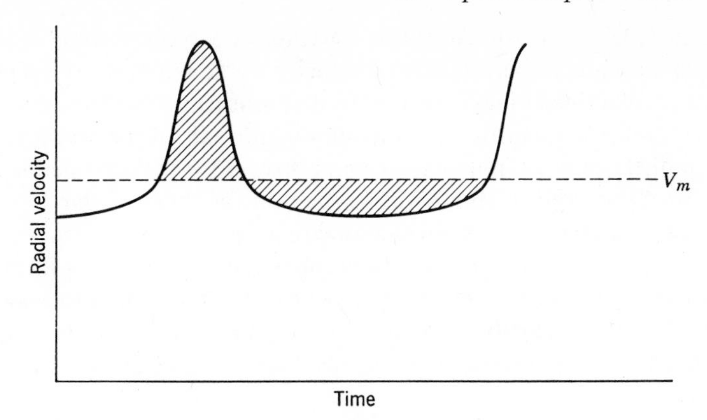

Figure 14.1 The velocity curve.

distance a star moves toward or away from the Sun is found by integrating the radial velocity over the appropriate time interval, it follows that V_m is found by constructing the straight line (the dashed line in the figure) such that the same area lies above as below it. Thus the shaded areas in Figure 14.1 are equal, and V_m in general is not the average of the extreme radial velocities.

The Orbital Elements. To see how the velocity curve is related to the orbital elements, consider Figure 12.3. There NN' is the line of nodes, which is the intersection of the orbital plane with the plane of the sky, the latter being the xy plane. The line of sight is along the z axis, and so it must be perpendicular to NN'. Let (r_1,θ_1) be the polar coordinates of star 1 relative to the center of mass. (Note Figure 12.1.) In the plane of the orbit, the component of r_1 along NN' is $r_1 \cos(\theta_1 + \omega)$, and its component normal to NN' is then $r_1 \sin(\theta_1 + \omega)$. Now, this is in the plane of the orbit, which is tipped by the angle i to the plane of the sky and by the angle $(\pi/2 - i)$ to the line of sight. It follows that the component of r_1 along the line of sight is $r_1 \sin(\theta_1 + \omega) \sin i$, and it is the time rate of change of this quantity that is the orbital radial velocity:

$$V_o = \sin i \left[r_1 \cos(\theta_1 + \omega) \frac{d\theta_1}{dt} + \sin(\theta_1 + \omega) \frac{dr_1}{dt} \right] \tag{14.2}$$

If equations (12.4) and (12.5) are then used to eliminate the above time derivatives, one obtains

$$V_o = K_1[\cos(\theta_1 + \omega) + e \cos\omega] \tag{14.3}$$

with

$$K_1 = \frac{2\pi a_1 \sin i}{P(1 - e^2)^{1/2}} \tag{14.4}$$

Equation (14.3) is the fundamental equation of a spectroscopic binary, and it relates the observed orbital radial velocity to the relevant orbital elements. If it is a double-lined binary, two such equations can be written, one for each star. In (14.4) the quantity a_1 is the semi-major axis of the absolute orbit of star 1 in distance units, not angular units. Except for the semi-major axis, the elements of the orbit of star 2 will be the same as those of star 1.

Since everything on the right side of equation (14.3) is constant except θ_1, the maximum value of V_o occurs when the first cosine term has its maximum value of unity; likewise, the minimum value of V_o corresponds to $\cos(\theta_1 + \omega) = -1$. The latter velocity is one of approach and thus is a negative quantity. It is customary to let A and B be the absolute values of these extreme orbital velocities, so

$$\begin{aligned} A_1 &= K_1(1 + e \cos \omega) \\ B_1 &= K_1(1 - e \cos \omega) \end{aligned} \tag{14.5}$$

It follows that

$$K_1 = \tfrac{1}{2}(A_1 + B_1) \tag{14.6}$$

Since A_1 and B_1 are obtained directly from the velocity curve, K_1 is easily found.

In equation (14.3) the quantity K_1 fixes the range of the velocity variation, and the second factor involving e and ω fixes the shape of this curve. One can show that all possible shapes of velocity curves can be produced by varying e and ω, and it is also possible to uniquely determine e and ω from the observed shape. Another orbital element that can be found from the velocity curve is the time of periastron passage T, for this is the time for which $\theta_1 = 0$. Then the orbital elements which can be found for a spectroscopic binary are P, e, ω, and T. Nothing whatever can be said about the value of Ω, since this element has no influence upon the velocity curve. As for a_1 and i, it is seen that the determination of K_1 and the independent evaluation of P and e means that the product $a_1 \sin i$ can be found, but a_1 and i cannot be separately found. This is reasonable, since an integration of the velocity curve will give the projection of the orbital size upon the line of sight, and this can be reduced to the projection of the semi-major axis $a_1 \sin i$. A large orbit nearly in the plane of the sky (small i) has the same velocity curve as a small orbit seen nearly edge-on (large i), the other elements being the same. Sometimes additional information is available, and then these elements can be separated. For example, if the system is also a visual binary of known distance, then a_1 is known and i follows. As is pointed out in Section 15, the inclination i can be found for eclipsing binaries, so again a_1 and i can be separately found.

Masses. For a two-lined binary both K_1 and K_2 can be observed. Then it follows from equation (14.4) that

$$\frac{K_1}{K_2} = \frac{a_1}{a_2} = \frac{\mathcal{M}_2}{\mathcal{M}_1} \tag{14.7}$$

The mass ratio is thus easily found for any two-lined spectroscopic binary.

As far as the masses themselves are concerned, the situation again depends on whether the lines of one or of both components are visible. In the latter case, $(a_1 \sin i)$ and $(a_2 \sin i)$ are both observable, so Kepler's third law can be written

$$a^3 \sin^3 i = (a_1 \sin i + a_2 \sin i)^3 = P^2(\mathcal{M}_1 + \mathcal{M}_2) \sin^3 i \tag{14.8}$$

In this case the products $(\mathcal{M}_1 \sin^3 i)$ and $(\mathcal{M}_2 \sin^3 i)$ are found, but not the individual masses themselves. This still represents an appreciable amount of information. One can calculate the average value of $\sin^3 i$ in order to obtain an idea of the correction needed for the unknown inclination. According to methods discussed in Section 1, one has

$$\begin{aligned} \overline{\sin^3 i} &= \frac{\int \sin^3 i \, d\omega}{\int d\omega} \\ &= \frac{1}{2} \int_0^{\pi} \sin^4 i \, di \\ &= \frac{3\pi}{16} = 0.59 \end{aligned} \tag{14.9}$$

In this result it is assumed that the orbits are oriented at random and that there are no observational selection effects. A selection effect is anything which would tend to make the observed sample nontypical, and it is seen that there is a selection effect working here. A system of large i would have a larger radial velocity variation than one of small i, other things being equal, and so it would have a better chance of being discovered as a spectroscopic binary. Thus the mean value of $\sin^3 i$ to be used among observed spectroscopic binaries is somewhat larger than the 0.59 found above. It has been suggested that $\frac{2}{3}$ is a better value. If this is correct, one may expect that the masses on the average are some 50% greater than the $(\mathcal{M} \sin^3 i)$ values found from observation.

The information obtained from a one-lined binary is considerably less. Only $(a_1 \sin i)$ is known. One can write

$$\begin{aligned} (\mathcal{M}_1 + \mathcal{M}_2)P^2 &= (a_1 + a_2)^3 \\ &= a_1^3\left(1 + \frac{a_2}{a_1}\right)^3 \\ &= a_1^3\left(1 + \frac{\mathcal{M}_1}{\mathcal{M}_2}\right)^3 \end{aligned}$$

or

$$\frac{a_1^3 \sin^3 i}{P^2} = \left(\frac{\mathcal{M}_2}{\mathcal{M}_1 + \mathcal{M}_2}\right)^2 \mathcal{M}_2 \sin^3 i \tag{14.10}$$

The quantity on the right side of equation (14.10) is known as the mass function, and it represents all that can be determined about the mass of a one-lined binary. If the two masses are equal, then the mass function is $\frac{1}{4}(\mathcal{M}_2 \sin^3 i)$, and a lower limit to the mass of the secondary (fainter) star could be assigned. Actually the fact that the binary is one-lined means that there is an appreciable luminosity difference between the two stars. Since star 1 is the brighter, $\mathcal{M}_2$ is probably less than $\mathcal{M}_1$ and so the mass function is probably less than the quantity mentioned above. Still, the mass function can yield important statistical information relating mass ratios, inclinations, and the masses of secondary components.

Spectroscopic vs. Visual Binaries. For a system to be a visual binary, its components must be far enough apart to be seen separately. Thus there is an observational selection effect which results in visual binaries having large orbits. It is not likely that a system will be discovered to be a spectroscopic binary unless there is a large radial velocity variation over a reasonably short period. Thus spectroscopic binaries tend to have short periods, which means that they have small orbits. This is a difference which illustrates how difficult it is to determine the true distribution of properties of binaries. Selection effects are so strong that average properties of binaries have little meaning.

The components of spectroscopic binaries are often so close together that each is distorted into a nonspherical shape, or one may see evidence of matter streaming from one star to the other, or the expansion which stars exhibit in certain phases of their evolution may be inhibited, or the like. In short, spectroscopic binaries are close binaries, and this can cause them to differ considerably from more distant binaries or from single stars. One should therefore take some care in applying data obtained from close binaries to stars in general. This is not a problem for most visual binaries.

The fact that spectroscopic binaries are close pairs makes them interesting from a different point of view. The effects mentioned above cause the orbits to be somewhat different from the simple two-body orbits described by Kepler's laws. These deviations from Kepler's laws are observed in some cases, and their interpretation is very complicated and challenging. Also, some effects are observed which are probably due to evolutionary changes in the stars themselves, effects that would not be noticed except in binaries of very short period. Many of these points are also relevant to the eclipsing binaries discussed in Section 15.

15. Eclipsing Binaries

An eclipsing binary is one in which the orbital inclination is sufficiently close to 90° that the two components periodically intersect each other's line of sight, causing an eclipse. The apparent brightness of the system diminishes during eclipse, and this is how such systems are discovered.

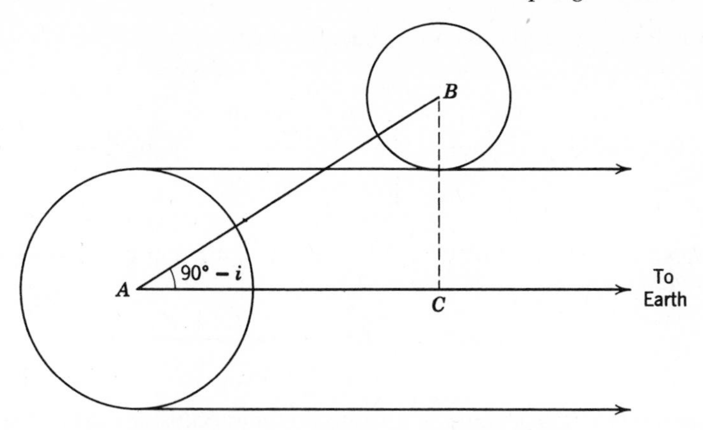

Figure 15.1 The limiting case for eclipses.

The probability that a given system will have eclipses as seen from the Earth can be easily calculated. Figure 15.1 shows two spherical stars which have an assumed circular orbit about each other. The view shows the orbit edge on, the plane of the diagram containing the normal to the orbit and the direction toward Earth. If a circular cylinder is constructed around one of the stars and extended toward Earth, then an eclipse will take place whenever the other star intersects this cylinder. The figure shows the limiting case. Angle BAC is the angle at which the orbital plane is tilted from the line of sight, and equals $(90° - i)$. Distance AB is a, the radius of the orbit, and BC is the sum of the radii of the two stars, $R_1 + R_2$. An eclipse will take place only if $90° - i$ is smaller than the limiting angle shown. Since the sine of the limiting angle is $BC/AB = (R_1 + R_2)/a$, an eclipse will occur if

$$|\sin(90° - i)| = |\cos i| < \frac{R_1 + R_2}{a} \tag{15.1}$$

The absolute value of the trigonometric functions is indicated because i can be anywhere in the range 0–180°, and so $\cos i$ can be positive or negative. (The decision whether i is greater or less than 90° is based on the direction of the orbital motion and on the right-hand rule. Thus if the orbital motion in Figure 15.1 is clockwise as seen from the Earth, then i is greater than 90°.)

The inclination is the angle between the line of sight and the normal to the orbit, so i can be identified with the angle θ of Figure 1.2 in which the z axis is toward Earth, and r is the normal to the orbit. The probability of any given range of i is proportional to the solid angle corresponding to that range, so the probability of eclipse is obtained by integrating over all solid angles which satisfy (15.1). If L and U are the lower and upper limits, respectively, of i which satisfy (15.1), then one has, from equation (1.2),

$$\text{Probability of eclipse} = \frac{\int_0^{2\pi} d\phi \int_L^U \sin i \, di}{\int_0^{2\pi} d\phi \int_0^{\pi} \sin i \, di} = \tfrac{1}{2}(\cos L - \cos U)$$

It is apparent from expression (15.1) and the figure that $\cos L = -\cos U = (R_1 + R_2)/a$, and so

$$\text{Probability of eclipse} = \frac{R_1 + R_2}{a} \tag{15.2}$$

Equation (15.2) expresses the obvious fact that eclipsing binaries tend to be composed of large stars with small orbits. It was pointed out previously that visual binaries tend to have large orbits, so the latter are not likely to be eclipsing binaries as well. As an example, two solar type stars separated by 10 AU have approximately $R_1 = R_2 = 10^{-3}a$, and equation (15.2) indicates that such a binary would only have about one chance in 500 of being inclined so that eclipses would be seen from the Earth. Most visual binaries have much greater separations than this.

The result is quite different for spectroscopic binaries, for an eclipsing system tends to have small a and large i, and this combination also makes for large orbital radial velocities. Most eclipsing binaries should also be spectroscopic systems, and this expectation is realized in almost all cases in which the relevant observations have been made.

As in the case of spectroscopic binaries, there are certain types of true variable stars which look very much like eclipsing binaries. Again, by careful analysis one can distinguish the true variables from the others.

The Light Curve. The observational data for an eclipsing binary consist of a relation between the apparent brightness of the system and time. A graph of this data, as in Figure 15.2, is known as the light curve of the binary. The light of the system is usually approximately constant between eclipses. At A the main or primary eclipse begins, and it lasts until D. A half period or so later the secondary eclipse occurs at E. At the latter the two stars have reversed their positions from the primary eclipse.

Usually both of the eclipses are partial, which means that only part of the eclipsed star is covered, but sometimes one of them is total. Figure 15.2 represents one of these total eclipses. The eclipse starts at A and becomes total at B. Between B and C the eclipsed star is completely covered, so the light from the system remains constant. The primary eclipse becomes partial again at C and ends at D. At E the large star is being eclipsed by the small one, so the secondary eclipse cannot be total.

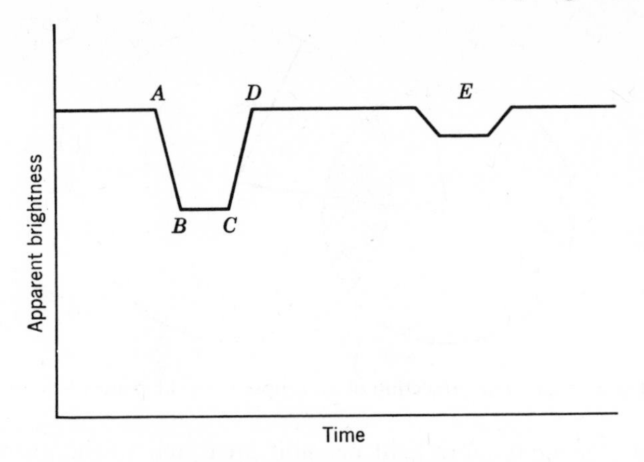

Figure 15.2 The light curve.

The Orbital Elements. Just as the shape of the velocity curve of a spectroscopic binary is fixed by e and ω, so the general shape of the light curve is also determined by these orbital elements. For a circular orbit ($e = 0$) or for an elliptical orbit with the long axis toward the observer ($\omega = 90°$ or $270°$), the minima will be equally spaced; otherwise, the secondary minimum will be displaced toward the preceding or the following primary minimum because of the nonuniform orbital motion. The lengths of the two eclipses will be equal for circular orbits or for elliptical orbits with the short axis toward Earth ($\omega = 0°$ or $180°$), but otherwise they will have different durations. In general an elliptical orbit will cause the orbital motion during the first half of an eclipse to be different from that during the second half, and this causes the eclipse to be nonsymmetric. If the orbit is sufficiently elliptical there may be no secondary eclipse. If the observations of these effects can be made accurately enough, then e and ω can be found. Otherwise one can sometimes use spectroscopic observations to obtain these elements as described in Section 14. In any event, the eccentricity and the longitude of periastron can be found, in principle, from the light curve alone.

The eclipsing binaries which have accurate eccentricities determined usually turn out to have nearly circular orbits. This is not surprising, since equation (15.2) shows that eclipsing systems tend to have stars which are a large fraction of the size of their orbits. When this occurs the eccentricity must be small, for otherwise the stars would collide with each other. The following discussion will be simplified by the assumption that the orbits are circular, although this assumption is usually not necessary in practice. For circular orbits, the elements ω and T are not defined.

For circular orbits, the area of one star which is blocked off during one eclipse equals the area of the other one blocked off during the other eclipse.

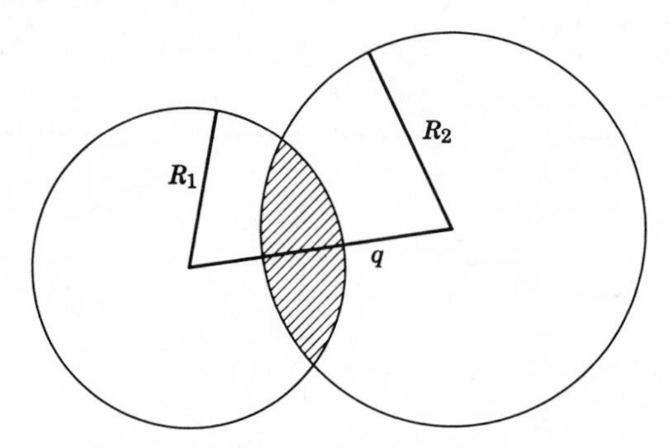

Figure 15.3 The projection of an eclipse onto the plane of the sky.

The hotter star emits more light per unit area, and so the primary eclipse must occur when the hotter star is behind the cooler one.

Figure 15.3 shows the projection of the stars onto the plane of the sky during one of the eclipses. Lines R_1 and R_2 are the stellar radii, and q is the distance between them projected onto the plane of the sky. The shaded area is the area covered during eclipse, and it can be on either star. It is apparent that this area is a function of the two stellar radii and q. If $\Delta F_p(t)$ is the amount by which the observed flux from the system is decreased during primary eclipse as a function of time t, then

$$\Delta F_p(t) = \frac{F_h A_{ec}(t)}{\pi R_h^2} \tag{15.3}$$

where $A_{ec}(t)$ is the area of the hot star which is covered at any time t, and F_h is the flux observed from the Earth from the hot star by itself. In this equation it is assumed that the star has a uniform surface brightness. If the light curve is compared at two different times, t_1 and t_2, during the primary eclipse, then

$$\frac{\Delta F_p(t_1)}{\Delta F_p(t_2)} = \frac{A_{ec}(t_1)}{A_{ec}(t_2)} = \frac{A_{ec}[R_1,R_2,q(t_1)]}{A_{ec}[R_1,R_2,q(t_2)]} \tag{15.4}$$

This equation simply means that the ratio of the light loss at two different times during the primary eclipse is equal to the ratio of the areas which are eclipsed, and these areas are functions of R_1, R_2, and the projected separations $q(t_1)$ and $q(t_2)$. An equation similar to (15.3) could be written for the secondary eclipse, and it would involve the flux and radius of the cool star.

Figure 15.4 shows two views of the binary orbit. In (a) is a view from above the orbit, and in (b) is shown the view from the Earth. The latter is simply the former tipped through the angle i, and A', B', and C' are the projected positions of points A, B, and C. While C is the center of the relative orbit, A is the position of the cool star at mid-eclipse, and B is the position of the cool

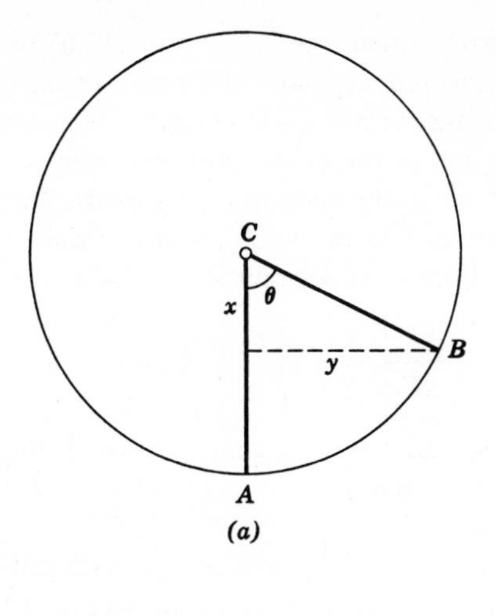

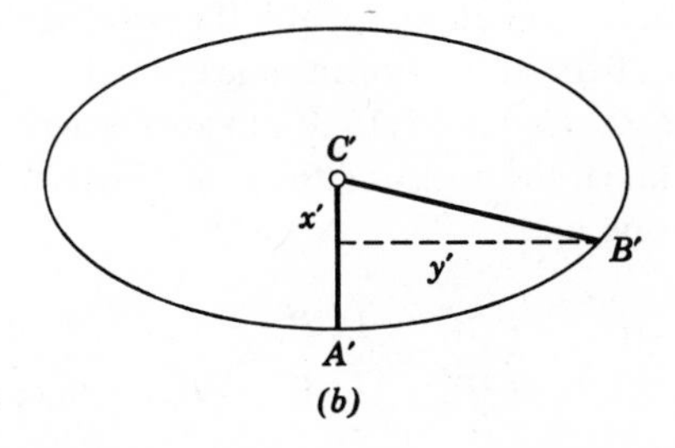

Figure 15.4 The orbit seen from above in (a) and seen from the Earth in (b).

star at an arbitrary time. Then $AC = BC = a$, and $B'C' = q$, the apparent separation of the two stars. If (x,y) are the coordinates of the point B as shown, then

$$x = a \cos \theta \qquad y = a \sin \theta \qquad \theta = \frac{2\pi}{P}(t - t_0) \tag{15.5}$$

In (15.5), the quantity t_0 is the time of mid-eclipse. The coordinates of point B' are (x',y'), where it is apparent that $x' = x \cos i$, $y' = y$. Then it follows that, since $q^2 = x'^2 + y'^2$, one has

$$q = a(\cos^2 i \cos^2 \theta + \sin^2 \theta)^{1/2} \tag{15.6}$$

This relates the apparent separation to the orbital elements a and i, plus the position angle θ. Since the period P is known, θ is known at any time from the third relation of (15.5).

It might appear from equations (15.4) and (15.6) that if the ratio of the light losses were measured at many different times, one could determine R_1, R_2, a, and i, but this is not quite correct. One cannot measure directly from the light curve any of the areas, only the ratio of the areas eclipsed. It is apparent that this ratio depends not on the absolute size of the system, but only on the relative sizes of R_1, R_2, and a. Equations (15.4) and (15.6) can then be combined into the schematic relation

$$\frac{\Delta F_p(t_1)}{\Delta F_p(t_2)} = f\left(\frac{R_1}{a}, \frac{R_2}{a}, i, t_1, t_2\right) \tag{15.7}$$

By making these measurements at many times during the eclipses (both eclipses can be used for this), one can find values of R_1/a, R_2/a, and i which best fit the light curve.

If the orbit is not circular, the above analysis still holds in general outline. Thus 5 of the 7 orbital elements discussed in Section 12 are found: P, e, ω, T, and i. Nothing can be said about Ω, and the orbit size a can be found only in terms of the radii of the stars, not in miles or AU.

Radiation Parameters. Equation (15.3) can also be written for the secondary eclipse, and if one takes the ratio of these at times such that the eclipsed areas are the same, one finds

$$\frac{\Delta F_p}{\Delta F_s} = \frac{F_h R_c^2}{F_c R_h^2} = \frac{L_h}{L_c}\frac{R_c^2}{R_h^2} = \frac{F_h\,(\text{surface})}{F_c\,(\text{surface})} \tag{15.8}$$

The subscript h or c stands for hot or cool, the L's are luminosities, the F's are fluxes measured from the Earth, and F (surface) is the flux at the surface of the star. The ratio of the radii is known from the above analysis, although one still has to determine whether the hot star is the larger or smaller one. Then measurements of the relative depths of primary and secondary eclipses will yield the relative luminosities and surface brightnesses as in equation (15.8). These quantities are known in whatever wavelength system the observations are made in. If the observations are made at enough different wavelengths so that bolometric quantities are known, then the relative effective temperatures follow from the above and equation (8.4).

Complications. There are three complications which are usually of some importance and which cause the analysis described above, and the equations that go with it, to need modification. These are limb darkening, the reflection effect, and ellipticity.

The intensity of light coming from the apparent center of a stellar disc is greater than that coming from near the edge or limb, hence the term limb darkening. The physical reason for this is discussed in Section 19. It results in a considerable complication, for the amount of light blocked off during an eclipse depends not only on the area being eclipsed, but also on which part of the stellar disc is affected. Thus equation (15.3) is not correct, and a correction factor for limb darkening must be included.

In principle the amount of limb darkening can be determined from the light curve. In practice one usually assumes that the intensity emitted at the angle θ to the normal to the stellar surface is given by

$$I(\theta) = I(\theta = 0)[1 - x(1 - \cos \theta)] \tag{15.9}$$

where x is the coefficient of limb darkening. Whereas $x = 0$ corresponds to a disc of uniform brightness, $x = 1$ is the other extreme, in which the brightness goes to zero at the limb. The value of x can be found from the light curve, although it is often assumed on the basis of the theory of stellar atmospheres. It does depend on the wavelength region of the observations.

The part of a star which is facing its companion is heated up by the radiation from this companion, and this is known as the reflection effect. It is important only for close binaries, and it is increased by the fact that most close binaries have equal rotation and revolution periods. It is the same part of the star that always faces its companion, and the hot spot can become quite appreciable, particularly on a cool star with a hot companion. The result of the reflection effect (it is more complicated than simple reflection) is that the cool star has nonuniform brightness in addition to that caused by limb darkening. The hot spot faces Earth around secondary minimum, so there will be a rise to a maximum just before and after secondary eclipse, and the total apparent brightness will not be constant outside of the eclipses.

Two close binary stars will distort each other out of the spherical shape by their gravitational fields. This effect is known as ellipticity because it is usually assumed that the stars will have an ellipsoidal shape. This has two effects on the light curve: (1) the area facing the observer varies with time, even outside of eclipses, and (2) the effective surface gravity (and hence surface brightness) varies across the surface of the star. The latter is sometimes known as gravity darkening. Ellipticity can cause the light curve to have strong variations even outside of eclipses.

The three effects mentioned above are far too complicated to be treated in an exact fashion, and the observational data are not accurate enough to warrant this anyway. Instead, it is assumed that these effects can be represented by relatively simple models or ideas, such as equation (15.9) for limb darkening. These models involve certain constants, such as the limb darkening coefficient, which must be evaluated to give the best possible fit. The unknowns to be found from the light curve, therefore, include these constants as well as the orbital elements and radiation parameters discussed previously. The solution has been greatly aided by the publication of numerous numerical tables which have been calculated for various values of these constants.

Eclipsing and Spectroscopic Binaries. A source of considerable information is an eclipsing system for which spectroscopic data from both components are also available. It is seen that the light curve supplies the value of the orbital inclination, which could not be obtained from the velocity curve by itself. Then the size of the orbit can be found as well as the individual masses. But the light curve gives the radii of the stars in units of the semi-major axis, so the radii in miles are also determined. If the luminosities can be determined, then the effective temperatures follow from equation (8.5).

Unfortunately, when spectroscopic data are available, they frequently do not agree with the data obtained from the light curve. The complications caused by close pairs mentioned in Section 14, such as mass loss from the stars and the possible enveloping of the system in a gaseous ring or shell, may be the source of the trouble. If a spectral line comes from material being ejected by one of the stars, it may not reflect accurately the orbital motion.

Eclipsing binaries are potentially the source of a tremendous amount of important information; however, in many cases some of these potential sources are only complications which limit the trustworthiness of the information which can be obtained. More accurate data and a better theoretical understanding of the effects involved are needed before astronomers can obtain maximum use of the information that is potentially available in binary stars. As an illustration, stars of the W Ursae Majoris type are examples of binaries whose components are so close together as to be essentially in contact with each other. Simple theories of ellipticity and the reflection effect are probably not adequate for them.

The investigation of a visual or a spectroscopic binary is largely dynamical in nature; that is, the laws of motion are prominent in the analysis. The eclipsing binary, however, is largely a geometric problem in that one tries to find how two stars can be placed in order to reproduce the light curve. The masses of the stars enter only if spectroscopic data can also be obtained.

Origin of Binaries. The subject of binary stars should not be left without at least a brief word about the problem of their origin. This is not an isolated point of minor concern, for probably at least one-half of all stars are in systems of two or more components. (The number of triple, quadruple, and higher multiplicity systems is surprisingly large.)

A certain set of circumstances, not well understood, will result in the formation of a binary system. These circumstances are certainly not very rare. Will a very small change in these circumstances cause three, four, or more stars to be physically connected instead of two? Are these circumstances essentially different from those that produce the star clusters containing hundreds or thousands of stars? How must things be different in order for most of the mass to be concentrated in one body, thus forming a single star or a star with a planetary system? The problem of how stars form out of the very tenuous gas and dust that exists in interstellar space is extremely difficult, and not very much is known about it today except of a rather general nature. Much progress has been made in recent years, and one can hope that the answers to questions such as those mentioned above will come in the relatively near future.

16. The Intrinsic Variables

The eclipsing binaries discussed in Section 15 are observed to vary in apparent brightness for geometric reasons; however, many stars actually have variable luminosities, and these are discussed in this section.

The most basic data about a variable star are contained in the light curve. This is usually plotted as some kind of apparent magnitude as a function of time. The different kinds of variable stars can be distinguished by the forms of their light curves, among other things. Some are quite regular, and the light curves look like distorted sine curves. These are periodic and repeat themselves quite accurately after one period. Others are known as semi-regular variables because there is a tendency for the light curve to approximately repeat itself after an approximately constant time interval; the light curve is not strictly periodic, however, and one can define only an average period. Different kinds of irregular variables are known in which the light curve fluctuates in an erratic fashion which shows little or no regularity. Finally, some stars stay at or near a constant luminosity for some time and then suddenly change their brightness, sometimes quite drastically. In the remainder of this section, examples of some of the more important kinds of variable stars are discussed.

Cepheids and Cluster Variables. Among the most important types of stars are the classical Cepheid variables, or just Cepheids. The group gets its name from the star δ Cephei, which is a variable of this type. The light curve of a Cepheid repeats itself quite regularly, and so the period is well defined. The periods range from about 2 days to about 2 months, and the total range of the light variations can be as large as 2 magnitudes for some stars and is barely detectable for others.

Many of the characteristics of Cepheids are closely correlated to the length of the period. For example, a Cepheid with a 10-day period will be very much like other 10-day Cepheids, but it will be quite different in many respects from those with much longer or much shorter periods. This observed fact allows a convenient method of obtaining the approximate properties of a Cepheid, for the period is much easier to determine than most of the other properties. The most important property that is found to be correlated with the period is absolute magnitude.

In 1912 Miss Henrietta Leavitt found that the large number of Cepheids appearing in the Magellanic Clouds, which are two nearby external galaxies, show a relation between period and apparent magnitude. The distances of the Magellanic Clouds are extremely great compared with their size, and to a very good approximation all of the stars in one of the clouds are thus at essentially the same distance from the Earth. This means that the observed period/apparent-magnitude relation indicates the existence of a period/absolute-magnitude or a period/luminosity relation for these variables. Once this relation is calibrated by the independent measuring of the absolute magnitudes of at least a few Cepheids, it can be applied to any Cepheid whose period is known.

Harlow Shapley was the first to calibrate this relation. Shapley's calibration was verified and generally accepted until the early 1950's, when Walter Baade indicated that a change was needed. Baade found that Cepheids of a given period are about 1.5 magnitudes brighter than had been previously thought. The discovery of this very large error meant that the Cepheids were actually about twice as far away as had been thought. The error was caused primarily by a confusion over different types of variables. The point is that no Cepheid is close enough for a trigonometric parallax to be obtained, and so statistical methods must be used; however, the early determinations included two other groups of variables which are now known to be quite different from the classical Cepheids, and an unfortunate weighting of the samples gave the large error. One of these groups is the Population II Cepheids, or W Virginis stars, and the other is the RR Lyrae, or cluster variables. When the latter groups were recognized as separate, the analysis could be applied to the classical Cepheids by themselves, and this resulted in the revision of the period/luminosity relation. Figure 16.1 shows this relation for the separate groups of variables, and it still has a rather large uncertainty.

Classical Cepheids are extremely luminous stars, and this makes them very

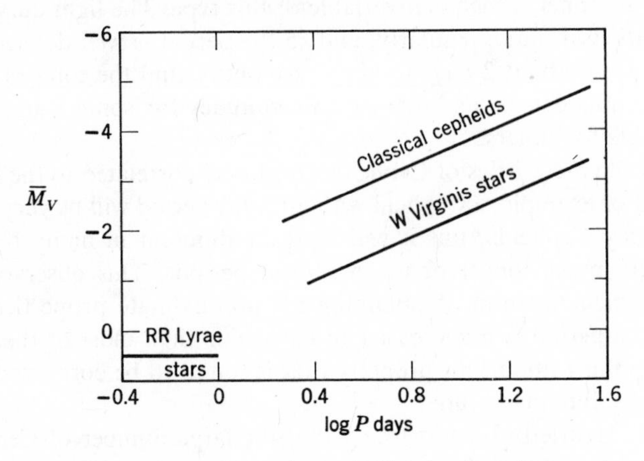

Figure 16.1 The period-luminosity relation.

important as distance indicators. Any group of stars which contains Cepheids can have its distance determined from these Cepheids, and the high luminosities mean that they can be observed at great distances. It is interesting that the relative error in finding the distance to a Cepheid is independent of its distance. This is in considerable contrast to the trigonometric parallax method, for which the relative error increases with distance. From equation (9.8), one finds

$$M = m + 5 - 5 \log r = m + 5 - 2.17 \ln r$$

where the symbol ln stands for the natural logarithm. If it is assumed that the apparent magnitude can be measured with arbitrary accuracy, then an uncertainty ΔM in the absolute magnitude will produce an uncertainty Δr in the distance of the star. If these are treated as small quantities, then one finds, by taking differentials of the above equation,

$$\frac{\Delta r}{r} = -0.46 \Delta M \qquad (16.1)$$

If the period/luminosity relation for Cepheids were accurate to within 0.25 in the absolute magnitude, an assumption that may be too optimistic, then the Cepheid distances would all be known with about a 10% accuracy. This also requires an accurate correction fŏr the effects of interstellar absorption, as discussed in Section 23.

Whereas W Virginis stars are of Population II, the classical Cepheids are Population I objects. The stellar populations are discussed in greater detail in Sections 25 and 26, but one of the important properties is the spacial distribution. Certain types of clusters of stars, certain regions of the Galaxy, and certain galaxies as a whole tend to be composed primarily of objects of one population or the other. As a result W Virginis stars and classical Cepheids do not occur together, and so the differences between the two groups were not completely realized until rather recently. The W Virginis stars seem to be about 1.5 magnitudes fainter than classical Cepheids of the same period.

The RR Lyrae variables are also of Population II, so a direct comparison with classical Cepheids to determine their relative luminosities is difficult. Because they are very abundant in certain star clusters, RR lyrae stars are also known as cluster variables. They have very short periods, from a few hours to about a day.

The W Virginis stars are rather rare and their properties not very well known, so their usefulness as distance indicators is limited. The RR Lyrae stars and the classical Cepheids compliment each other nicely, since they are of opposite populations; however, it is unfortunate that the RR Lyrae stars are much fainter than the Cepheids, and so they can be used only for relatively nearby objects. The luminosities of RR Lyrae variables seem to be independent of their periods, and the mean visual absolute magnitude is around +0.5.

These types of stars have radial velocities which vary with the same period as the light variations. It was once thought that Cepheids were eclipsing binaries, but this is now known to be incorrect. For one thing, the minimum brightness occurs at or near maximum radial velocity, i.e., when the velocity is largest away from the observer, while the orbital radial velocity must be zero during an eclipse. These are pulsating stars which periodically expand and contract. The physical conditions in a star change during the pulsation cycle, and a variable luminosity is one of the results.

The Cause of Pulsations. Why do some stars pulsate while others do not? Most stars do not have large-scale mass motions. They are in hydrostatic equilibrium, which means that the pressure forces pushing outward at any point exactly balance the gravity force pulling in at the same point.

Suppose that somehow a star were expanded to a size that would be larger than its equilibrium size. The gravity force would become smaller because of the greater distances, but the pressure force would decrease by an even greater amount. [The force of gravity is proportional to the inverse square of the distance, while pressure, which is proportional to density times temperature, must fall off as temperature divided by the cube of the distance. Since temperature itself decreases as the star expands, the pressure must decrease more rapidly than gravity for an expanding star. See, e.g., equation (21.35) for one special type of expansion.] As a consequence the gravity force would dominate, and the star would start to contract back toward its equilibrium size again.

When this equilibrium size is reached, the forces again balance, but the mass is moving rapidly inward and will overshoot the equilibrium position. Further contraction will cause the pressure to become dominant, so the contraction will eventually come to a halt and expansion will then start, etc. In general these pulsations will not continue indefinitely because various dissipative processes are also taking place, so the motions will overshoot the equilibrium size by an amount that becomes smaller each time.

It is convenient to consider these pulsations in terms of energy. There is a certain amount of energy involved in the pulsation process. Part of this is in the form of the kinetic energy of the pulsation motions, part in the form of excess gravitational potential energy, and part in the form of excess thermal or internal energy. At the moment of greatest size, the pulsation energy is all in the form of excess potential energy. As contraction starts, this is converted into the kinetic energy of contraction, and this continues until the equilibrium size is reached. As contraction continues, the kinetic energy is transformed to excess internal energy, and eventually the contraction stops. This transfer of energy goes in the opposite direction during expansion.

If there were no dissipation, there would be just enough excess internal energy during contraction to push the star back out to the same state of expansion it previously reached, so the pulsations would go on indefinitely. In practice there always is some dissipation. Some of the pulsation energy is lost in each cycle by radiative or convective processes, so there is generally not enough excess internal energy to push the star out as far as it had been the previous cycle, and the pulsations are damped out. The situation is very much like a bouncing ball, in which there is an exchange between the gravitational potential energy, the kinetic energy, and the internal energy of deformation of the ball. The frictional forces will dissipate the energy, so the ball will eventually stop bouncing unless some mechanism continues to feed energy into the bouncing process. The fact that pulsating stars are observed must mean that there is some mechanism that feeds energy into the pulsation process fast enough to replace the energy dissipated.

A. S. Eddington considered two possible mechanisms for keeping the pulsations going. One depends on the energy sources in the star, such as the nuclear reactions which are known to take place near the center. The point is that the increases in temperature and density that occur during compression also increase the energy generation rate. This provides more internal energy right at the time it is needed, so it can help offset the dissipative losses and give the star a stronger outward push. It has been shown, however, that the pulsations do not extend appreciably into the central regions where the nuclear reactions take place, so this mechanism is not a sufficient cause.

Eddington's second mechanism is a sort of valve action. Energy is constantly flowing outward in a star. If it were possible to dam up some of this energy during the compression and release it during expansion, then this dammed up energy could help offset the dissipative losses and give the compressed star the push needed to maintain the pulsations. This normally does not occur, for stellar material is usually more transparent to radiation at higher temperatures, i.e., during the compressed phases; however, in a region where an abundant element such as hydrogen or helium is being ionized, the heat capacity and absorption work in the correct direction. The regions of ionization of hydrogen and helium are thin shells in the outer parts of stars, and these thin zones are apparently responsible for stellar pulsations.

S. A. Zhevakin, J. P. Cox, R. F. Christy, and others have shown in detailed theoretical work that this mechanism actually is sufficient. The reason that all stars do not pulsate is that the size and position of the ionization zones are quite critical: if these zones are too deep in the star, the amplitude of the pulsations in them will be too small, and if they occur too far out they will not involve enough mass. Also, in many stars convection is a very important means of carrying energy through the ionization zones, and it is unlikely that the valve action can be very important in such cases.

The Analysis of Pulsating Stars. A great deal of information can be obtained from the combined light and velocity curves of pulsating stars. First, however, one must correct the observed radial velocity V in order to obtain the true pulsation velocity V_p. (The radial velocity has already been corrected for the Earth's motion.) The quantity V is the average over the visible hemisphere of the star of the radial velocity of each element of surface area, weighted by the contribution of that element to the flux observed from the star. The flux is given by equation (8.6), and so one has

$$V = \frac{\int_0^{\pi/2}\int_0^{2\pi} V(\theta,\phi)I(\theta,\phi)\cos\theta\sin\theta\,d\theta\,d\phi}{\int_0^{\pi/2}\int_0^{2\pi} I(\theta,\phi)\cos\theta\sin\theta\,d\theta\,d\phi} \tag{16.2}$$

where (θ,ϕ) are spherical angles in a coordinate system whose origin is the center of the star and whose z axis points toward the Earth. The quantity $V(\theta,\phi)$ is the radial velocity of a point on the star's surface whose position is (θ,ϕ). The quantity $V(\theta,\phi)$ is the sum of two motions: V_m, the radial velocity of the star as a whole, and $(V_p \cos\theta)$, the radial component of the pulsation velocity. If one makes the substitution $\mu = \cos\theta$, then equation (16.2) becomes

$$V = V_m + V_p \frac{\int_0^1\int_0^{2\pi} I(\mu,\phi)\mu^2\,d\mu\,d\phi}{\int_0^1\int_0^{2\pi} I(\mu,\phi)\mu\,d\mu\,d\phi} \tag{16.3}$$

If one further assumes that the star has azimuthal symmetry so that the intensity does not depend upon ϕ, and if one uses the standard limb-darkening formula of equation (15.9), then (16.3) reduces to

$$V = V_m + \frac{4 - x}{6 - 2x} V_p \tag{16.4}$$

As before, x is the coefficient of limb darkening and has a value between 0 and 1. Its value can be obtained from the theory of stellar atmospheres, so it can be considered known; however, like many other "known" quantities, it may introduce an appreciable uncertainty.

Equation (16.4) is of the same form as equation (14.1) for spectroscopic binaries. As in the latter equation, the periodic term involving V_p must bring the material back to the same position, with respect to the center of the star, after one period; therefore, V_m must be such as to bisect the area of the velocity curve, and this allows V_m and V_p to be found.

According to equation (8.5), the luminosity of a star is proportional to the product $(R^2T_e^4)$, where R is the radius and T_e the effective temperature.

Suppose a pulsating star is observed at two times, t_1 and t_2. Then

$$\frac{L_1}{L_2} = \left(\frac{R_1}{R_2}\right)^2 \left(\frac{T_e(1)}{T_e(2)}\right)^4 \tag{16.5}$$

If this is changed to magnitudes, one finds

$$m_{\rm bol}(2) - m_{\rm bol}(1) = 5 \log \frac{R_1}{R_2} + 10 \log \frac{T_e(1)}{T_e(2)} \tag{16.6}$$

The magnitudes in (16.6) are bolometric, since the radiation in all wavelengths is being considered. The complications of obtaining bolometric magnitudes from the observations are discussed in Section 17, and it is here assumed that the required bolometric corrections are available.

Another topic discussed in Section 17 is the relation between effective temperature and color for stars. If this relation is known, then equation (16.6) provides a means of determining the ratio R_1/R_2. It is important that only the ratio $T_e(1)/T_e(2)$ is required, as this is much more accurately determined than either temperature itself. Thus the light curve allows the ratio of the radius of a pulsating star at any two times to be found.

The velocity curve allows the pulsation velocity V_p to be found from equation (16.4), and one obviously has

$$R_1 - R_2 = \int_{t_1}^{t_2} V_p \, dt \tag{16.7}$$

The sign convention used is that radial velocities are positive away from the observer, so V_p is positive for a contraction of the star. Equations (16.6) and (16.7) then fix both R_1 and R_2, so the stellar radius is determined for any time. If the effective temperature is known, then the luminosity can also be found.

The color of a Cepheid, like its other characteristics, varies throughout its period. There are two times each period when the color is a given value, once during the expansion and once during the contraction, and it happens that these times do not correspond to times of equal luminosities. A. J. Wesselink chose for t_1 and t_2 two such times of equal color. Then, by virtue of the assumption that equal colors go with equal effective temperatures, the effective temperature terms drop out of equation (16.6), and the procedure is continued as described above.

There are many modifications to this type of procedure, which was first suggested by W. Baade, but there are some important uncertainties involved. For example, the regions giving rise to a spectral line may change throughout the period, so the observed radial velocity may not represent the true mass motions of the pulsation. Further, the radius of a line-forming region may be significantly different from the continuum radius that appears in equations (16.5) and (16.6).

Another characteristic of note for pulsating stars is the period-density relation. This relation is usually given in the form

$$P(\rho/\rho_o)^{1/2} = Q \tag{16.8}$$

where P is the period in days, ρ the average mass density of the star, and ρ_o the same for the Sun, and where Q is called the pulsation constant. Theory indicates that a relation of the form of equation (16.8) might hold, and observations indicate that the value of Q is approximately constant for different pulsating stars; thus it can be used to estimate the mean density of stars of known periods.

A typical value of Q is perhaps 0.05 day, which indicates that if the relation were valid for all stars, the Sun would have a pulsation period of about an hour. Variables with long periods must have low mean densities. Periods of more than a year are not uncommon. With $\rho_o = 1.41$ g/cm^3, a star with $P = 1$ yr would have a mean density of about 3×10^{-8} g/cm^3. With a fantastically low density like this, it must be difficult to determine where the star ends and where the interstellar medium begins. One can see that the concept of radius for such a star is somewhat ambiguous.

J. D. Fernie has found evidence in favor of a period-radius-mass relation of the form

$$P = kR^2\mathcal{M}^{-1/2} \tag{16.9}$$

With R and $\mathcal{M}$ in solar units, he finds $k = 0.00675$ day for at least some types of pulsating stars. If this relation is correct with k a constant for all stars, then Q of equation (16.8) is not constant, but is proportional to the square root of the radius of the star. Whether or not (16.9) is a more fundamental relation than (16.8) remains to be seen.

Long-Period Variables. There are many more types of pulsating stars. Long-period variables are an important type. The name, naturally, comes from the periods, most of which are in the range 0.5–2.0 yr. These are also called Mira variables after the star Mira. These stars are not strictly periodic. The light curve of a long-period variable only very roughly repeats itself after a time interval that fluctuates about a mean value. These are very cool stars, and they show a large range in visual magnitude.

The star Mira can be anywhere between apparent visual magnitude 2 and 5 at maximum light, while at minimum it is somewhere between 8 and 10. The period averages 332 days, and its effective temperature varies from about 1900°K to about 2600°K, according to the measures of E. Pettit and S. B. Nicholson. If the distance of Mira is 50 pc, then its absolute visual magnitude at maximum is around 0 or -1, so Mira variables are rather luminous stars.

Mira is one of the few stars whose angular size has been determined. In the 1920s F. G. Pease found an angular diameter of 0.″056. Using 50 pc as the distance, one finds that Mira is some 300 times the linear size of the Sun. This very large size is consistent with the very low mean density predicted by the period-density relation.

The radius and effective temperature data allow one to determine the luminosity of Mira. The result is that at maximum light Mira is about $(300)^2 \times (2600/5800)^4 = 3600$ times as luminous as the Sun, using 5800°K for the effective temperature of the Sun.

The low temperature of Mira has several important effects. For example, the absolute bolometric magnitude of the Sun is about +4.8, so that of Mira is

$$M_{\text{bol}}\,(\text{Mira}) = M_{\text{bol}}\,(\text{Sun}) - 2.5 \log \frac{L\,(\text{Mira})}{L\,(\text{Sun})}$$

$$= 4.8 - 2.5 \log 3600$$

$$= -4.1$$

Thus Mira is nearly 9 bolometric magnitudes brighter than the Sun. In absolute visual magnitudes, however, one finds M_v (Mira) = 0, approximately, and for the Sun $M_v = 4.8$, so Mira is only about 5 visual magnitudes brighter than the Sun. The reason for this difference is that the Sun has a moderate temperature, so that most of the energy it radiates is in the visual region; therefore, there is little difference between the visual and the bolometric magnitudes of the Sun. Mira has such a low temperature that very little of its energy is radiated in the visual region, most of it being in the infrared; therefore, Mira is much brighter in the bolometric than in the visual magnitude system.

Another illustration of this temperature effect is in the brightness range of Mira. Suppose that the changes in radius can be neglected (spectroscopic observations show that this is a good approximation) and that the star radiates as a black body (this is a poor approximation). Then if the visual magnitude system is taken to be the same as the monochromatic magnitude at 5500 Å, which is nearly correct, then the above temperatures and equation (2.2) indicate that

$$\frac{L_V\,(\text{max})}{L_V\,(\text{min})} = \frac{B_\lambda(2600°\text{K})}{B_\lambda(1900°\text{K})} = 41$$

and so the range in visual magnitudes is

$$\Delta m_V = 2.5 \log 41 = 4.0$$

The range in bolometric magnitudes is given by equation (16.6):

$$\Delta m_{\text{bol}} = 10 \log \frac{2600}{1900} = 1.4$$

Thus the bolometric range is less than $1\frac{1}{2}$ magnitudes. It is seen that observations in the visual give a very distorted view of these low-temperature objects.

There are many other types of pulsating stars, such as the β Canis Majoris or β Cephei stars, RV Tauri stars, δ Scuti or dwarf Cepheids, semi-regular and irregular red stars, and perhaps several more. A more detailed discussion of the properties of pulsating stars can be found in some of the references listed at the end of the chapter.

Novae and Supernovae. In addition to the pulsating variables there are also the eruptive variables. The two best-known kinds are novae and supernovae.

A nova is a star that suddenly flares up in brightness by perhaps 10 to 12 magnitudes, the rise usually taking place in a matter of hours or days. Absolute visual magnitudes at maximum are around -7 to -8. Maximum brightness lasts only a very short time, and it is followed by a slow and often irregular decline. After months or years the star reaches essentially its pre-outburst faintness.

Both the general appearance of the spectrum and the doppler shifts of the lines indicate that a nova ejects matter with high velocities, although the amount of matter thrown off is only a very small fraction of the total mass of the star. An expanding gaseous shell is often seen around a former nova, and a comparison of the observed rate of angular expansion with the measured doppler shifts can yield the distance of the star.

Recurrent novae are similar to ordinary novae, except that they are much fainter and they repeat the outburst. The frequency of the explosions seems to be inversely related to the maximum brightness. It has been suggested that all novea are recurrent with very long periods.

Many, perhaps all, novae are members of close binary systems, and this could be a clue to the origin of the explosions. One suggestion is that a resonance might occur between the orbital period and the natural pulsation period of the star, causing an instability in the star. Another suggestion is that matter may stream from one star to the other. If the latter is a degenerate star (see Section 5) that has lost all of its hydrogen, this accumulation of hydrogen-rich material on its surface could eventually lead to an outburst which could repeat periodically.

Supernovae look very much like novae but on a much larger scale. Absolute magnitudes at maximum are around -15, so these objects are comparable in brightness to an entire galaxy. Material is ejected with very large velocities, and, in contrast to ordinary novae, an appreciable fraction of the mass of the star may be blown off. Studies of other galaxies indicate an average of one supernova per galaxy per several hundred years, although they may be more frequent in some galaxies than in others.

The Crab nebula in Taurus is now known to be the material ejected by a supernova which was observed in the year 1054 AD. The observed rate of expansion of the material and the observed position both agree with this identification. Most of the energy given off by the Crab nebula is synchrotron radiation, which is emitted by very-high-speed electrons being accelerated by a magnetic field. Other supernovae remnants have also been discovered in our Galaxy.

The amount of energy released in a supernova explosion makes it clear that the star as a whole is affected in a major way. A possible source of this kind of explosion is described in Section 22.

PROBLEMS

1. A visual binary has a period of 100 yr, and the semi-major axis of the relative orbit is 1″.00. If one of the stars is identical to the Sun and has an apparent visual magnitude of +8.0, what is the mass of the other star?

2. An eclipsing binary has a light curve similar to that of Figure 15.2. The period is 100 days, each eclipse lasts 0.812 day, and the light remains constant at mid-eclipse for 0.172 day. Spectroscopic data show that the brighter component has a radial velocity which varies between +2.9 and +57.1 km sec^{-1}, while the fainter star varies between −8.3 and +68.3 km sec^{-1}. The combined light of the system has $m_{\mathrm{bol}} = 5.545$ normally, $m_{\mathrm{bol}} = 5.990$ during the primary minimum, and $m_{\mathrm{bol}} = 5.790$ during the secondary minimum. The distance is 50 pc. If the orbit is circular and if the plane of the orbit is in the line of sight ($i = 90°$), determine as many characteristics of the stars as possible.

3. It is shown in Section 21 that if a pulsating star changes in such a way that it is always a simple scale model of itself, then $T_e = \text{constant}/R$. [See equation (21.36.)] Suppose this is valid for a pulsating star having a period of 10 days and for which m_{bol} varies between 8.0 and 8.5. If the radial velocity of the star indicates that the radius has a total range of 10^{12} cm, estimate the mass and radius of the star.

REFERENCES

There are many works, including any general astronomy text, which contain discussions of binary and variable stars. See also References 1–4 of Chapter I, and the following:

1. Aitken, R. G. *The Binary Stars*, Dover, New York, 1964.
2. Hynek, J. A. (Ed.). *Astrophysics*, McGraw-Hill, New York, 1951.
*3. Struve, O., and Zebergs, V. *Astronomy of the 20th Century*, Macmillan, New York, 1962.

Stellar masses and the mass-luminosity relation are covered in Reference 4 of Chapter I and in the following:

* This is on a popular or semi-popular level.

4. Eggen, O. J. *Astron. J.*, **70**, 19, 1965.
5. Popper, D. M. *Ann. Rev. Astron. Astrophys.*, **5**, 85, 1967.

Pulsating stars are discussed in the following:

6. Christy, R. F. *Ann. Rev. Astron. Astrophys.*, **4**, 353, 1966.
7. Zhevakin, A. *Ann. Rev. Astron. Astrophys.*, **1**, 367, 1963.
8. Rosseland, S. *The Pulsation Theory of Variable Stars*, Dover, New York, 1964.

III

Astrophysics

With the previous chapters as background, the reader is now prepared for a systematic investigation of stellar astronomy. This chapter develops the facts and physical principles which are thought to be important in an understanding of individual stars and their environment, the interstellar medium. Sections 17 and 18 give an observational introduction to astrophysics, and the next three sections present the theory which is needed to interpret the observations. In Section 22 theory and observation are brought together in the study of the most fundamental topic in stellar astronomy, the life histories of stars. The chapter is brought to a close with the description of the interstellar medium in Section 23.

17. Colors and Temperatures of Stars

The apparent brightness of a star is usually measured in some magnitude system. The most common systems are those introduced by H. L. Johnson in the early 1950s, and they are called the ultraviolet, blue, and visual systems, or U, B, and V. Each of these systems has a rather narrow response function $\phi(\lambda)$, hence they differ little from monochromatic magnitudes. The average wavelength of each system does depend some upon temperature, but over a wide temperature range the average wavelengths of U, B, and V are very close to 3700 Å, 4450 Å, and 5500 Å, respectively.

The color of a star is defined as the difference between two magnitude systems for that star. Thus the three magnitudes (U,B,V) define two independent colors for stars, the $(U - B)$ color and the $(B - V)$ color. These colors are extremely important measures. One reason is that they are rather easy to make, but more to the point, a star's color may be expected to reveal something about its physical properties. The material covered in Section 2 might lead one to expect that the color of a star should be related to its temperature, and this is correct.

Black-Body Colors. Stellar energy distributions do appear rather similar to those of black bodies, at least in general form; therefore, it is worthwhile to calculate the colors of black bodies in order to see if this can help in the interpretation of actual colors of stars.

As is shown in Section 9, the apparent visual magnitude of a star can be written

$$V = C_V - 2.5 \log F_V \tag{17.1}$$

$$F_V = \int_0^\infty F_\lambda \phi_V(\lambda)\, d\lambda \tag{17.2}$$

The fact that the visual magnitude system is being used here as the example is of no special significance; any other system could be used without changing the procedure. In the above, F_λ is the monochromatic flux measured at the Earth and corrected for the Earth's atmosphere, C_V is the visual system constant, and ϕ_V is the visual response function. Since ϕ_V is a rather narrow function, one can write approximately

$$F_V = \Delta\lambda_V F_{\lambda_V} \tag{17.3}$$

where λ_V is the average visual wavelength (about 5500 Å) and $\Delta\lambda_V$ is the average width of ϕ_V (about 900 Å). If this is put into equation (17.1), one has

$$V = C'_V - 2.5 \log F_{\lambda_V} \tag{17.4}$$

where the constant term involving $\Delta\lambda_V$ has been included in the new constant C'_V.

So far nothing has been said about the radiation, but it is now assumed that it is black-body radiation of temperature T. Then equation (2.4) indicates that the flux in the outward direction at the surface of the star is $\pi B_\lambda(T)$, where the latter is the monochromatic Planck function given by equation (2.1) or (2.2). But equation (8.7) shows that the flux measured at any point is inversely proportional to the square of the distance to the star, so one has

$$F_\lambda(r) = \frac{R^2}{r^2} \frac{2\pi hc^2}{\lambda^5} \frac{1}{e^x - 1} \tag{17.5}$$

where $x = hc/\lambda kT$, R is the radius of the star and r is its distance. If this is substituted into equation (17.4) with all of the constants being absorbed into a new C''_V, one has

$$V = C''_V + 12.5 \log \lambda_V + 2.5 \log (e^{x_V} - 1) - 5 \log \frac{R}{r} \tag{17.6}$$

A corresponding equation will hold for any other magnitude system. The apparent magnitude depends on (R/r), the angular size of the star as seen from the Earth; however, all magnitudes depend on this in exactly the same way, so this term will drop out when any color is determined. Using the U and

B magnitudes, one finds the following expressions for the colors:

$$(B - V) = C_{BV} + 12.5 \log \frac{\lambda_B}{\lambda_V} + 2.5 \log \frac{e^{x_B} - 1}{e^{x_V} - 1}$$
$$(U - B) = C_{UB} + 12.5 \log \frac{\lambda_U}{\lambda_B} + 2.5 \log \frac{e^{x_U} - 1}{e^{x_B} - 1} \qquad (17.7)$$

Everything in the above equations is known except the two constants C_{BV} and C_{UB}. These constants are rather difficult to determine because the magnitudes are defined in terms of actually observed stars, rather than in terms of an absolute energy distribution. One would have to compare in detail the energy distribution of a star of known color and a black body. This has been done by C. R. Lynds, H. Arp, and others, but here a much simpler procedure will be used. The quantities C_{UB} and C_{BV} will be defined so that a black body of the same temperature as the Sun will have the same colors as the Sun. If stars were black bodies, this would be the same as the more complicated procedure. Since stars are not black bodies, the colors found here will differ by a constant from colors on the correct (U,B,V) system. To make this difference explicit, primes will be used on the magnitudes, thus $(B' - V')$. [Because the average wavelengths and widths of the magnitude systems are not quite constant, as has been assumed here, the differences between the colors on the (U,B,V) and the (U',B',V') systems will not be exactly constant; they will depend slightly on temperature.]

For the Sun,

$$T_e \text{ (Sun)} = 5800°\text{K} \qquad U - B = +0.10 \qquad B - V = +0.62 \qquad (17.8)$$

In order to reproduce these colors from equations (17.7), one must have $C_{BV} = +0.60$, $C_{UB} = -0.10$. Equations (17.7) then give the black-body colors on the (U',B',V') system as shown in Table 17.1.

Note that the colors become more negative as the temperature rises. At

TABLE 17.1. BLACK-BODY COLORS

T (°K)	(B′ − V′)	(U′ − B′)	T (°K)	(B′ − V′)	(U′ − B′)
3000	1.68	1.27	10,000	0.17	−0.37
4000	1.13	0.67	12,000	0.06	−0.47
5000	0.80	0.32	14,000	0.00	−0.55
5800	0.62	0.10	16,000	−0.05	−0.60
6000	0.58	0.09	20,000	−0.11	−0.67
8000	0.32	−0.19	∞	−0.32	−0.90

higher temperatures, more energy is radiated at the shorter wavelengths, and so the U' region of the spectrum increases in importance with respect to the B' region, and likewise for the B' and V' regions. But magnitudes become smaller as the source becomes brighter, so the observed effect can be understood. It is interesting that the colors become less sensitive at the higher temperatures. These colors are useful only at those temperatures for which a fairly large fraction of the energy is radiated near the wavelengths of the magnitude systems. At the highest temperatures, one would obtain more useful information by using magnitude systems with wavelengths much shorter than those given; likewise, for very low temperatures one should use magnitudes far out into the infrared.

The question now arises: How accurate is the above color-temperature relation when applied to real stars? If the $(B - V)$ color of a star is $+0.17$, can one be sure that it has an effective temperature of 10,000°K? There is one easy way to make a check on this point: find out if the relation between $(B' - V')$ and $(U' - B')$ for black bodies is the same, except for a constant shift, as the $(B - V) - (U - B)$ relation observed in stars. This is known as a color-color relation for obvious reasons, and Figure 17.1 illustrates this. The solid curve is the black-body relation taken from Table 17.1, and the dashed curve is that observed for certain types of "normal" stars. The black-body curve has been adjusted so that the two lines intersect at the position of the Sun. The black-body curve is nearly a straight line, while the stellar color-color curve deviates quite significantly from a straight line. This is

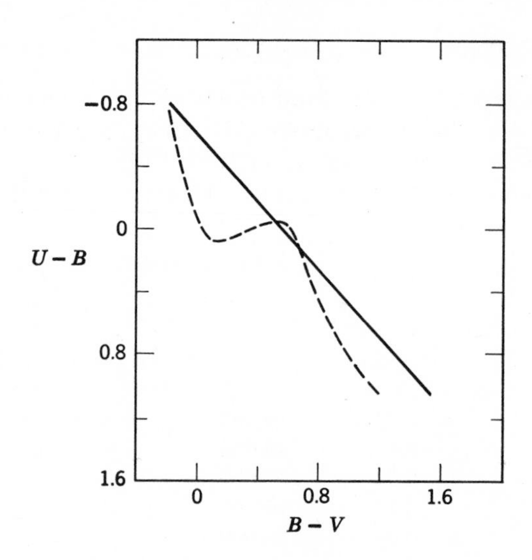

Figure 17.1 The color-color relation for black bodies (solid line) and stars (dashed line).

particularly true at high temperatures. This means that one must use considerable caution in applying the black-body color-temperature relation to the stars. [It happens that most of the difference between the two curves in Figure 17.1 is due to the $(U - B)$ colors; the black-body $(B - V)$-temperature relation is a fairly good approximation to that of the stars.]

Calibration of the Color-Temperature Relation. The above does not mean that stellar colors cannot yield effective temperatures; it means that black-body colors are not completely reliable indicators of stellar colors. One still expects a strong correlation between color and T_e for stars, but some other means of calibrating it (i.e., of finding the numerical values in the relation) must be found. Once this relation has been calibrated, the effective temperature of any star can be found simply by measuring its color.

This procedure brings up two separate problems. First, it must be possible to determine directly the effective temperature of at least a few stars, so that the T_e-color relation can be calibrated. Second, the color of a star may be affected by other quantities as well as by T_e, and thus different types of stars may have different T_e-color relations. Astronomy consists to a large extent of calibrating observations, and these two related problems occupy a significant part of an astronomer's efforts.

There are essentially two methods of determining the effective temperature of a star. One is based on studies of the structure of the outer layers of stars, and this general approach is discussed in Section 19. The other is basically an observational procedure and will be described now.

From the material covered in Sections 8 and 9 it is easily shown that the apparent bolometric magnitude of a star can be expressed as

$$m_{\text{bol}} = C_{\text{bol}} - 2.5 \log F$$

$$= C_{\text{bol}} - 2.5 \log \left(\frac{L}{4\pi r^2}\right)$$

or

$$m_{\text{bol}} = C_{\text{bol}} - 5 \log \frac{R}{r} - 2.5 \log \sigma T_e^4 \tag{17.9}$$

Determining the effective temperature of a star can be reduced to the problem of measuring its bolometric magnitude and its angular size.

Measuring m_{bol} is not at all straightforward. Most observatories are not equipped to measure easily the radiation in all wavelengths, and even if they were, the Earth's atmosphere absorbs certain wavelengths so completely that to make an accurate correction for it is very difficult. The atmosphere is opaque to radiation below about 3000 Å, so bolometric magnitudes for very hot stars must be based on theory or must be measured from above the atmosphere.

Astronomers usually do not express their results in bolometric magnitudes, but prefer to use the visual magnitude and a quantity called the bolometric correction, or *BC*. The *BC* is just the difference $(V - m_{\mathrm{bol}})$, and so it is a sort of color of the star. (Sometimes the *BC* is defined as the negative of this.) The point is that m_{bol} is very difficult to measure, so most of the time an astronomer will measure only *B* and *V*. The *BC*, like other colors, depends primarily upon the temperature of the star; therefore, the relatively few stars that do have measured values of m_{bol} can be used to calibrate a $BC - (B - V)$ relation. Approximate values of *BC* and m_{bol} are then found for any star from its measured values of *B* and *V*.

Figure 17.2 shows the $BC - (B - V)$ relation for a group of "normal" stars. The *BC* is actually defined so that its minimum value for these stars is zero, and it is seen that this minimum occurs for stars with $(B - V) = 0.45$. This then defines the constant C_{bol} appearing in equation (17.9).

The *BC* is a measure of the ratio of the total energy radiated by a star to that which it radiates in the visual region of the spectrum. For very hot or very cool stars, this ratio is quite large, since they have such a small part of of their energy in the visual. For moderate temperatures, most of the energy is radiated in the visual and the *BC* is least. Although stars are not black bodies, temperature effects in stars are very pronounced, and this general type of reasoning does apply to them.

A comparison of the energy distribution of the Sun with that of other stars indicates that $BC = 0.07$ for the Sun. With $M_V = +4.84$ (from the work of J. Stebbins and G. E. Kron), the absolute bolometric magnitude of the Sun

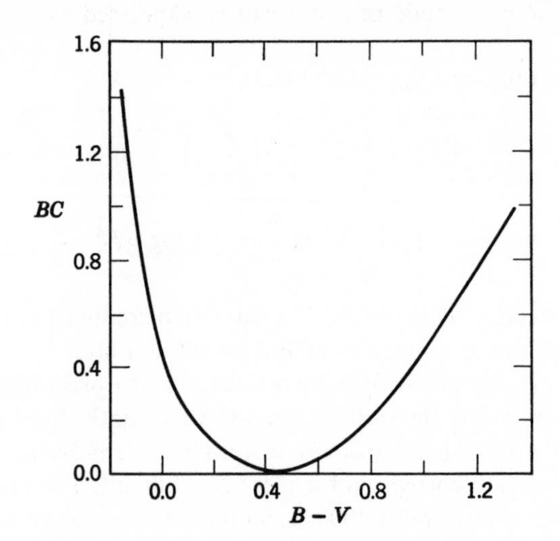

Figure 17.2 Bolometric correction/color relation.

is +4.77. Slightly different values from these are often quoted. For any star of luminosity L, one has

$$M_{bol} = 4.77 - 2.5 \log \frac{L}{L\,(\text{Sun})} \tag{17.10}$$

The second part of determining the effective temperature of a star is measuring (R/r), the apparent size of the star as seen from the Earth. According to Section 15, it is possible to find the radii of certain eclipsing binaries, and this can be done only if the distance to the system is also known. For such a binary system, the effective temperature can immediately be found; however, these systems tend to be close binaries, and these are not typical stars. One must show some care in using these stars to calibrate a color-temperature relation to be used for stars in general.

A second method is the direct measurement of (R/r), using the stellar interferometer. The interferometer is an instrument which was invented by A. A. Michelson, and it used the interference phenomenon of light waves to measure extremely small angles, angles too small to be seen or photographed directly. Michelson and F. G. Pease, in the 1920s, measured (R/r) for several stars, and their effective temperatures can be found. But these are all unusually large stars, as one might expect, so they are not of much help for calibrating the color-temperature relation for normal stars. More recently R. H. Brown, R. Q. Twiss, and others have applied a variation of the Michelson interferometer to the determination of the angular diameters of some other stars. Results for the two stars Sirius and Vega have been published, and they have provided important calibration points for the color-temperature relation. The method of Brown and Twiss is being applied to other stars of bright apparent magnitudes.

Figure 17.3 shows the result of all of this effort. This color-temperature relation does not differ very much from the black-body relation of Table 17.1 for moderate temperatures, but the differences become considerable for very hot and very cool stars. The curve in Figure 17.3 is also less reliable for these temperature extremes. One can now use Figure 17.3, or a revised version if improved data become available, to find the effective temperature of any star of known $(B - V)$ color, if one is convinced that the star is sufficiently normal. Section 18 will help to clarify the question of what stars are normal. Of course, a separate calibration curve must be determined for any group of abnormal stars, if enough data are available, so the analysis is not restricted to normal stars.

18. Stellar Spectra and H-R Diagrams

The spectra of most stars show dark lines superposed on a continuous background. In view of Kirchhoff's spectroscopic laws mentioned in Section 3,

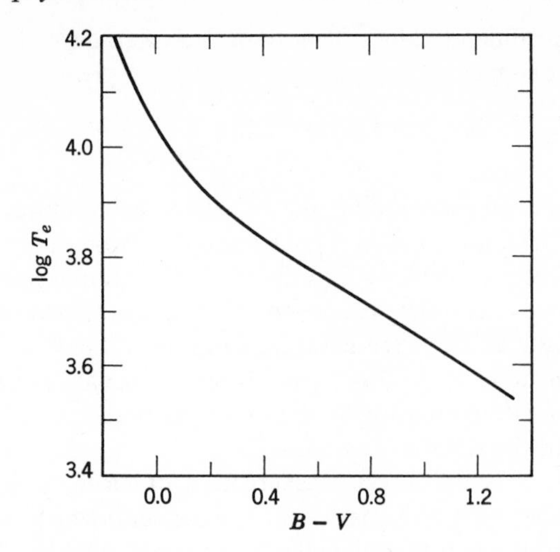

Figure 17.3 Effective temperature/color relation.

this suggests that stars consist of a hot source of continuous radiation surrounded by a cooler, tenuous gas that produces the absorption lines. The lines are somehow formed by transitions between different bound levels of the atoms in the gas.

In 1802 the dark lines in the solar spectrum were discovered by William Wollaston, but a systematic study of them was first made by Joseph Fraunhofer in the 1820s. Many of the stronger lines are still known by the letters that Fraunhofer gave to them, and absorption lines are often called Fraunhofer lines. Gustav Kirchhoff formulated his spectroscopic laws in 1859, laws which gave a qualitative explanation of the different types of spectra. The first identification of chemical elements in stars using spectral lines was made by William Huggins in 1864.

Some spectra are very complicated, showing thousands of lines without any obvious order, while others are quite simple in appearance. Sometimes emission lines are observed, but most of the lines are in absorption. It is not difficult to identify the lines, at least in principle. Laboratory spectra of different elements and compounds are observed at various temperatures and electron pressures, and they are compared with the stellar spectra. Coincidences in relative wavelength and strength provide the identification. In practice there can be some complications, but most of the more conspicuous lines have been identified. This immediately tells what some of the elements are that stars are made of, but it requires a highly complicated analysis to find out the amounts of the different elements in the stars. This type of analysis of the chemical abundances is discussed in Section 20.

It is the presence of the very narrow lines in the spectra of stars that makes possible the measurement of accurate radial velocities. The wavelength observed in the star is compared with that measured in the laboratory for the same line, and if the difference is due to the doppler effect, then the radial velocity is found.

Spectral Classification. The appearance of stellar spectra may be expected to be related to a number of the properties of a star, such as composition, temperature, and the like. One would like to set up some objective means of describing the appearance of the spectra, and then calibrate these descriptions in terms of the relevant physical quantities, as in the color-temperature calibration discussed in the last section. This requires that a classification be set up which depends only on the observed properties of the spectra.

Angelo Secchi made one of the first classifications of stellar spectra in 1863, but it was the Harvard group, notably Miss Antonia Maury, Miss Annie Jump Cannon, and E. C. Pickering, who brought spectral classification almost to its present state. The object in classifying spectra is to find an arrangement of them such that all of the features change as smoothly as possible as one moves from one end of this arrangement to the other. This step is obvious when colors are being considered. A color is simply a number, and when the stars are arranged in numerical order of their colors, they have been placed in a smooth color sequence or classification. A spectrum is a much more complicated thing, and it is by no means obvious that a single sequence of them will be sufficient.

A classification scheme in common usage divides stars into seven groups designated by the letters O, B, A, F, G, K, and M (mnemonic device: Oh Be A Fine Girl, Kiss Me). A few other letters are also used, but most stars are included in these seven. Each letter group except O is then divided into about 10 subdivisions which are indicated by numbers from 0–9. The O group only runs from 5–9. Thus a B5 star is roughly halfway between B0 and A0 in the general appearance of its spectrum. The spectral features of different stars change rather smoothly from early O to late M. (The term *early* is often used to mean occurring in the spectral sequence closer to O5, and *late* means closer to M9. Thus an early A star is one around A0–A2, while a late A star is one whose spectral type is about A7–A9.) This classification was used by the Harvard group for stars published in the Henry Draper Catalogue, and it is known as the Draper classification.

A scheme such as the Draper classification differs in two important respects from the color classification already discussed. First, the latter is a continuous sequence, since the color of a star can be any of a continuous set of numbers, while the Draper system is discrete. A star can be in one of only a finite number of spectral slots allotted to them, and so these slots must be chosen with care. For example, if there is too great a difference in appearance between an F2 and an F3 star, then information will be lost by forcing an intermediate star into one of the other of these classes. If there is too small a difference between them, on the other hand, the specification of F3 may really mean anywhere from F1 to F5, and the precise classes lose their meaning. It is because of these points that certain numbers are not used and certain others have been introduced. For example, the difference between an O9 and a B0 spectrum has been found to be too large, so the intermediate class O9.5 has been introduced.

The second point of difference between color and spectral classes has already been mentioned: while a color sequence is one-dimensional, it may not be possible to put all stars on a one-dimensional spectral sequence. Certain lines in a spectrum may indicate one class while other lines in the same spectrum indicate a different class. Sometimes one sees a classification such as A0p, which shows that the star is approximately A0 but that there is something peculiar about it. To uniquely classify all stars, a multi-dimensional scheme may be necessary. Spectral classification is much more complicated than color classification, but spectra are the source of much more information than colors.

In spite of the above complications, most stars can be put very nearly into a one-dimensional sequence based on the appearance of their spectra. This means that, if certain lines indicate one spectral class, then other lines in the same spectrum will usually correspond to the same class. This very important fact will be discussed further. Again, not all stars can be uniquely classified in this way.

In the multi-dimensional classifications there are two general points of view: one can accurately measure a few features or one can roughly estimate a large number of them. D. Barbier, D. Chalonge, and L. Divan have developed a three-dimensional classification of the former kind. Of the same type is a classification of Bengt Strömgren, who uses narrow filters and makes the measures photoelectrically. This may then seem more like a three-dimensional color classification than a spectral classification, but—in constrast to most color systems—the filters are chosen to be sensitive to selected individual spectral features. The point of these types of classifications is that a few features are found which are very sensitive to the physical quantities responsible for variations in stellar spectra, and these features are measured very accurately. One can then form a set of continuous sequences into which all stars can fit.

The other type of classification is one in which simple eye estimates of a large number of lines are made. In this method stars of abnormal properties are less apt to be overlooked, and these systems are potentially the source of more information, since more features are examined; however, some stars may have ambiguous classifications. The Yerkes or MK system of W. W. Morgan and P. C. Keenan is of this type, and since it is one of the most common in use today, it will be described in greater detail.

The MK system is two-dimensional, and one of these dimensions is essentially the same as the old Draper classes, running from O5 to M9 and containing nearly 70 discrete classes. The second dimension is not nearly as elaborate, and it contains only six groupings called luminosity classes. These are designated Ia, Ib, II, III, IV, and V, although intermediate classes such as Iab, IV–V, and the like are some times used. The complete spectral description is given by the combination of both the spectral and luminosity classes, for example, as B0 III, G2 V, and so on. It should be remembered that this classification into a two-dimensional spectral-luminosity class is made on the basis of the apperance of the spectrum alone, without regard to the physical interpretation of this appearance.

Most stars fit uniquely into one of the several hundred spectral-luminosity classes, but there are some important exceptions. The fact that the spectral dimension has about 10 times as many slots as the luminosity dimension indicates that one dimension is nearly enough by itself. This does not mean that the difference between O and M stars is necessarily greater than that between Ia and V stars; it only means that the spectra are more sensitive to whatever physical quantities are varying along the spectral dimension than to those that vary along the luminosity dimension.

A brief description of the appearance of the different spectral types, observed in the blue-visual region of the spectrum, will now be given. Only a few of the more conspicuous features are mentioned, and Figure 18.1 illustrates the relative strengths of some of these features.

O: There are very few lines in the visible spectrum, and these are mostly ionized helium and highly ionized silicon, nitrogen, and so on. The Balmer lines of hydrogen are easily visible.

B: Ionized helium is replaced by neutral helium which rises to a maximum and then disappears at late B. Lines from much the same elements as in O stars, but of lower ionization stages. The Balmer lines become very strong at late B.

A: Balmer lines of hydrogen completely dominate at early A. Singly ionized metals fairly strong, the number of lines visible increases toward late A.

F: Balmer lines are much weaker, but still strong. Ionized metals are getting weaker as neutral metals increase. Spectra become more complicated toward later types.

G: Ionized calcium lines H and K are strongest as the Balmer lines are still weakening. Spectra filled with ionized and neutral metal lines.

K: The strongest lines are those of neutral metals, and a few molecular bands of titanium oxide are present at late K. The Balmer lines are still visible.

M: The molecular bands of titanium oxide dominate the spectra, with the neutral metals being very strong.

The above descriptions have to do only with the spectral classes. The differences between the luminosity classes are much smaller. Generally, the

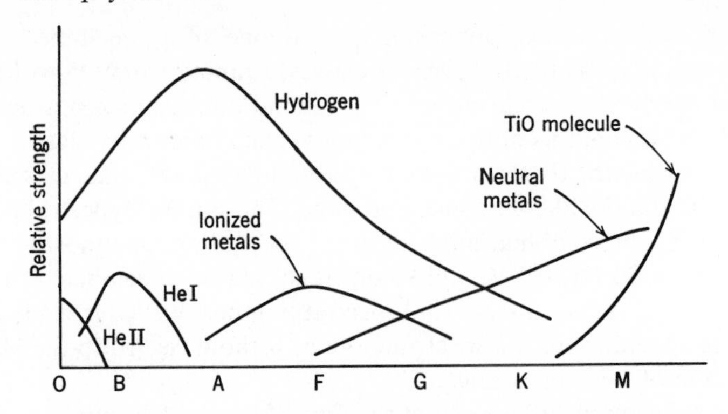

Figure 18.1 Approximate line strengths for different spectral types.

stronger lines become much sharper toward class Ia, and more lines are usually visible for these than for V stars at the same spectral class. There are many other details that define the luminosity classes. Figure 18.2 shows examples of some of the spectra.

Calibration with Physical Quantities. Before the spectral types can be calibrated with the relevant physical quantities, these quantities must be identified. What changing physical conditions are responsible for the observed changes along the spectral and luminosity classes? It was once thought that differences in chemical composition are the main cause: the O and B stars are unusually rich in helium, A stars are almost pure hydrogen, and so on. But M. N. Saha showed in the 1920s that the spectral differences are easily understood in terms of variations in excitation and ionization conditions in the stars. In view of the discussion in Section 4, this means that the spectral-luminosity sequences are temperature and electron-pressure sequences.

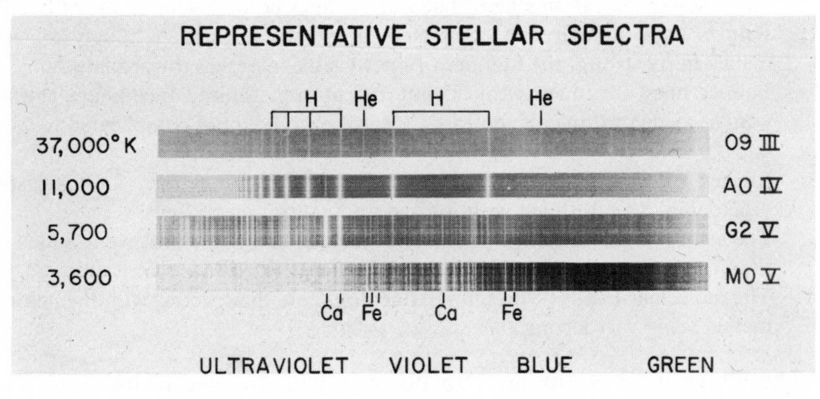

Figure 18.2 Examples of MK spectrum types. (Kitt Peak National Observatory.)

Helium is very difficult to excite, so only the hottest stars show strong helium lines. The Balmer lines of hydrogen arise from the level $n = 2$. It takes a rather large amount of energy to excite hydrogen to this level, but too much energy will ionize it; therefore, these lines reach maximum prominence at moderately high temperatures and fall off for both higher and lower temperatures. Neutral metals have very low ionization potentials, so their lines show only in the coolest stars, and the same reasoning applies to molecules. This does not mean that abundance is unimportant. Helium lines will not appear in a star that does not have any helium; however, even if helium is present, the conditions of temperature and electron pressure must be in the proper ranges for the helium lines to be visible.

The appearance of a spectral line depends on temperature, electron pressure, and the abundance of the element in question. There may be other quantities involved, such as magnetic field or rotation of the star, but these will be assumed to be unimportant for the time being. This then suggests that a complete classification scheme for stellar spectra must have three dimensions. This does correspond to the Barbier-Chalonge and the Strömgren systems, but the fact that a one-dimensional classification is nearly satisfactory indicates that one of the quantities is much more important than the others, or at least that the three physical effects do not have large independent variations.

By using the color-temperature relation discussed in Section 17, one can show that the spectral sequence is essentially a temperature sequence. The O and B stars have blue colors (small or negative values of $B - V$), while M stars are quite red (large positive values of $B - V$), and the corresponding temperature differences are sufficient to explain almost all of the spectral variations from O to M stars. This means not that the abundances and the electron pressure are constant along the spectral sequence, but only that the changes they produce are of secondary importance.

It may be surprising that the chemical abundances, perhaps the most obvious cause, turn out to be relatively unimportant. This is partly because variations in the chemical composition are smaller than one might have supposed, and partly because conditions are often such that fairly large differences produce only small changes in the appearance of the Fraunhofer spectrum. It is known that there are large variations in composition among certain stars, and this effect is a major contributor of the objects which do not fit uniquely into the two-dimensional MK system; however, most stars do fit into this system, and for these the effects of abundance variations are very small. One expects, therefore, that the physical quantities most important in causing the different spectral-luminosity classes are temperature and electron pressure.

The two-dimensional MK system can be calibrated with temperature and

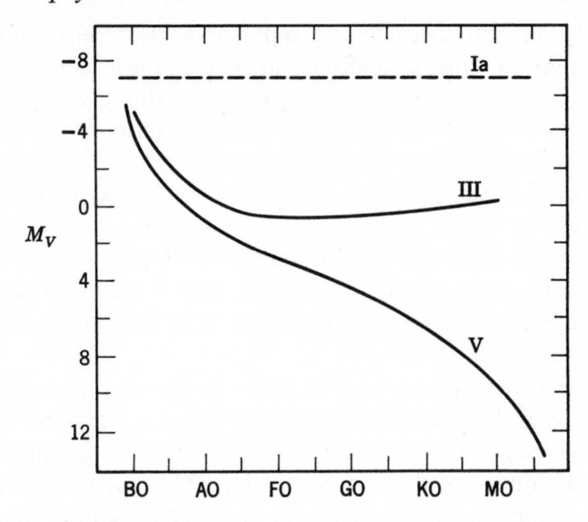

Figure 18.3 An H-R diagram for the MK classes.

electron pressure, but this does not mean that there is a one-to-one correspondence between them. In fact, a spectral class is determined mainly by its average degree of ionization. Then if either temperature or electron pressure varies with luminosity class (if neither does, then what causes the difference?), the other must also in order to keep the average ionization degree approximately constant. It is still true that the spectral sequence is primarily a temperature sequence and the luminosity sequence primarily an electron-pressure sequence.

H-R Diagrams. Shortly after 1910 Einar Hertzsprung and Henry Norris Russell independently discovered a correlation that astronomers have been very busy with ever since. This correlation is usually given in the form of a diagram called the Hertzsprung-Russell or H-R diagram. An H-R diagram is a plot in which a star is represented by a point. The abscissa is some quantity which is related to the temperature of the star, such as log T_e, $(B - V)$, or spectral type, and the ordinate is some measure of the luminosity of the star, such as log L or some kind of absolute magnitude. Traditionally these coordinates are arranged so that temperature increases to the left and luminosity increases upward. This tradition is so strong that many astronomers would not recognize an H-R diagram plotted in a different way. Figure 18.3 is an H-R diagram for stars which are normal with respect to the MK system.

Figure 18.3 does not show individual stars, but only lines which indicate constant luminosity classes. Class Ia is indicated with a dotted line, since the absolute magnitudes are quite uncertain. Classes Ib, II, and IV are intermediate between the ones shown. It now becomes obvious why the luminosity classes are so-called: apparently the electron pressure is closely related to the luminosity of a star.

The H-R diagrams are important for a number of reasons, and one obvious reason is that, if a star is normal in the MK system, then its absolute magnitude can be found from its spectral-luminosity classification, if the appropriate calibration has been made. Knowing both the absolute and apparent magnitudes, one can find the distance to the star. Parallaxes found in this way are known as spectroscopic parallaxes, and these are extremely useful for stars which are too far away for their trigonometric parallaxes to be measured. Of course, trigonometric parallaxes or some other independent method must be used to calibrate the absolute-magnitude/spectrum relation.

The temperature-spectrum relation is most easily calibrated by means of the color-temperature relation which was discussed in Section 17. The electron pressures will be considered again in Section 19, but for the present it is far more important that the luminosity of a star can be found from its spectrum than that the value of P_e might also be found.

It is interesting that, while a correlation between the observed features of a spectrum and the temperature, electron pressure, and composition of a star would be expected, one also finds the spectrum to be related to the luminosity of a star. But luminosity and effective temperature determine radius, so the radius of a star is also related to the appearance of the spectrum. Perhaps mass, age, in fact all intrinsic properties of a star are related in some fashion to the apperance of the spectrum. If this is correct, then it is only necessary to perform the required calibration in order to determine any property of a star from its spectrum. This would mean that if two stars had identical spectra, the stars would be identical in all respects.

Suppose that this idea is correct. Under the assumption that such effects as magnetic fields, rapid rotation, membership in close binary systems, and the like can all be ignored, there is good reason to believe that the appearance of a spectrum depends on the values of two physical quantities plus chemical composition. These two quantities have been identified with temperature and electron pressure, but there may be two other quantities that are even more fundamental and which in turn fix T_e and P_e. In other words, all properties of a star may be fixed by the values of two quantities plus composition. The theoretical basis for this statement is a fundamental part of the study of the structure and evolution of stars, and the two most basic physical quantities relating to a star will later be identified with the mass and the age of the star.

Figure 18.3 does not indicate the relative numbers of stars of different kinds. In each unit volume of space, at least near the Sun where most of the data come from, nearly all of the normal stars are on the V sequence, many more at late types than at early types. There is another concentration of stars along the

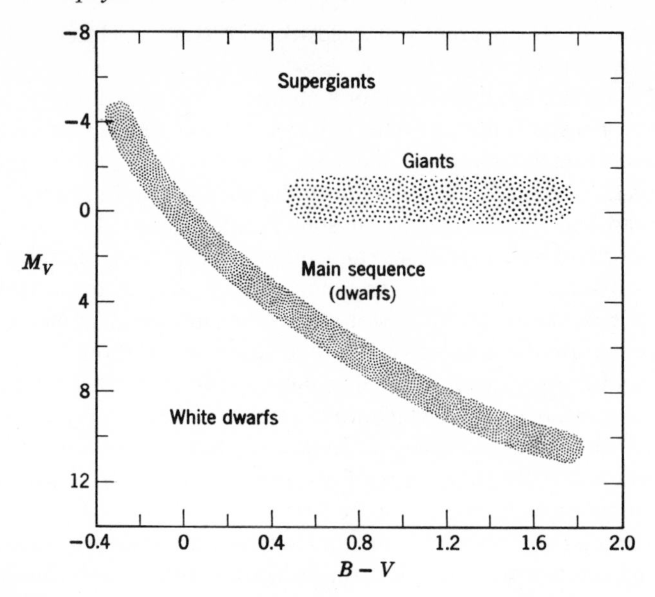

Figure 18.4 An H-R diagram for all stars. Only some of the more important groupings are shown.

III sequence, and the other regions of the H-R diagram are rather sparsely populated. Ia stars are extremely rare, and a rather large number of them are known only because they can be seen for very great distances.

One of the disadvantages of Figure 18.3 is that not all stars can be plotted on it, only those stars that have a unique classification on the MK system. Of course, one could force a star into some classification, but this would not provide a true physical picture, and the calibrations would not necessarily be valid for it. From what has already been stated, one expects the spectral classes to be closely related to color, and each star has a unique color; therefore, an H-R diagram in which, for example, M_V is plotted against $(B - V)$ will provide a position for all stars, no matter how peculiar they are. Such a diagram is Figure 18.4 The coordinates could have been log L versus log T_e, or any of a number of other quantities, but color and absolute visual magnitude are the most common because they are usually the easiest to measure. The main features of Figures 18.3 and 18.4 are the same, but the latter has some new details owing to the presence of the "peculiar" stars.

Stars which are in special parts of the H-R diagram are given special names. Thus the large group running from upper left to lower right is called the main sequence, and main sequence stars correspond to luminosity class V in the MK system. The nearly horizontal branch contains the giants, class III among MK stars. The very brightest stars at the top are known as supergiants, and correspond to luminosity classes Ia and Ib. In view of the latter names, main-sequence stars are often called dwarfs. A special group of stars that is below the main sequence, yet that has fairly high temperatures, is called the white dwarfs. The white dwarfs are shown in Figure 18.4, but they do not appear in Figure 18.3 because of their peculiar nature.

One physical quantity that can be found from the position in the H-R diagram is the radius, since L is proportional to $R^2T_e^4$. If a plot of M_V versus $(B - V)$ is all that is available, then the calibration of these into L and T_e is also necessary. It is assumed that the bolometric corrections and the color-temperature relations are known. Lines of constant radius on an H-R diagram run from lower right to upper left, L and T_e increasing together in such a way that (L/T_e^4) remains constant. These lines are nearly parallel to the main sequence, but not quite so steep. This means that radius changes only very slowly along the main sequence, earlier stars having slightly larger sizes. For stars off the main sequence, however, the radius can vary by a tremendous amount.

A cool star has very little energy being radiated per unit area, a hot star has much. Stars in the upper right corner of the H-R diagram, the cool giants and supergiants, have a large luminosity in spite of low energy output per unit area; therefore, they must have a very large surface area, and so they have large radii. On the lower left part of the H-R diagram, the region of the white dwarfs, one finds the opposite situation. The stars have a very low luminosity in spite of a large energy output per unit area; these stars must have very small surface areas, and so they have small radii. As examples of these extremes, consider a white dwarf with $T_e = 10{,}000°K$ and $M_V = 10$, and a cool supergiant with $T_e = 3000°K$, $M_V = -7$. If the bolometric corrections are assumed to be 0.6 for the white dwarf and 2.0 for the supergiant, then the absolute bolometric magnitudes turn out to be 9.4 and -9. Equation (17.10) then indicates that these stars have luminosities of 1.4×10^{-2} and 3.2×10^5 times that of the Sun. Since the effective temperature of the Sun is 5800°K, the radius of the white dwarf turns out to be $0.04R_o$, and the radius of the supergiant is $2100R_o$, where R_o is the solar radius. These values are close to the extreme ranges of all stars, and they must not be thought of as typical in any sense.

One can also use the binary-star data to check masses of stars in different parts of the H-R diagram. It is found that the "normal" stars that do satisfy the usual mass-luminosity relation are none other than the main-sequence stars. Giants have their own mass-luminosity relation, and, presumably, so do other types of stars. One finds that mass varies along the main sequence only slowly, and even away from the main sequence the mass variations among stars are not very great. The largest masses known are probably no larger than somewhere between 50–100 solar masses, while S. S. Kumar has calculated that objects with masses less than about 0.05–0.1 solar mass probably never become true stars.

O. J. Eggen has found evidence for different mass-luminosity relations for stars of slightly different chemical abundances, even though the spectra may appear very nearly the same. This emphasizes the fact that spectra are much more sensitive to some parameters than to others. Still, stellar spectra are potentially the sources of essentially any information on stars if the appropriate calibrations have been made. The theoretical approach to these calibrations will be stressed in the next several sections.

19. Continuous Radiation of the Stars

When light is traveling in a material medium, it will eventually be absorbed. The same photon may reappear in a different direction, in which case it has been scattered. The energy of the absorbed photon may help to heat the material or it may reappear in the form of photons of different frequencies. These latter cases are nonscattering or true absorption processes, although both scattering and non-scattering processes are usually included when one speaks of absorption.

If a photon has a frequency that corresponds to an allowed transition that many atoms can make, then the probability of absorption for that photon is large. The photon will not travel very far on the average before it is absorbed, and it has a short mean free path. If very few atoms are in energy states from which a given photon can be absorbed, then it will travel a long distance before it is absorbed, and it has a large mean free path. The mean free path of a photon is inversely related to the absorbing power of the material.

The light which a star emits originates in the nuclear reactions which take place in the deep interior of the star. The photons are created near the center, and they are absorbed and emitted a very large number of times. Eventually they work their way out to the surface and are emitted by the star. The atmosphere of a star is defined as those layers which are near enough to the surface that a typical photon, emitted in the outward direction, has a fairly good chance of escaping from the star before it is absorbed again. The atmosphere consists of the outer few mean free paths of typical photons which the star emits. In other words, the atmosphere of the Sun goes about as deep as one can see into the Sun. Stellar atmospheres do not have sharp boundaries, and it is not correct to picture a star as having a gaseous atmosphere over a solid or liquid surface. Stars are gaseous throughout.

Stellar atmospheres are usually quite thin and contain a negligible fraction of the mass of the star; nevertheless, they are extremely important. For, by definition, the radiation leaving the star was last absorbed and emitted somewhere in the atmosphere. As a result, the character of this radiation is fixed by conditions in the atmosphere, not by conditions in the deep interior where it originated. The total rate of energy loss by a star is determined by the interior, but the frequency distribution is changed by the nonscattering processes that take place as the energy works its way out. An understanding of the frequency distribution of the emitted radiation, including the character of the line spectrum, therefore, requires a knowledge of the physical conditions in the atmosphere.

Emission and Absorption. The radiation which is emitted by matter can be described by an emission coefficient j_λ. This is defined so that $j_\lambda\, d\lambda\, d\omega$ is the energy with wavelength between λ and $\lambda + d\lambda$ which is emitted per second by unit volume into the element of solid angle $d\omega$, and the dimensions of j_λ are energy per second per volume per wavelength interval per solid angle. This emission includes radiation scattered into the given direction from other directions. Then $j_\lambda\, d\lambda\, d\omega\, dx\, dy\, dz$ is the energy per second emitted by the volume element $dx\, dy\, dz$ into $d\omega$, as illustrated by Figure 19.1. If $d\omega$ is directed along the x-axis, then one can divide this expression by $dy\, dz$, by $d\omega$, and by $d\lambda$, to obtain energy per second per unit area, per solid angle, and per wavelength interval, and this is the contribution of the volume element to the intensity in this direction:

$$dI_\lambda = j_\lambda\, dx \tag{19.1}$$

A volume element contributes to the intensity in any direction an amount equal to the emission coefficient times the length of the element in that direction.

Matter also absorbs radiation energy. If σ_λ is the absorption coefficient for radiation of wavelength λ, measured in cm^{-1}, then the intensity will be decreased on traveling the distance dx by an amount [cf. equation (10.1)]

$$dI_\lambda = -\sigma_\lambda I_\lambda\, dx \tag{19.2}$$

The net change of intensity is the excess of emission over absorption:

$$dI_\lambda = (j_\lambda - \sigma_\lambda I_\lambda)\, dx \tag{19.3}$$

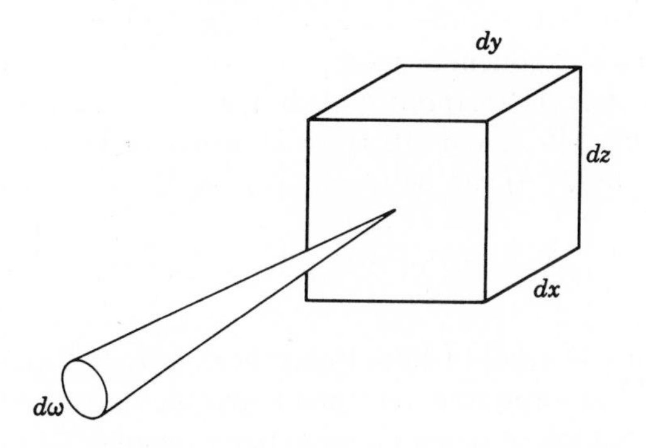

Figure 19.1 Emission from a volume element.

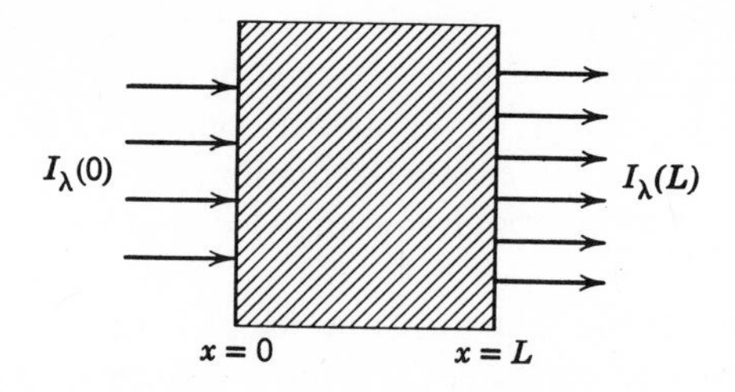

Figure 19.2 Transfer of radiation through a uniform slab.

This is known as the equation of transfer, and it determines how radiation is transferred through any material of given emission and absorption coefficients.

Consider a uniform slab of thickness L, as in Figure 19.2. In this slab j_λ and σ_λ are constants. Equation (19.3) can be written

$$dI_\lambda + \sigma_\lambda I_\lambda \, dx = j_\lambda \, dx$$

Multiply this by $e^{\sigma_\lambda x}$:

$$e^{\sigma_\lambda x}(dI_\lambda + \sigma_\lambda I_\lambda \, dx) = d(e^{\sigma_\lambda x} I_\lambda) = j_\lambda e^{\sigma_\lambda x} \, dx$$

Integrate this between the limits 0 to L:

$$I_\lambda(L)e^{\sigma_\lambda L} - I_\lambda(0) = \int_0^L j_\lambda e^{\sigma_\lambda x} \, dx$$

or

$$I_\lambda(L) = I_\lambda(0)e^{-\sigma_\lambda L} + \int_0^L j_\lambda e^{-\sigma_\lambda (L-x)} \, dx \tag{19.4}$$

The intensity emerging from the slab is composed of two terms. The first term consists of the part of the radiation incident at $x = 0$ which is not absorbed by the slab, and the second term consists of energy which is emitted within the slab. Note that the latter is formed by the emission term of equation (19.1) cut down by the usual exponential absorption $\exp - [\sigma_\lambda(L - x)]$ between the point at x and the edge. This exponential absorption is the same as that which appears in equations (10.2) and (10.4) for atmospheric extinction.

The integration in (19.4) can be readily performed, and the result is

$$I_\lambda(L) = I_\lambda(0)e^{-\sigma_\lambda L} + \frac{j_\lambda}{\sigma_\lambda}(1 - e^{-\sigma_\lambda L}) \tag{19.5}$$

There are two limiting cases of importance here, $\sigma_\lambda L \ll 1$ and $\sigma_\lambda L \gg 1$. In the former case the absorption is very small, and the slab is essentially transparent, while in the latter case it is essentially opaque. These are usually known as the optically thin and the optically thick cases, respectively. It is easily seen that equation (19.5) reduces to the following for these two extremes:

$$I_\lambda(L) = I_\lambda(0) + j_\lambda L \qquad (\sigma_\lambda L \ll 1)$$

$$I_\lambda(L) = \frac{j_\lambda}{\sigma_\lambda} \qquad (\sigma_\lambda L \gg 1) \tag{19.6}$$

The first equation of (19.6) follows directly from (19.1) in the absence of absorption.

In the optically thick situation, material very far from the edge obviously will not contribute appreciably to the emitted radiation. Only the matter within about one mean free path or so of the boundary can be very important for this. Since the mean free path is equal to the reciprocal of the absorption coefficient, the form of the second equation of (19.6) can be understood. The ratio of the emission coefficient to the absorption coefficient is quite important in radiation theory, and it is known as the source function.

Formulation of the Problem. The absorption lines in a stellar spectrum appear superposed on a continuous background. Because of the extreme complexity of the general problem, the continuous radiation and the lines are studied separately in practice. The continuum is discussed in this section, and the lines are considered in Section 20.

It has already been noted that stars radiate like black bodies to a fairly good approximation. In this approximation, the atmosphere is assumed to be a region of uniform properties. The effective temperature may be found by determining the temperature of a black body which gives the best fit with the observed continuum. It is noted in Section 17 that this approximation is of limited accuracy, and it is often desired to improve upon it.

The assumption that temperature is constant for all parts of the atmosphere is obviously artificial, and one might improve on it by allowing conditions to vary with depth in the atmosphere. Since the black-body assumption is not very bad, perhaps T could increase inward, and each layer could radiate like a black body at its own temperature. The emitted radiation would then be the sum of black-body radiation of different temperatures, since there is a range of depths which contribute to it. This approximation is a considerable improvement, but it is also a considerable complication. The effective temperature is no longer the temperature of the atmosphere, but it is only some average over the layers of importance. This approximation is known as local thermodynamic equilibrium because the absorbing and emitting properties are assumed to be the same as in thermodynamic equilibrium under the same conditions. This differs from TE in that conditions change from one point to another. Local thermodynamic equilibrium will now be considered in some detail.

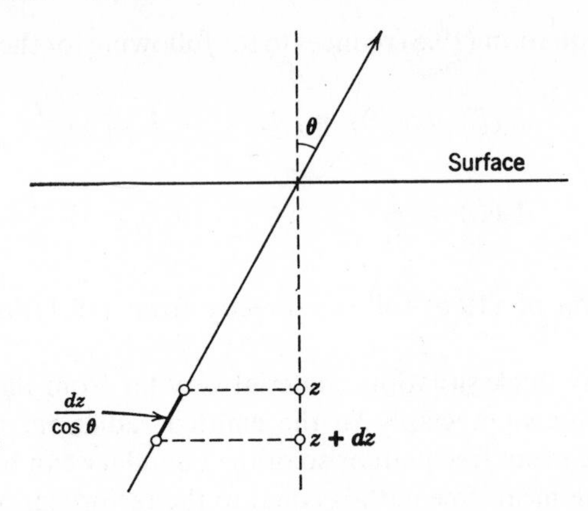

Figure 19.3 Geometry of a plane-parallel atmosphere.

Equation (19.4) suggests that the intensity of radiation emerging from a star is the integral through the atmosphere of the emission coefficient multiplied by the exponential absorption of the layers above. There is no incident radiation, and here both j_λ and σ_λ will change with depth. Let z be the distance of a point below the surface of a star and θ the angle between a given direction and the direction radially out of the star. If the atmosphere of the star is sufficiently thin, as is usually the case, then the curvature can be ignored and the atmosphere can be considered flat or plane-parallel as illustrated in Fig. 19.3. Let the optical depth of any point z, for radiation of wavelength λ, be defined as

$$\tau_\lambda(z) = \int_0^z \sigma_\lambda(z')\, dz' \tag{19.7}$$

This is the same as equation (10.3), and $\exp\,[-\tau_\lambda(z)]$ is the exponential absorption term for the material directly above the depth z. Then the intensity emerging from the star at an angle θ to the normal is given by

$$I_\lambda(0,\theta) = \int_0^\infty j_\lambda(z) \exp - [\tau_\lambda(z)/\cos\theta]\, dz/\cos\theta \tag{19.8}$$

This integration should only extend through the star, but this is essentially infinity as far as the thickness of the atmosphere is concerned.

So far nothing has been assumed about the nature of the absorbing and emitting processes. In strict thermodynamic equilibrium, conditions do not change from one point to another, and the material must be optically thick; thus the second equation of (19.6) would be valid, and the intensity would equal j_λ/σ_λ, the source function. But in TE, the intensity of radiation is given by the Planck function, so one has

$$j_\lambda = \sigma_\lambda B_\lambda(T) \tag{19.9}$$

This relation is valid in TE, but the atmosphere is assumed to emit radiation as if it were in TE; therefore, the emission coefficient at any depth z is equal to the absorption coefficient at that depth times the Planck function of the local temperature. If this is substituted into equation (19.8), the result is

$$I_\lambda(0,\theta) = \int_0^\infty B_\lambda(T)e^{-\tau_\lambda(z)/\cos\theta}\sigma_\lambda(z)\,dz/\cos\theta \tag{19.10}$$

The integration variable in this can be changed from z to optical depth by means of equation (19.7). If this is done, one obtains

$$I_\lambda(0,\theta) = \int_0^\infty B_\lambda(T)e^{-\tau_\lambda/\cos\theta}\,d\tau_\lambda/\cos\theta \tag{19.11}$$

Equations (19.10) and (19.11) are equivalent to each other, and which one is to be used is only a matter of convenience. In any case the angle θ is a constant with respect to the integration, this being a result of the assumption that curvature can be neglected.

The above relations could be used to calculate the intensity in any direction coming from a star if the temperature and absorption coefficient were known at each depth in the atmosphere. These quantities are usually not known, however. One might suppose that the problem could be turned around so that observations of the intensity could be used to determine how T and σ_λ vary with depth. It will shortly be shown how this is possible in a limited fashion, but intensities are not observable for most stars. As is pointed out in Section 8, stars other than the Sun are so far away that only a flux can be measured, not an intensity. If $F_\lambda(r)$ is the monochromatic flux of a star measured at a distance r from the star, then

$$F_\lambda(r) = \left(\frac{R}{r}\right)^2 F_\lambda(R) = \left(\frac{R}{r}\right)^2 \int I_\lambda(0,\theta)\cos\theta\,d\omega \tag{19.12}$$

where R is the radius of the star and r is its distance, and the integration over solid angle is to be taken over the outward hemisphere. The quantity $F_\lambda(r)$ can be observed, at least in principle, and it can be related to the structure of the atmosphere from the previous relations. Equations (19.11) and (19.12) together yield

$$F_\lambda(r) = 2\pi\left(\frac{R}{r}\right)^2 \int_0^\infty B_\lambda(T)E_2(\tau_\lambda)\,d\tau_\lambda \tag{19.13}$$

where E_2 is the second exponential-integral function:

$$E_2(x) = \int_0^1 e^{-(x/y)}\,dy \tag{19.14}$$

The quantity $E_2(x)$ is rather similar to the ordinary exponential function. It has a value of 1.0 at $x = 0$, and it goes to zero for large x like (e^{-x}/x). If equation (19.12) is integrated over all wavelengths, the result is

$$F(r) = \left(\frac{R}{r}\right)^2 F(R) = \left(\frac{R}{r}\right)^2 \sigma T_e^4 \qquad (19.15)$$

In this σ is the radiation constant, not the absorption coefficient.

The problem is to determine the structure of the atmosphere so that equations (19.13) and (19.15) give results that are in agreement with observations, if these observations exist. In practice what is usually known is only the relative flux distribution $(F_\lambda/F_{\lambda_o})$, where λ_o is some standard wavelength. (Some observers insist on calling this ratio the absolute flux distribution; they use the term relative flux to mean the flux in one star relative to that in another star at the same wavelength.) In calculating the relative flux it is seen from equation (19.13) that the highly uncertain quantity (R/r) drops out, and only the shape of the wavelength distribution is important. Also, the effective temperature T_e is not usually known with very great accuracy from observations, so the integrated flux in equation (19.15) does not make a very good comparison between theory and observation.

There is one further constraint on the problem: energy must be conserved. Since all of the energy originates in the deep interior of the star, the total energy incident on the bottom of the atmosphere per second must equal the rate at which the star as a whole radiates energy from its surface. The integrated flux at any distance r from the center, therefore, must be $(L/4\pi r^2)$, if r is large enough to include all of the energy generation regions. The quantity L is the luminosity of the star. Since the atmosphere is very thin, by previous assumption, all points in the atmosphere are essentially at distance R from the center. The integrated flux throughout the atmosphere is then $(L/4\pi R^2)$, which is a constant:

$$F(z) = \int_0^\infty \left[\int I_\lambda(z,\theta) \cos\theta \, d\omega \right] d\lambda = \sigma T_e^4 = \text{constant} \qquad (19.16)$$

where $I_\lambda(z, \theta)$ is the intensity at depth z in the direction of θ, and it is given by an expression similar to equation (19.10) but somewhat more general.

The problem is now to determine the structure of the atmosphere so that energy is conserved, i.e., so that equation (19.16) is satisfied, and so that the observed relative energy distribution is obtained. This structure consists of the details of how temperature and any other relevant physical quantity vary with depth in the atmosphere. One is trying to construct a theoretical model of a stellar atmosphere, a model that will satisfy all of the physics considered important, and that will reproduce the observed radiation of the star.

The Absorption Coefficient. One of the requirements of obtaining a satisfactory model atmosphere is that the absorbing properties of the stellar material be known. The absorption coefficient σ_λ depends on how many atoms and molecules are in the quantum states from which absorption of photons of wavelength λ is permitted, and it also depends on how likely such transitions are. For example, a hydrogen atom in the excited level $n = 3$ can absorb a photon of wavelength 5000 Å with a bound-free transition; therefore, σ_λ for 5000 Å will depend on how many H atoms there are in the level $n = 3$, and it will also depend on the probability that this transition will take place.

The transition probabilities must be either calculated by quantum mechanics or measured in the laboratory. In either case they are often quite uncertain because the required calculations and measurements are very difficult to carry out. Transition probabilities are among the important sources of uncertainty in stellar atmospheres, but the situation is constantly improving.

The number of atoms in a given quantum mechanical level will depend on the abundance of the element in question and on the fraction of these atoms in the given level. The latter can be found as a function of temperature and electron pressure by means of the excitation and ionization equations (Section 4), at least if the atmosphere is close enough to thermodynamic equilibrium for these equations to be valid. In principle, therefore, knowledge of the transition probabilities will allow one to calculate the absorption coefficient as a function of temperature, electron pressure, and the chemical composition of the star. This can be represented by the schematic relation

$$\sigma_\lambda = \sigma_\lambda(T, P_e, \text{composition}) \tag{19.17}$$

It is known that most stars are composed mainly of hydrogen and helium, so it might be imagined that these elements are responsible for most of the absorption. It happens that these elements are rather difficult to excite, however, so neither H nor He is very important for moderate- to low-temperature stars, such as the Sun. Helium is important only for O and B stars and, perhaps, early A stars. Hydrogen is not completely negligible even for the Sun, but it does not become a strong absorber of important wavelengths until spectral type F and earlier.

A measure of the importance of hydrogen is made possible by the so-called Balmer discontinuity. A photon of wavelength λ 3647 or shorter has enough energy to ionize an H atom from the level $n = 2$, while photons of wavelength longer than this do not. This causes the absorption coefficient of hydrogen to have a discontinuous jump at λ 3647, and this jump should show on the observed energy distribution curves if hydrogen absorption is very important. A jump in F_λ is observed in stars at λ 3647, and it is called the Balmer discontinuity because it occurs at the limit of the Balmer series. Specifically, the

Balmer discontinuity is defined as the quantity D, where

$$D = \log \frac{F_{\lambda^+}}{F_{\lambda-}} \tag{19.18}$$

The fluxes in equation (19.18) are for the long- and short-wavelength sides of λ 3647. The quantity D is one of the classification criteria for some spectral classification systems. As expected, D is maximum for early A stars and falls of for higher temperatures, owing to too much ionization of hydrogen, and at lower temperatures, owing to too little excitation. For the Sun D is small enough to show that hydrogen absorption is not very important. Of course a strong Lyman discontinuity occurs in the Sun at λ 912, arising from the absorption edge at the limit of the Lyman series of lines; however, this region of the spectrum is not important in the total energy balance of the solar atmosphere.

In 1939 R. Wildt suggested that the negative hydrogen ion may be a major source of continuous absorption in moderate-temperature stars like the Sun. The H^- ion consists of a proton surrounded by two electrons, and H. Bethe and E. A. Hylleraas had shown nearly 10 years earlier that this system could exist. Although the two electrons will repel each other because of their like charges, each one of them can see enough of the nucleus with its positive charge to hang on and make a stable configuration. The ionization potential of H^- is only 0.75 eV, and when it is ionized a neutral H atom plus a free electron will result. The calculations of S. Chandrasekhar in the mid 1940s of the transition probabilities (both bound-free and free-free) of H^- showed that it is the dominant source of continuous absorption in the Sun.

A very important by-product of the domination by H^- of the absorption in the Sun follows from the fact that H^- absorption does not depend very strongly on wavelength. This means that for the Sun and similar stars σ_λ may to a fairly good approximation be taken as independent of wavelength. An atmosphere with absorption independent of wavelength is called a gray atmosphere, and a gray atmosphere represents a tremendous simplification. For such an atmosphere τ_λ does not depend upon λ, and this subscript can be left off. One can show that in this case equation (19.16) reduces to

$$T_e^4 = 2\int_\tau^\infty T^4 E_2(\tau' - \tau)\, d\tau' - 2\int_0^\tau T^4 E_2(\tau - \tau')\, d\tau' \tag{19.19}$$

This holds for all values of τ, and this equation can be solved to yield the dependence of temperature upon optical depth once and for all. One does not need to know anything about the atmospheric composition, pressures, or the like. An approximate solution of equation (19.19), which is known as the Eddington approximation, is given by

$$T^4 = \tfrac{3}{4} T_e^4(\tau + \tfrac{2}{3}) \tag{19.20}$$

Because equation (19.20) is a good approximation to the solution for the gray problem, and because real atmospheres of moderate temperatures are nearly gray, it follows that the temperature distribution in many stars is rather close to (19.20). The Eddington approximation is a great improvement over the black-body assumption and is quite sufficient for many purposes. If one wants the temperature as a function of geometric depth z, or if one wants pressures, or the like, one must consider the problem in more detail.

At very high temperatures hydrogen and helium are largely ionized, and there is a high concentration of free electrons. This causes electron scattering or Thomson scattering to be an important part of the continuous absorption. Electron scattering under these conditions is independent of wavelength, but this also causes the material to have an emission different from that of a black body, and so a modified treatment is necessary.

For very cool stars molecular absorption is important. Although the visible region of the spectrum is dominated by titanium oxide absorption, only a very small part of the energy radiated by these stars is in the visible. The infrared absorption is what is important, and this is probably dominated by water vapor and carbon monoxide.

Another major source of absorption is the cumulative effect of all of the absorption lines, called the blanketing effect by E. A. Milne. This does not mean that all of the lines need be included in detail, but some sort of statistical method of taking them approximately into account is necessary for accurate studies of the continuum. The fact that very hot stars show few lines in the visible region is deceiving; there are many lines in the ultraviolet region of the spectrum of hot stars, and the blanketing effect is quite important even for them.

There are other sources of continuous absorption which may be of importance at some special wavelength and temperature regions, but a complete description of them will not be given here. All sources which are thought to be needed are added together so that a general relation of the form of equation (19.17) is obtained. This is usually in the form of a series of numerical tables instead of an actual equation, but it does not matter. The important point is that σ_λ can be obtained whenever λ, T, P_e, and the composition are given.

The absorption coefficient calculated as given above may be rather considerably in error. Uncertainties in the transition probabilities and errors in the assumed excitation and ionization conditions are possible sources. Also, one cannot be sure that an important contributor to the absorption coefficient has not been left out. The abundance of H^- in the solar atmosphere is extremely small, yet it dominates the absorption coefficient. Its discovery followed from the realization that the absorption coefficient calculated without it is very obviously in error. If it had been only a 20–30% contributor to the absorption coefficient, its importance might be unknown even today. There are many other sources of uncertainty in stellar atmospheres which would tend to mask errors of about this size in σ_λ.

Obtaining a Model Atmosphere. After the above digression on the absorption coefficient, attention now reverts to the evaluation of the integrals appearing in equations (19.13) and (19.16). The absorption coefficient is needed to do this, but σ_λ involves composition and electron pressure as well as temperature. This means two things: (1) composition must be given before a model atmosphere can be calculated; and (2) equations describing the physical state of the material must also be introduced so that the electron pressure can be determined.

In order to describe the physical state of the matter, it is usually assumed that it is not in a state of motion. This means that the various forces acting on the material cancel out. Now, there are two types of forces that are usually important: gravitational forces directed downward and pressure forces directed upward. The material inside a cylinder of base area A and height δz has a volume $A\,\delta z$ and a mass $\rho(z)A\,\delta z$, where ρ is the mass density. If g is the acceleration of gravity, then $g\rho A\,\delta z$ is the gravitational force acting on this mass element in the downward or positive z direction. The top of this element is at a depth z, and the bottom is at depth $(z + \delta z)$. The pressures at these two points are $P(z)$ and $P(z + \delta z)$, and if δz is small enough, the latter can be evaluated by the first-order terms of the Taylor expansion:

$$P(z + \delta z) = P(z) + \frac{dP}{dz}\,\delta z$$

Pressure is force per unit area, so the buoyant force pushing upward on the element is the area times the excess pressure at the bottom, or $A\,\delta z(dP/dz)$. The condition that the pressure and gravity forces are equal then leads to

$$\frac{dP}{dz} = g\rho(z) \tag{19.21}$$

Equation (19.21) is known as the equation of hydrostatic equilibrium. The pressure in this relation must be the total pressure, and for very hot stars the pressure due to radiation becomes quite appreciable and must be included. This is another complication of very-high-temperature stars. For very thin atmospheres the acceleration of gravity does not vary appreciably, and g may be considered a constant equal to its surface value. This is related to the mass and radius of the star by

$$g = \frac{G\mathcal{M}}{R^2} \tag{19.22}$$

where G is the gravitational constant.

In practice equation (19.21) is rewritten in a different form as follows:

instead of σ_λ, one generally uses the mass absorption coefficient κ_λ, whose units are cm^2/g. These two absorption coefficients are related by

$$\sigma_\lambda = \kappa_\lambda \rho \tag{19.23}$$

and thus the optical depth satisfies the relation

$$d\tau_\lambda(z) = \sigma_\lambda(z)\,dz = \kappa_\lambda(z)\rho(z)\,dz \tag{19.24}$$

Equations (19.21) and (19.24) together yield

$$\frac{dP}{d\tau_\lambda} = \frac{g}{\kappa_\lambda} \tag{19.25}$$

Everything that was stated about σ_λ is also true of κ_λ, so one can also write the schematic relation

$$\kappa_\lambda = \kappa_\lambda(T, P_e, \text{composition}) \tag{19.26}$$

Note that in equation (19.25) optical depth has replaced geometric depth as the independent variable, as in the case of equation (19.11).

The situation seems to be getting more complicated, and this is a good time to see what has been accomplished so far. Integrals of the form of equations (19.11) and (19.16) need to be evaluated in order to check whether the model conserves energy and satisfies observations. The absorption coefficient is needed to evaluate these integrals, and so the electron pressure must be introduced to the problem, and to this end the total pressure has been brought in through the equation of hydrostatic equilibrium. In other words, there are certain quantities that have to be found as functions of depth, and thus far there are four of them: T, P, P_e, and κ_λ. According to mathematical theory, there must be four independent equations in order for these four quantities to be found as functions of optical depth. So far only two equations are available, (19.25) and (19.26), and hence two more equations must be found.

One more equation comes from the ionization conditions. From the law of partial pressures as discussed in Section 5, the ratio of electron to total pressure must be the same as the ratio of the number of free electrons to the total number of free particles. If N_e is the number of free electrons per unit volume and N the total number of free particles per unit volume, then

$$\frac{P_e}{P} = \frac{N_e}{N} \tag{19.27}$$

But the ratio (N_e/N) can be found from the ionization equation as applied to all of the different atomic species as a function of T, P_e, and composition. Thus equation (19.27) can be written in the schematic form

$$P = P(T, P_e, \text{composition}) \tag{19.28}$$

This relation will be in the form of some numerical tables, and it represents the third of the four equations needed to determine the structure of the model atmosphere.

The fourth independent relation is supplied by equation (19.16). (The equation of state will not help. It does add a new equation, but it also adds a new unknown function, namely, the density.) Equation (19.16) must be a part of the physics which the model should obey; however, this relation is much too complicated a function of the physical conditions to be used directly, so it is used only indirectly. The actual procedure is that a relation between temperature and optical depth is assumed, and this assumed $T(\tau_\lambda)$ function is the fourth and last equation needed. These equations are then solved to give T, P, P_e, and κ_λ as functions of optical depth. The details of how this is done will not be considered here, but it should be noted that this can be rather involved in itself. When the depth dependence of these variables has been found, one has a tentative model atmosphere. The model must still satisfy equation (19.16), and this is the check on whether the assumed $T(\kappa_\lambda)$ relation is correct.

The integration of the model atmosphere equations yields all of the physical variables one needs in order to determine integrals like those appearing in equations (19.11) and (19.13). This means that the entire radiation field can be calculated, and in particular the energy conservation check can be made. If the orignally assumed $T(\tau_\lambda)$ relation is in error, then equation (19.16) will not be satisfied; one must then adjust this relation and try again. The construction of an accurate model atmosphere is an iterative scheme.

A great deal of effort has been put into the two problems of how to make a good first guess on $T(\tau_\lambda)$ and how to correct this first guess when it is not good enough. One possible answer to the first question is offered by the gray temperature distribution. Actually the Eddington approximation (19.20) is often used and is far better than an arbitrary guess. Certain mean absorption coefficients have been introduced by S. Rosseland and by S. Chandrasekhar which make the use of the gray solution an even better first approximation.

The second problem, that of improving a temperature-optical depth relation which is not accurate enough, has been the subject of many investigatons. A method due to M. Krook and E. H. Avrett is probably the most satisfactory in solving this problem, and essentially arbitrary accuracy in obtaining a temperature distribution which will satisfy equation (19.16) is now possible.

Semi-Empirical Models. There is another way to determine the temperature/optical-depth relation, but it works only for the Sun. It involves measuring the limb darkening, defined as the ratio $I_\lambda(0, \theta)/I_\lambda(0, 0)$. The numerator is the intensity coming from the surface at angle θ to the normal, while the denominator is the intensity observed at the center of the disc. As has already been mentioned, this can be observed only for the Sun, since every other star is too far away to show a measurable disc.

A photon emitted at an optical depth τ_λ has to survive for $\tau_\lambda/\cos\theta$ mean free paths to escape from the surface, as equation (19.11) indicates. Obviously, the deeper layers make a greater contribution to radiation leaving at small θ values than at large ones. Since the deeper layers generally have higher temperatures and, therefore emit more radiation, the intensity is greatest for small θ values, i.e., near the center of the disc.

As a simple example of how observations of limb darkening can help determine the temperature distribution, suppose that the limb-darkening formula used for eclipsing binaries, equation (15.9), is valid for monochromatic radiation. Then this, plus equation (19.11), would indicate that

$$I_\lambda(0,0)(1 - x_\lambda + x_\lambda \cos\theta) = \int_0^\infty B_\lambda(\tau_\lambda) e^{-\tau_\lambda/\cos\theta} \frac{d\tau_\lambda}{\cos\theta}$$

where x_λ is the limb-darkening coefficient for wavelength λ. The Planck function has been explicitly indicated as a function of depth, since it depends upon temperature, which itself varies with depth. It is easily verified that the above will be satisfied if

$$B_\lambda(\tau_\lambda) = I_\lambda(0, 0)(1 - x_\lambda + x_\lambda \tau_\lambda) \tag{19.29}$$

If one can measure the intensity at the center of the disc and the limb-darkening coefficient at some wavelength, then this equation gives the value of the Planck function at any depth. Since this is a known function of temperature [cf. equation (2.1)], the temperature can be found as a function of τ_λ. This temperature distribution can then be used as a basis for calculating the structure of the model atmosphere by the methods already discussed. This is sometimes called the semi-empirical method, since it uses observations more directly than the other methods.

Equation (19.29) is of limited accuracy because only a crude limb-darkening formula was used. However, A. K. Pierce and J. H. Waddell used a more accurate expression in their semi-empirical model of the solar atmosphere, and a much more accurate temperature distribution resulted.

If the form used above for the limb darkening is assumed for integrated rather than monochromatic intensity, then the temperature distribution turns out to be of the same form as the Eddington approximation (19.20). That is, an equation like (19.29) is obtained, except that the quantities are all integrated. When the integrated Planck function is expressed in terms of temperature through equation (2.10), the result is

$$T^4 = \frac{\pi}{\sigma} I(0,0)(1 - x + x\tau) \tag{19.30}$$

When this is compared with equation (19.20), it is seen that a gray atmosphere in the Eddington approximation has an integrated coefficient of limb darkening of 0.6.

Convection. Equation (19.16) is correct only if all of the energy passing through the atmosphere is carried by radiation; however, it may happen that convection becomes important, and this relation must then be modified. Energy still must be conserved, and so the total flux is still constant. In this case the total flux is the sum of the flux carried by radiation, F_{rad}, and that carried by convection, F_{conv}. Then

$$F_{\text{rad}}(z) + F_{\text{conv}}(z) = \sigma T_e^4 = \text{constant} \tag{19.31}$$

If convection is important, then this relation replaces equation (19.16). The radiative flux is still calculated in the usual way, but now one needs to know how to obtain the convective flux. This is one of the important uncertainties in model-atmosphere theory, since a completely satisfactory theory of convective energy transport does not now exist. A form of the so-called mixing-length theory of convection as developed by L. Prandtl, L. Biermann, E. Vitense, and others is generally used. It is fortunate that in most stellar atmospheres convection does not have a very strong influence, but in the general case equation (19.31) must be used.

Comparison with Observation. The theoretical models allow one to calculate any property relating to stellar atmospheres, but the accuracy of these models is sometimes difficult to assess. For certain types of stars the relevant observations are not available, and model atmospheres are the only source of information. As a general rule models are quite reliable for stars whose energy is radiated mainly in the easily observed visible region of the spectrum, for the major errors of these models were easily detected and corrected. For very high or very low temperatures, however, the models are certainly less reliable. This situation is rapidly improving, since observations at very long and very short wavelengths are now becoming available.

The important bolometric corrections discussed in Section 17 can be found from model atmospheres. The monochromatic flux $F_\lambda(R)$ is calculated, so the relative amount of energy under the visual filter can be determined. This, of course, does not depend on the size or the distance of the star. Any color of a star, such as its $(B - V)$, can also be found. A radiative quantity that depends on the size of the star, such as L or M_v, cannot be calculated from a model atmosphere alone; additional information about R must also be available.

This brings up the important question of the physical features of a model atmosphere. If the procedure for constructing a model is studied, it will be noted that certain information must be given before the model can be calculated. These are the input parameters of a model atmosphere, and they are chemical composition, surface gravity, and effective temperature. The composition enters the absorption coefficient and the ionization conditions. The surface gravity enters the equation of hydrostatic equilibrium, and the effective temperature is involved in the energy balance equation (19.16) or (19.31).

If the parameters of a model atmosphere are given, then a model can be constructed, and that model will be unique. Of course, one can make a numerical error, or an incorrect theory of convection can be used, or an important source of absorption can be left out, so it is possible to have two models with the same parameters but having different characteristics. Nature, however, does not make these mistakes, so the theory indicates that if two stars differ from each other in some atmospheric respect, i.e., if their spectra are not identical, then the stars must differ from each other in one or more of the parameters of T_e, g, or composition. It is not a coincidence that this is similar to the statement that was suggested in Section 18, namely, that the properties of a star depend upon the values of two quantities plus chemical composition.

The properties of a model atmosphere turn out to be very sensitive to the value of T_e, much less so to g and the composition. This means that effective temperatures of real stars can be found rather accurately by comparing the the energy distributions of stars with those of models, but the values of surface gravity and composition found in this way are less certain. Apparently the luminosity classes differ from each other primarily in gravity, as far as the atmospheric parameters are concerned, and it has already been noted that temperature produces far greater differences in spectra than does luminosity. Composition effects are worth a more detailed examination.

Blanketing. It happens that composition effects produced through the ionization relation (19.28) are quite small, and those produced through the continuous absorption coefficient are also rather small. It is not surprising that the absorption lines are the features most sensitive to composition. Most of the information on this subject comes from detailed line study, which will be considered in Section 20; however, the gross effects of the lines on model atmospheres through the blanketing effect is quite appreciable.

A number of stars are observed to be somewhat below the main sequence in H-R diagrams in which color is the abscissa. Because of the terminology of the other parts of the H-R diagram, these stars became known as subdwarfs. J. W. Chamberlain, L. H. Aller, and others began to accumulate data which indicated that subdwarfs differ from normal stars in composition, and the explanation of subdwarfs came in terms of blanketing. The point is that most lines in moderate temperature stars come from the metals group, from elements such as iron, aluminum, calcium, and potassium. It is easily verified by an examination of the solar spectrum that these lines are much more numerous toward shorter wavelengths. Although normal stars are composed mainly of hydrogen and helium, they have enough of the heavier elements to undergo a rather large blanketing effect. Blanketing distorts the energy distribution of the emitted radiation and causes less energy to be radiated at those wavelengths at which the lines are most prominent. Since most of the lines occur at short wavelengths, blanketing causes less light to be radiated in the blue and ultraviolet regions than would be radiated in the absence of lines.

Subdwarfs have considerably smaller abundances of the heavy elements than normal stars, so their radiation is less distorted by blanketing. This causes their colors to be bluer than normal stars of the same temperature, so it would be more precise to say that they are to the left of the main sequence rather than below it in a color-magnitude diagram.

Subdwarfs not only have a smaller $(B - V)$ than normal stars of the same temperature, they also have a smaller $(U - B)$ than normal stars of the same $(B - V)$. The ultraviolet excess of a star is defined as the difference between its $(U - B)$ color and that of a normal star of the same $(B - V)$. Subdwarfs have large ultraviolet excesses, and they lie well above normal stars in color-color diagrams such as Figure 17.1. When it has been properly calibrated, the ultraviolet excess can be used to determine the approximate heavy-element content of a star, and it can give a quickly obtained measure of how normal the star's composition is.

The Chromosphere and Corona. The Sun is the only star whose atmosphere can be studied in great detail, and its atmosphere is usually divided into three regions. The inner atmosphere is the region of main concern in this section, and it is known as the photosphere. The photosphere is the source of the bulk of the energy the Sun gives off. The outer part of the atmosphere of the Sun consists of two regions known as the chromosphere and the corona. The chromosphere is some 10,000 km thick and lies just above the photosphere. It is the region in which the solar spectrum changes from absorption lines on a strong continuum to strong emission lines on a weak continuum. The corona is the extreme outer part, which gradually blends with the interplanetary medium. The radiation of both the chromosphere and the corona is completely lost in the bright glare of the photosphere, so the former are visible only in special conditions, such as during total eclipses of the Sun or from above the Earth's atmosphere, or with special instruments.

The density falls off as expected as one proceeds outward through the solar atmosphere, but the temperature does a strange thing. Temperature drops as one goes outward in the photosphere, reaching a minimum value of about 4200–4500°K in the upper photosphere or the lower chromosphere; then it rises again through the chromosphere until it reaches about one or two million degrees in the corona. These high temperatures are indicated by the spectra, as emission lines of very highly ionized elements are observed. Also, the radio emission from the Sun indicates these high temperatures. This unexpected behavior of the temperature in the outer regions where the radio emission originates causes the Sun to have limb brightening at radio frequencies instead of the usual limb darkening.

This temperature rise must mean that energy of some sort is continuously being dumped into the outer atmosphere. Where does it come from? Attempts to explain it in terms of external factors, such as an infall of meteors, are unsuccessful, so the source must be internal. The lower regions of the solar photosphere are in turbulent motion, and it was stated above that the resulting convective transport of energy does not strongly affect the structure of the photospheric layers; however, there must be some sort of disturbance which is generated in this convective region and which is propagated upward, eventually to be dissipated in the outer atmosphere. The details of this have not yet been satisfactorily worked out. The amount of energy involved in this process is quite small compared with the total solar luminosity, but it is still of great importance. The hot corona is apparently continuously evaporating, giving rise to a stream of matter coming out from the Sun and known as the solar wind. This undoubtedly has major effects on solar-terrestrial relations.

There is no reason to believe that the Sun is unique in the possession of an outer atmosphere, so other stars are expected to have chromospheres and coronas also. Many stars do show emission lines and other features indicating an extended atmosphere, and perhaps these are just stars with unusually large chromospheres and coronas. Many stars are observed to be losing matter, and in some cases the mechanism may be similar to that of the solar wind but on a larger scale. O. C. Wilson and others have also obtained some evidence of chromospheric activity in stars from the properties of some of their strongest lines, notably the H and K lines of ionized calcium. The point is that the centers of the strongest lines originate very high in the atmosphere, well above the photosphere and in the chromosphere.

One would like to construct models of the chromosphere and the corona in much the same fashion by which one obtains models of the photosphere; procedures are rather different, however, for two reasons. First, conditions are so much different that many of the simplifying assumptions which are valid for the photosphere are no longer valid for the chromosphere and the corona. Second, the above-mentioned uncertainty in the energy source means that one does not have an energy-balance equation such as (19.31) as a constraint on the problem.

As an example of the differences in conditions, the outer atmosphere is very far from local thermodynamic equilibrium. This means that the material does not radiate like a black body at a unique temperature, and it also means that the usual excitation and ionization equations need modification. The correct relations are much too complicated to be used in their complete generality, and the validity of the relations used in practice is often open to debate. In addition, the idea of a spherically symmetric outer atmosphere in hydrostatic equilibrium is probably much oversimplified. Magnetic fields are probably important in the structure of these regions.

The lack of an accurate energy-balance equation means that a semi-empirical approach is necessary, just as the semi-empirical model photosphere method makes the use of the energy-conservation equation unnecessary. It does represent a major lack of understanding of the physics of the problem, and it also makes extrapolation of information from the easily observed Sun to other stars much less dependable.

20. Line Formation

To a certain approximation the analysis of line radiation is the same as that of continuous radiation. To the extent that this is true, a line is distinctive only in that it corresponds to a wavelength or frequency for which the absorption coefficient is larger than usual. Radiation at such a frequency must arise very high in the atmosphere, or it would not have a very good chance of escaping before being reabsorbed. Likewise, radiation at frequencies of low absorption are last emitted in the deeper parts of the atmosphere. If temperature increases with depth, and if the material emits like a black body, then radiation at frequencies of high absorption is emitted, on the average, in regions of low temperature, less energy is emitted, and absorption lines are the result. According to this explanation the absorption lines are caused by the large absorption and by the inward increase of temperature; without either one of these two effects, the absorption lines could not occur. (If the temperature decreased inward, for example, the lines would be in emission, but this does not mean that the observed emission lines necessarily indicate such a temperature reversal.)

The actual situation may be much more complicated than this, because line emission is less likely to be like that of a black body than continuous emission. A photon absorbed in a bound-bound transition has a good chance of being scattered, i.e., being emitted in the same downward transition. This does not necessarily have to be at exactly the same frequency as the absorption; if it is, then the scattering is said to be coherent; otherwise, it is noncoherent. In either case, if scattering is important, then significant deviations from local thermodynamic equilibrium (LTE) will occur. Also, it does not follow that all nonscattering processes which can take place will necessarily tend toward LTE. This complicated problem has been the subject of many theoretical investigations, notably by R. N. Thomas and J. T. Jefferies; however, the observations seem to indicate that, except for the central regions of the strongest lines, solar absorption lines are formed under conditions very close to LTE.

The Line-Absorption Coefficient. The details of the bound-bound absorption coefficient differ in a significant way from that of a bound-free or a free-free coefficient. In all cases the abundance of the absorbing element, the percentage of the atoms in the lower level, and the relevant transition probability are involved. But the bound-bound transitions are also influenced by a quantity, the broadening function, which is not important for continuous absorption. Line absorption is very sensitive to tiny changes of frequency, so effects which cause these small changes increase considerably the range of frequencies or wavelengths over which the line absorption is appreciable. It is because of this broadening that lines are not infinitely narrow, but have a finite width.

The line-absorption coefficient per atom is usually written in the form

$$a_\nu = (1 - e^{-h\nu_0/kT}) \frac{\pi e^2}{mc} f\phi_\nu \tag{20.1}$$

This is the absorption coefficient in cm^2/(atom in the lower level). (Frequency units are used here because this practice is more common, but wavelength units could just as easily have been used.) The absorption coefficient in cm^{-1}, as used in equation (19.2), is simply (20.1) multiplied by the number of absorbing atoms per unit volume, while the mass absorption coefficient κ_λ or κ_ν defined in equation (19.23) is a_ν multiplied by the number of absorbing atoms (i.e., the number of atoms in the lower level of the transition being considered) per gram of the matter.

The first term on the right side of equation (20.1) is the correction for induced emission. The effective number of absorptions is the total number minus the number of induced emissions, since the latter act very much like absorptions in reverse, and $(1 - e^{-h\nu_0/kT})$ corrects for this effect. The next term contains atomic constants (e = electronic charge, m = electronic mass, c = speed of light), and f is known as the oscillator strength or f value of the line. It is essentially the transition probability for the line, and it is a constant that must be calculated by means of quantum mechanics or measured in the laboratory. The final quantity, ϕ_ν, is the broadening function. It is defined so that if an absorption in the line in question does take place, then $\phi_\nu \, d\nu$ is the probability that the frequency of the absorbed photon lies between ν and $\nu + d\nu$, assuming equal intensities for all frequencies. It is apparent that the integral of ϕ_ν over all frequencies is equal to one. The value of ϕ_ν is large near ν_0, the frequency of the line center, and it falls off very rapidly for larger and for smaller frequencies.

The three most important types of broadening are the doppler effect, natural broadening, and pressure broadening. Magnetic fields produce what is known as the Zeeman effect on spectral lines, and this causes a type of broadening which must also be considered in certain special cases.

Doppler broadening is caused by the doppler effect (see Section 11) of moving atoms. Some atoms happen to be moving toward the observer during the moment of absorption or emission, and some are moving away. Even if atoms could absorb or emit only the frequency they see as ν_0, the observer would see a finite range of frequencies being absorbed or emitted. Anything that gives a spread of velocities to the atoms contributes to the doppler broadening. The most important source is usually the thermal motions due to the temperature of the gas, but often other sources are also acting. Many early type stars have rapid rotation, and if such a star is observed other than pole on, some of the observed radiation comes from areas of the star that are approaching and some from parts that are receding from the observer. This causes a rotationally broadened line. Also, some stars, notably supergiants, have a large amount of turbulence in their atmospheres, and these turbulent motions also broaden the lines because of the doppler effect.

The broadening function is easily calculated for any assumed velocity distribution, for ϕ_ν is just the probability that an atom has a radial velocity such that the frequency absorbed at ν_0 is observed at ν. If the velocity distribution is Maxwellian (see Section 5), then one has

$$\phi_\nu \text{ (doppler) } d\nu = \frac{1}{\pi^{1/2}} e^{-[(\nu-\nu_0)/\Delta\nu_D]^2} \frac{d\nu}{\Delta\nu_D} \qquad (20.2)$$

where $\Delta\nu_D$ is the doppler width in frequency units:

$$\Delta\nu_D = \frac{\nu_0}{c}\left[\frac{2kT}{M} + (\text{turbulent velocity})^2\right]^{1/2} \qquad (20.3)$$

where T is the temperature and M is the mass of the absorbing atom. As the temperature or the turbulent velocity, or both, rises, the doppler width increases and the probability that absorption can take place far from the line center also increases. The form of (20.3) is valid only if the turbulent velocities also follow a Maxwellian distribution. Rotational velocities are not Maxwellian, and (20.3) does not hold for them. The doppler width is the frequency shift that corresponds to the most probable velocity, as can be seen from equations (11.2) and (5.2).

Natural broadening is an inherent property of the energy levels, and thus the lines would have finite width even if the other mechanisms were not operative. It can be understood in terms of the quantum mechanical uncertainty principle. If an atom spends a time interval t, on the average, in a certain level, then it is not possible to know the exact value of the energy of this level. This limitation is due not to faulty measuring devices but to the nature of matter. The uncertainty of the energy is at least of the order of $(h/2\pi t)$, where h is Planck's constant. This means that an atom in a certain level is likely to have energy anywhere in the interval $E_0 - (h/2\pi t)$ to $E_0 + (h/2\pi t)$, where E_0 is the mean energy of the level. This gives a fuzzyness to the levels and broadens the lines. Most excited levels have lifetimes of the order of 10^{-8} sec or so, and with t this short the natural width can be appreciable. Metastable levels are excited levels which have very small transition probabilities to all lower levels, and so they have long lifetimes; consequently, metastable levels are quite sharp—as, of course, is the ground level.

The broadening function for this can be shown to be

$$\phi_\nu\,(\text{natural})\,d\nu = \frac{\gamma_N\,d\nu}{4\pi^2(\nu-\nu_0)^2+(\gamma_N/2)^2} \tag{20.4}$$

in which $\gamma_N = \gamma_1 + \gamma_2$, the quantities γ_1 and γ_2 being the reciprocals of the mean lifetimes of the two levels. The shorter the lifetimes, the larger γ_N and the greater the likelihood of absorptions far from the line center. The quantity γ_N is known as the natural-damping constant.

The energy levels of an atom are distorted by the electric fields of nearby particles, and this gives rise to pressure or collisional broadening. These effects are produced even by neutral particles if they come close enough. Pressure broadening is extremely complicated, and in practice some form of approximate theory is used. There are several types of pressure broadening, and it is an interesting paradox that hydrogen, the simplest atom, has the most complicated pressure broadening. This is due to the special quantum mechanical degeneracies of hydrogen that more complicated atoms do not have. Most atoms have a form of pressure broadening for which the broadening function is the same as that for natural broadening [equation (20.4)]. Pressure broadening has its own damping constant, γ_c which is the frequency of the disturbing collisions.

The total broadening function is found by combining the three effects discussed above. For those cases in which pressure broadening is of the same form as equation (20.4), one can write the result as follows:

$$\left.\begin{aligned} &\gamma = \gamma_N + \gamma_c \qquad a = \frac{\gamma}{4\pi\,\Delta\nu_D} \qquad v = \frac{\nu-\nu_0}{\Delta\nu_D} \\ &H(a,v) = \frac{a}{\pi}\int_{-\infty}^{\infty}\frac{e^{-y^2}\,dy}{a^2+(v-y)^2} \end{aligned}\right\} \tag{20.5}$$

Then

$$\phi_\nu\,d\nu = \frac{1}{\pi^{1/2}\,\Delta\nu_D}\,H(a,v)\,d\nu = \frac{1}{\pi^{1/2}}\,H(a,v)\,dv \tag{20.6}$$

This particular form of the broadening function is usually called the Voigt function. It is seen to be a complicated function of the total damping constant γ, the physical conditions through the doppler width, $\Delta\nu_D$, and the frequency.

In lines of astrophysical interest, the dimensionless quantity a is usually quite small. When this is true, the expression (20.6) reduces to the doppler

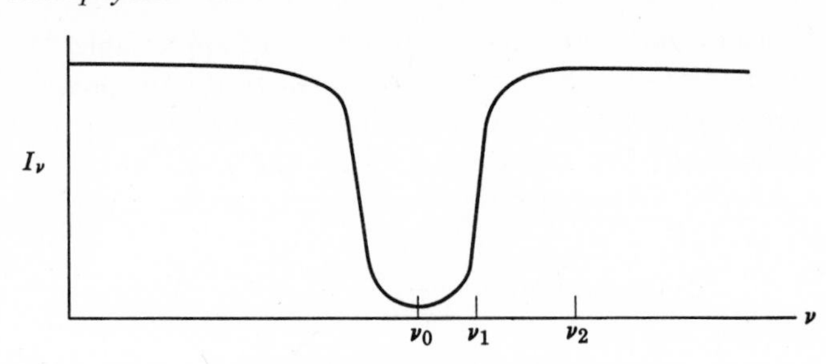

Figure 20.1 A line profile.

formula (20.2) very close to the line center, and it takes the form of equation (20.4) far from the line center. The position of the transition region between the central doppler core and the damping wings can be found approximately by equating equations (20.2) and (20.4), and the result is that the transition occurs at the frequency $\nu - \nu_0 \approx 3\ \Delta\nu_D$, or about three doppler widths from the line center. For typical conditions in the Sun and for an element of atomic weight 25, one doppler width is $3 \times 10^9\ \text{sec}^{-1}$ in frequency units, or 0.04 Å in wavelength units.

Line Profiles and Equivalent Widths. Equations (20.1) and (20.6) determine the line-absorption coefficient as a function of the abundance of the element and the physical conditions. If LTE is a satisfactory approximation for the line, then equation (19.10) or (19.11), which is just as valid for a line as for the continuum, can be used to evaluate the intensity at various frequencies within the line. In order to carry this out, the physical conditions must be known at all depths in the atmosphere, and this can be assumed known from a model atmosphere investigation as described in Section 19. A plot of intensity vs. frequency for any assumed abundance of the relevant element is known as the line profile, and Figure 20.1 is an example.

In the figure, ν_2 is a frequency far enough from the line center that line absorption is essentially zero, so I_{ν_2} is the same as the continuum intensity in the neighborhood of the line. The frequency ν_1 corresponds to the transition region between the doppler core and the damping wings of the line. It should be remembered that the frequency scale is such that continuum quantities, such as the Planck function B_ν, are for all practical purposes constant for the tiny frequency interval across the line.

The shape or profile of an absorption line is a very complicated function of a large number of quantities and conditions; therefore, it is the potential source of a greal deal of information on conditions in a star. Unfortunately, there is a large amount of uncertainty in much of the theory, and accurate observations of line profiles are difficult to obtain. As a result, much of this potential information is not available in practice, and the uncertainties are often too large to make much of it very useful. This situation has improved considerably in recent years.

It is a common procedure to study the total strength of a line rather than its detailed shape. The strength is measured in units called equivalent width, which is the width the line would have, for the same energy taken out, if the intensity in the line were zero:

$$W_\nu = \int_0^\infty \frac{I_c - I_\nu}{I_c}\, d\nu \tag{20.7}$$

The integral in (20.7) should extend only over the line, but the limits can be extended from 0 to ∞ if I_c is kept constant as the continuum value in the neighborhood of the line.

The point in using equivalent widths is that they are much easier to measure accurately, they retain a fairly high sensitivity to abundance, and they are much less sensitive to some of the uncertainties in the theory than the line profiles. On the other hand, much potential information is lost when only equivalent widths are used and the profiles ignored.

The Curve of Growth. The abundance of an element in a star is obtained by calculating a number of profiles or, more likely, equivalent widths, each with a different assumed abundance. A comparison with observations then tells which abundance gives the best fit. A large number of lines is used, if possible, in order to reduce the scatter and improve the accuracy.

The general relation between the equivalent width of a line and the number of absorbing atoms producing it is known as the curve of growth. Most abundance investigations use some form of a curve of growth analysis, and it is worthwile to consider a simplified example in some detail.

Consider a uniform slab of thickness L with continuum radiation incident on one side. The slab contains atoms which produce line absorption, but it is assumed for simplicity that there is no emission in the slab. Then if I_c is the incident intensity for frequencies in the neighborhood of the line and in the direction normal to the slab, the intensity emerging from the other side is

$$I_\nu = I_c e^{-\tau_\nu} \tag{20.8}$$

with

$$\tau_\nu = NLa_\nu = NLa_0\phi_\nu \tag{20.9}$$

$$a_0 = (1 - e^{-h\nu_0/kT})\frac{\pi e^2}{mc} f \tag{20.10}$$

where N is the number of absorbing atoms per unit volume. Equation (20.7) and the above then show that the equivalent width of the line is given by

$$W_\nu = \int_0^\infty (1 - e^{-NLa_0\phi_\nu})\, d\nu \tag{20.11}$$

In this example use will be made of the approximation mentioned in connection with equation (20.6), namely, ϕ_ν is given by the doppler expression (20.2) for $\left|\dfrac{\nu - \nu_0}{\Delta\nu_D}\right| < 3$, and with the dimensionless damping constant a very small, equation (20.4) shows that

$$\phi_\nu = \frac{\gamma}{4\pi^2(\nu - \nu_0)^2} = \frac{a}{\pi\,\Delta\nu_D v^2} \tag{20.12}$$

for $|(\nu - \nu_0)/\Delta\nu_D| > 3$. The quantities a and v are defined in (20.5).

At one extreme one can consider a very weak line, so weak that $NLa_0\phi_\nu \ll 1$ for all frequencies. In equation (20.11) one can then use the approximation $e^{-x} = 1 - x$ for small x and obtain, for the equivalent width,

$$W_\nu = NLa_0\int_0^\infty \phi_\nu\, d\nu = NLa_0 \tag{20.13}$$

For these very weak lines the equivalent width is proportional to the number of absorbing atoms.

If the number of absorbing atoms per unit volume N is increased sufficiently, then the line must become strong. The opposite extreme to that considered above occurs when the absorption in the line center is very large, $NLa_0\phi_\nu \gg 1$ near the line center $\nu = \nu_0$. Far enough from the line center, however, the absorption will again be small. Let the two frequencies $\nu_0 \pm \Delta$ be the transition frequencies, i.e., the frequencies for which $NLa_0\phi_\nu = 1$. Then virtually all of the contribution to the equivalent width comes from frequencies close to the center for which the absorption is large, so

$$W_\nu \cong \int_{\nu_0-\Delta}^{\nu_0+\Delta} (1 - e^{-NLa_0\phi_\nu})\, d\nu$$

Since the absorption is large throughout this range of frequencies, the second term in the above integrand is small compared with the first; therefore,

$$W_\nu \cong \int_{\nu_0-\Delta}^{\nu_0+\Delta} d\nu = 2\Delta \tag{20.14}$$

The strong lines can now be divided into two types according to how far from the center the absorption remains strong. For moderately strong lines, the absorption becomes weak while still within the doppler core ($\Delta < 3\Delta\nu_D$), whereas for the very strong lines the absorption remains large out into the damping wings ($\Delta > 3\Delta\nu_D$). This means that for the moderately strong lines

$$NLa_0\phi_{\nu_0\pm\Delta}\,(\text{doppler}) = 1$$

and for the very strong lines

$$NLa_0\phi_{\nu_0\pm\Delta}\,(\text{damping}) = 1$$

Equations (20.2), (20.12), and (20.14) then give

$$W_\nu = 2\Delta\nu_D\left[\ln\left(\frac{NLa_0}{\pi^{1/2}\,\Delta\nu_D}\right)\right]^{1/2} \tag{20.15}$$

for the moderately strong lines, and

$$W_\nu = \frac{1}{\pi}\,(NLa_0\gamma)^{1/2} \tag{20.16}$$

for the very strong lines.

Figure 20.2 is a curve of growth in the form of a plot of log W_ν versus log (NLa_0). The product NL is the number of absorbing atoms per unit area along the line of sight, and a_0 is proportional to the transition probability. For very small values of NLa_0, the line is quite weak, and equation (20.13) is valid. This is called the linear part of the curve of growth, since W_ν is proportional to N. As more absorbing atoms are added (or as the transition probability increases), the weak-line approximation becomes invalid, and the equivalent width increases less rapidly. This turning over of the curve is called saturation, and the physical reason for it is that essentially all of the available radiation has already been absorbed, so adding more atoms does not appreciably increase the equivalent width. Equation (20.15) becomes valid in this region.

As mentioned above, not enough atoms move fast enough to cause appreciable absorption through the doppler effect more than about three doppler widths from the line center. As equation (20.3) shows, three doppler widths may correspond to a large or a small frequency range, depending on temperature and the turbulent motions. This means that the region where saturation sets in depends on T and the turbulent velocities; higher temperatures mean

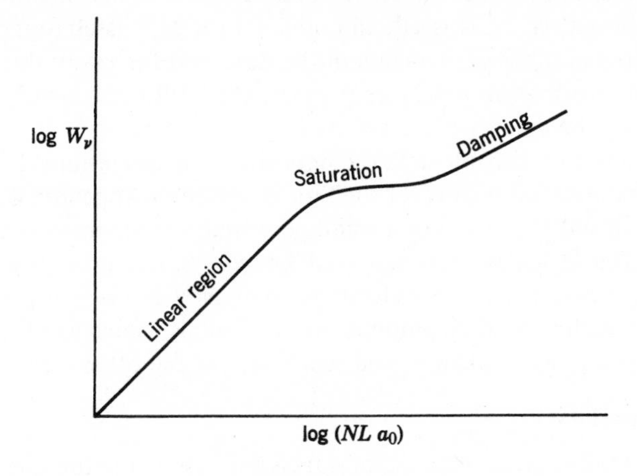

Figure 20.2 A curve of growth.

that the linear region lasts longer and that saturation occurs at a larger value of the equivalent width.

Beyond about three doppler widths the absorption is due primarily to natural and pressure broadening. If the number of absorbing atoms (or the transition probability) becomes large enough, the amount of absorption far out can start being important. The line develops what are known as damping wings, and equation (20.16) becomes valid. It is seen that, in the damping part of the curve of growth, the equivalent width is proportional to the square root of the abundance of the relevant element. The region where this is valid depends on the size of the damping constant a. The larger the damping constant, the sooner the transition occurs between equations (20.15) and (20.16), that is, between saturation and the wings.

There are a number of ways in which curves of growth can be applied. It is common to construct a theoretical curve based on an assumed model of line formation, and an observational curve based on observed equivalent widths of different lines and experimental data on the lines. A comparison of the two curves then determines something about the conditions of line formation and abundances. One of the big advantages of curve-of-growth procedures is that the results are not very sensitive to errors in the assumed model of line formation. For example, the simplified picture assumed above for the description of curves of growth is not at all physically realistic, yet it could be used in the analysis of stellar atmospheres, and the resulting abundances would not be unreasonable. In the analysis of stars other than the Sun, the flux F_ν rather than the intensity I_ν must, of course, be used.

Chemical Abundances. It was not until the 1920s that atomic theory and excitation and ionization conditions were sufficiently well understood that accurate abundance analyzes could be made. Important investigations were made by A. Unsöld, H. N. Russell, and others. In 1929 Russell found abundances for a large number of elements in the Sun, and for many this was the only source of information until recent years. One still occasionally sees the Russell mixture of elements being referred to.

A tremendous number of stellar composition investigations have been carried out, and detailed results for the Sun are given in Appendix C. For the more or less normal stars, all but a small percentage of the mass is hydrogen and helium. The H/He ratio is not well known, but is probably between 3/2 and 3/1 by mass. The heavier elements are mostly carbon, nitrogen, oxygen, and neon, with very small amounts of the "metals," such as silicon, iron, sodium, potassium, and calcium, and just traces of the other elements.

21. Stellar Interiors

The study of the interior structure of stars is quite similar to the study of their atmospheres. One attempts to apply the known laws of physics to stellar interior conditions in order to come up with a theoretical model of a whole star which is compatible with observations.

In some respects the interior represents a more complicated situation than the atmosphere. For example, it is obvious that variations of the mass and the distance from the center cannot be neglected. On the other hand, the physics of stellar interiors is simplified by the absence of any doubt that local thermodynamic equilibrium is a good approximation, and for most purposes the assumption of complete thermodynamic equilibrium is extremely good.

Physical Conditions in the Interior. The Sun is very nearly a spherical star, and it is quite stable for long periods of time. This suggests that spherical symmetry and hydrostatic equilibrium may be good approximations. There are certainly stars known for which this is not true, but such special cases should be considered only after the much simpler examples are fairly well understood.

Spherical symmetry means that physical conditions are functions of only one variable, namely, the distance r to the center of the star. Hydrostatic equilibrium means that an equation similar to (19.21) is valid for the pressure gradient, but it must be modified for the interior. This is easily done if one notes that the acceleration due to gravity, g, is not a constant. In analogy with equation (19.22), one has

$$g(r) = G\frac{\mathcal{M}(r)}{r^2} \tag{21.1}$$

where $\mathcal{M}(r)$ is the mass that is inside the sphere of radius r, and thus $\mathcal{M}(R) = \mathcal{M}$, the total mass of the star. The pressure equation is then

$$\frac{dP}{dr} = -G\frac{\mathcal{M}(r)\rho(r)}{r^2} \tag{21.2}$$

The minus sign means that pressure increases as radius decreases, since r increases in the opposite direction from the position variable z of Section 19.

As before, $P(r)$ is the total pressure supporting the material above r, and it must include the pressure due to radiation as well as the gas pressure. The radiation pressure is quite important for massive stars, but it is small for moderate- and low-mass stars like the Sun. It will here be assumed that radiation pressure can be ignored.

The mass distribution follows directly from the assumption of spherical symmetry. If one moves outward by an amount dr, the mass interior to the point in question increases by the amount of mass in the spherical shell of thickness dr. This shell has a volume of $4\pi r^2\,dr$, so

$$\frac{d\mathcal{M}(r)}{dr} = 4\pi r^2\rho(r) \tag{21.3}$$

This type of equation was not needed in Section 19, since it was assumed that essentially the total mass of the star is interior to the entire atmosphere.

Equations (21.2) and (21.3) are two equations for the three unknown functions $P(r)$, $\mathcal{M}(r)$, and $\rho(r)$. As in the case of the atmosphere, one more equation which does not introduce a new unknown function must be found before the solution can be obtained. Exactly as before, this extra relation will come from energy-conservation principles. Before going into this matter, however, it is instructive to see what can be determined from the equations already at hand. The Sun will be used as a typical example.

The mass of the Sun is 2.0×10^{33} g, and its radius is 7.0×10^{10} cm. This corresponds to an average density of 1.4 g cm^{-3}. In the photosphere the density is 10^{-7} g cm^{-3} or less, and toward the center it must get well above the average value. If it is assumed that the density is a constant throughout the Sun, then the concentration of mass toward the center will be ignored, and this assumption should yield a lower limit to the internal pressures.

If the density is taken as a constant equal to $\bar{\rho}$, then equation (21.3) can be immediately integrated to give

$$\mathcal{M}(r) = \tfrac{4}{3}\pi r^3 \bar{\rho} \tag{21.4}$$

which is obvious for constant density. Substitute this into equation (21.2) and integrate from r to R, using the fact that the pressure goes to zero at the surface:

$$P(r) = \tfrac{4}{3}\pi G \bar{\rho}^2 \frac{R^2 - r^2}{2} = \frac{G\bar{\rho}\mathcal{M}}{2R}\left(1 - \frac{r^2}{R^2}\right) \tag{21.5}$$

In particular, the pressure at the center of the star is

$$P_c = P(0) = \frac{G\bar{\rho}\mathcal{M}}{2R} \tag{21.6}$$

Putting in numerical values for the Sun, one finds that $P_c = 1.3 \times 10^{15}$ dyne cm^{-2}. As stated above, this should be an underestimation of the central pressure in the Sun, but it does give a rough idea of the pressures to be expected in the solar interior. This pressure is about 10^9 atmospheres, and material cannot retain the cohesion of a solid or a liquid under this tremendous pressure. The assumption that stars are gaseous throughout is then justified.

Typical temperatures to be expected in stellar interiors can be found from the equation of state, (5.8). Since this is only to be a crude calculation, one is justified in assuming that the material is pure hydrogen. Anticipating rather high temperatures, one can also assume that the hydrogen is completely ionized, and thus that the matter consists of equal numbers of protons and free electrons. This means that the average mass per free particle m is essentially one-half of the proton mass, or about 0.8×10^{-24} g. When the pressure and mean density found above are put into equation (5.8), the temperature turns out to be 5.6×10^6 °K. It is difficult to tell at present whether this should be an overestimate or an underestimate, but temperatures of millions of degrees Kelvin are to be expected in stellar interiors.

Energy Transport. Now that a rough idea of the conditions in the interior have been established, attention will revert to the problem of obtaining an extra relation so the equations of stellar structure can be solved. Let $L(r)$ be the luminosity at r, that is, the total net energy leaving the sphere of radius r per second. Then $L(r) = 4\pi r^2 F(r)$, where—as usual—F is the flux. In the atmosphere, across which both $L(r)$ and r are constant, this reduces to the relation $F =$ constant. In the interior, both $L(r)$ and r are varying. In the central regions where the energy is being generated, $L(r)$ does vary, and this means that a relation which determines how the luminosity varies will be needed, but this will come up again later in this section.

Now, $L(r)$ will consist of parts due to conduction, convection, and radiation:

$$L(r) = 4\pi r^2 [F_{\text{cond}}(r) + F_{\text{conv}}(r) + F_{\text{rad}}(r)] \tag{21.7}$$

This is analogous to equation (19.31). The higher densities in the interiors make conduction of greater importance than in the atmospheres, but it is usually quite small (though not necessarily negligible) except in certain special conditions. It will not be considered further here.

The radiation term in equation (21.7) can be handled very much as in Section 19 for the atmosphere. In fact, it can be shown that if the Rosseland mean absorption coefficient is used, then the radiative relations in the interior are the same as those valid in a gray medium, which is a medium in which the absorption coefficient does not depend on frequency. The Rosseland mean is defined as follows:

$$\frac{1}{\kappa_R} = \frac{\displaystyle\int_0^\infty \frac{1}{\kappa_\nu} \frac{dB_\nu}{dT} d\nu}{\displaystyle\int_0^\infty \frac{dB_\nu}{dT} d\nu} \tag{21.8}$$

where B_ν is the monochromatic Planck function. It is mentioned in Section 19 that the use of the Rosseland mean helps to make the gray $T(\tau)$ relation a good approximation; in the interior, this approximation becomes almost exact, and the Eddington approximation to the gray solution, equation (19.20), is also nearly exact. Generalizing equation (19.20) in an obvious way, one has

$$\frac{dT^4}{d\tau_R} = -\frac{1}{\kappa_R \rho} \frac{dT^4}{dr} = \frac{3}{4} \frac{F_{\text{rad}}(r)}{\sigma}$$

or

$$L_{\text{rad}}(r) = -\frac{16\pi\sigma}{3} \frac{r^2}{\kappa_R \rho} \frac{dT^4}{dr} \tag{21.9}$$

The same general remarks made in Section 19 about the absorption coefficient hold here for the Rosseland mean κ_R. If all the sources of absorption have been identified and if all of the transition probabilities are known, then κ_R can be calculated as a function of T, P_e, and composition. The σ in equation (21.9) is again the Steffan-Boltzmann constant.

Equation (21.9) shows that the radiation contribution to the luminosity is determined largely by the temperature gradient, or rather the gradient of T^4. This is a very important point, for one generally thinks of the luminosity as being determined by the nuclear energy sources. Of course, after a long enough time the temperature gradient will adjust itself so that energy is carried off at the same rate it is produced, but the energy sources have only this indirect effect on the luminosity. The instantaneous luminosity is determined only by the temperature and the other quantities appearing in equation (21.9).

The convection term of equation (21.7) is quite different. One can use some theory of convection to compute $L_{\text{conv}}(r)$ as a function of conditions in the manner mentioned in Section 19 for the atmosphere, but for the interior the uncertainties in the convection theories are usually of no importance. The reason for this is that the densities and internal energies are so great in interiors that even very slight motions of the material have a large influence upon $L(r)$. Convection is so efficient, in fact, that it is not a question of how much convection is there, but only a question of whether there is any. If there is none, then radiation is dominant, and equation (21.9) holds for the total luminosity. If there is any, then it is dominant, and the radiative contribution can usually be neglected.

The condition for convection to occur may be examined as follows. Suppose a small mass element is in equilibrium with its surroundings at pressure P and density ρ. Then let it suddenly find itself displaced downward by the amount δr, owing to a random perturbation. Its new surroundings have a pressure $P + \delta P$ and a density $\rho + \delta\rho$, and the excess pressure of the surroundings will cause the element to contract until its pressure is also $P + \delta P$. If this contraction takes place so rapidly that there is no appreciable energy exchange between the element and the surroundings,then the contraction will be what is known as an adiabatic change. The new density ρ' of the element will be given by

$$\rho' = \rho + \left(\frac{d\rho}{dP}\right)_{\text{ad}} \times \delta P$$

where the subscript ad indicates an adiabatic change.

If the element is subjected to a force which tends to push it back up to the level it originally came from, then random motions are damped out, and convection will not occur. But if the element feels a force which tends to move it further down, away from its original position, then these random motions will be enhanced, and convection will occur. The direction of the force on the element depends on whether it is more or less dense than the new surroundings, and it is apparent that convection will occur if $(d\rho/dP)_{ad} \times \delta P$ is greater than $\delta\rho$. Since δP and $\delta\rho$ are very small, it follows that convection will occur if

$$\left.\frac{d\rho}{dP}\right)_{act} < \left.\frac{d\rho}{dP}\right)_{ad} \tag{21.10}$$

The subscript act stands for the actual value of the quantity that occurs in the star. The expression (21.10) indicates that convection will occur at a point in a star if the density at that point is changing less rapidly than the adiabatic rate. If density does not increase with depth fast enough, the lower layers will not be able to support the upper layers without becoming unstable to convective motions.

It is more common to express the condition for convection in terms of temperature rather than density. Whenever the density, temperature, and pressure of a perfect gas change, equation (5.8) indicates that these changes must be related by

$$d\rho = \frac{1}{k(T/m)}\left[dP - \frac{P}{(T/m)}\,d\left(\frac{T}{m}\right)\right] \tag{21.11}$$

If this is substituted into (21.10), and one notes that the minus sign changes the "less than" to "greater than," the result is

$$\left.\frac{d(T/m)}{dP}\right)_{act} > \left.\frac{d(T/m)}{dP}\right)_{ad} \tag{21.12}$$

as the condition for convection to occur. If the average mass per free particle m is not changing appreciably, then this takes the more familiar form

$$\left.\frac{dT}{dP}\right)_{act} > \left.\frac{dT}{dP}\right)_{ad} \tag{21.13}$$

If the temperature is changing more rapidly than the adiabatic value, then convection will occur. The adiabatic gradient $(dT/dP)_{ad}$ can be calculated as a function of the local conditions and the composition, so one can check at any point in a model star to see whether (21.13) is satisfied.

In principle one can go ahead and calculate $L_{conv}(r)$ as a function of the local conditions whenever (21.12) or (21.13) is satisfied, but this is not generally done in practice. The reason is that the amount of energy carried by convection is proportional to the actual temperature gradient minus the adiabatic one raised to the 3/2 power; however, in interiors the quantity that multiplies this is so large that the gradient need be only a very tiny amount in excess of the adiabat in order for convection to carry the entire luminosity of the star. This means that in convective regions of a star, conditions follow essentially adiabatic relations. One has, for example,

$$P = \text{constant} \times \rho^{\gamma} \tag{21.14}$$

for an adiabatic change. Here γ is essentially equal to the ratio of the specific heat at constant pressure to that at constant volume, which is 5/3 for a perfect gas which is completely neutral or completely ionized. In other words, in a convection zone of a star one can forget about the luminosity, knowing that convection will take care of however much energy transport is needed. The adiabatic relation (21.14) then replaces equations (21.7) and (21.9).

What happens if the temperature gradient is very much greater than the adiabat? Then convection will drain energy out of the deeper layers at a tremendous rate, causing the deeper layers to cool and lowering the gradient to essentially the adiabatic value. This would give the star a very large luminosity for a short period of time.

Equation (21.9) can give one an idea of when to expect convection to be important. For example, if the mean absorption coefficient κ_R is very large, a large temperature gradient is needed to drive all of the energy by radiation. If κ_R becomes too large, the inequaltity (21.13) will become satisfied and convection will occur. Also, if the energy sources are very strongly concentrated toward the center of the star, then most of the luminosity is produced in a very small volume and a large gradient is again needed to carry off this energy by radiation. Thus a convection zone may occur near the center of some stars.

Convection also occurs in the outer regions of many stars. This hydrogen convection zone as it is called is caused by the combination of two effects. The first is a very large absorption coefficient, due primarily to H^- in the cooler layers and to neutral H in the deeper layers. The second effect is the lowering of the adiabatic gradient due to the ionization of hydrogen and, to a smaller degree, helium. It is apparent from (21.13) that a small value for the adiabatic temperature gradient favors convection. The hydrogen convection zone extends very deep in stars of low mass, but for massive stars it is quite thin and unimportant.

It should be emphasized that most of the above remarks hold only for stellar interiors. In the atmosphere, convection is not very efficient, and rather large motions may produce very little convective energy transport. The temperature gradient in the atmosphere can be considerably greater than the adiabat, and the amount of convective transport must be calculated as described in Section 19.

Energy Sources. The equations of stellar structure could now be solved if the luminosity were known at all positions in the star. In order to find $L(r)$, however, the energy sources must be identified and their properties must be determined. The general situation will not be considered, but it will be assumed for simplicity that the star in question has been in existence for a long enough time that an equilibrium has been reached between the energy sources and the temperature gradient. Thus $L(r)$ will exactly equal the energy produced per second by the matter within the sphere of radius r.

It is usual to define ε as the energy produced per second by one gram of stellar material. Then the equilibrium assumed above means that

$$L(r) = \int_0^r 4\pi r^2 \varepsilon(r)\rho(r)\, dr \tag{21.15}$$

or in differential form,

$$\frac{dL(r)}{dr} = 4\pi r^2 \varepsilon(r)\rho(r) \tag{21.16}$$

It must now be determined how ε can be calculated.

A star like the Sun has many different forms of energy stored in it, and it must be determined which ones are the most important sources of the luminosity. The first form of stellar energy one might think of is the thermal or internal energy E_{th}. This is the kinetic energy of the particles which gives rise to the temperature, and it has already been noted in connection with equation (21.9) that this is the direct source of the luminosity; however, one would like to know where the thermal energy came from and whether it is being constantly replenished.

If U is the thermal energy per gram matter, then

$$E_{\text{th}} = \int_0^{\mathcal{M}} U\, d\mathcal{M} = \bar{U}\mathcal{M} \tag{21.17}$$

where $\bar{U}$ is the average value over the whole star. Using equation (5.13), one finds

$$E_{\text{th}} = \tfrac{3}{2} k \left(\frac{\overline{T}}{m}\right) \mathcal{M} \tag{21.18}$$

If this is applied to the Sun and the previously determined typical values ($m = 0.8 \times 10^{-24}$ g, $T = 5.6 \times 10^6$ °K) are used, one finds

$$E_{\text{th}}(\text{Sun}) \approx 4 \times 10^{48} \text{ erg} \tag{21.19}$$

This is only a very roughly determined value and it can be in error by a factor of 2 or 3, but it should not be wrong by a factor of 10.

The present luminosity of the Sun is about 3.9×10^{33} erg sec^{-1}. This means that if the thermal energy of the Sun is not being replenished, the Sun can keep shining at its present rate for about 10^{15} sec $= 3 \times 10^7$ yr. This thermal time scale of a few tens of millions of years was once thought to be quite long, but there is now much evidence, both astronomical and geological, that the Sun has been shining much as it now is for times of the order of 10^9 to 10^{10} years. The thermal energy in the Sun is, therefore, only about 1% of the energy the Sun is known to have radiated over its lifetime, so the thermal supply must be continually replenished from some other source.

A second source is the gravitational energy E_g, which is the potential energy of the matter due to the attraction of gravity. When an electron drops to a lower quantum state in an atom, the atom loses potential energy of the electrostatic attraction, and this lost energy can appear as excess kinetic energy or as energy in a radiated photon. In exactly the same fashion, if a star contracts to a smaller size, gravitational potential energy is lost, and this may be an important supply to the thermal energy.

Suppose that one is building a star by bringing in the material from an infinite distance. If $\mathcal{M}(r)$ is the mass which has already been brought in and squeezed into the sphere of radius r, then the potential energy lost by bringing an additional gram mass from infinity to r is given by [see equation (3.5)]

$$\Delta PE = -\int_\infty^r F\,dr = +G\mathcal{M}(r)\int_\infty^r r^{-2}\,dr = -\frac{G\mathcal{M}(r)}{r} \tag{21.20}$$

The integral of this over the total mass of the star is the gravitational potential energy E_g of the star:

$$E_g = -\int_0^{\mathcal{M}} \frac{G\mathcal{M}(r)}{r}\,d\mathcal{M} = -4\pi G\int_0^R \mathcal{M}(r)\rho(r)r\,dr \tag{21.21}$$

Equation (21.21) can be integrated in the same approximate fashion that led to the result (21.19), but it is more instructive to approach this in a different way. From equations (21.17) and (5.13), the thermal energy is

$$E_{\text{th}} = \frac{3}{2}\int_0^{\mathcal{M}} \frac{P}{\rho}\,d\mathcal{M} = 6\pi\int_0^R P(r)r^2\,dr$$

The second of the above relations follows from (21.3). If this is integrated by parts, the result is

$$E_{\text{th}} = 6\pi\left[\tfrac{1}{3}P(R)R^3 - \frac{1}{3}\int_0^R r^3\frac{dP}{dr}\,dr\right]$$

The pressure vanishes at the surface of the star, so $P(R) = 0$. Using equation (21.2) for (dP/dr), one obtains

$$E_{\text{th}} = 2\pi G\int_0^R \mathcal{M}(r)\rho(r)r\,dr = -\tfrac{1}{2}E_g \tag{21.22}$$

Equation (21.22) is a special form of what is known as the virial theorem, which is a statement of how the energy of a body is distributed among its various forms as a function of the body's relevant properties. [It is interesting that equations (3.6) and (3.7) indicate that for an electron in a circular orbit around a proton, $E_{kin} = -\frac{1}{2}E_{pot}$, in exact agreement with (21.22).] Equation (21.22) indicates the very important fact that as long as a star is in hydrostatic equilibrium, its thermal energy is directly proportional to the negative of its gravitational energy. Now it seems reasonable that stars were once very large, tenuous gas clouds which somehow contracted to their present sizes. In the very early stages before gravitational forces had much effect on this contraction, the gravitational energy was nearly zero, and the thermal energy must also have been nearly zero (unless an external energy source existed). As the contraction started, the gravitational energy started to decrease, and, by equation (21.22), the thermal energy started to increase. The stars began to heat up. Of course, hydrostatic equilibrium cannot hold rigorously for a contracting star, but it will be a very good approximation if the contraction does not proceed too rapidly. During this contraction, as the gravitational energy is decreased by the amount ΔE, the internal energy is increased by the amount $\frac{1}{2}\Delta E$. Since energy is conserved, the remaining amount, $\frac{1}{2}\Delta E$, must be the quantity that is radiated away by the star during this time interval.

This conversion of gravitational energy into thermal energy through contraction is a plausible mechanism for the stars to have become hot, but it still does not satisfy the total energy requirements for the Sun. The gravitational time scale is of the same order as the thermal time scale, namely, a few tens of millions of years, so the main supply of solar energy must come from some other source.

In the last century H. von Helmholtz and Lord Kelvin suggested that gravitational contraction is the source of the solar energy. Although this is now known to be incorrect, the physics of this mechanism is still valid. Thus gravitational contraction does play a vital role during several stages in the life of a star, including the very early stages in which the star is heating up, but it is not the main energy source for most of the star's lifetime.

The nuclear energy sources of a star provide the third great supply. Nuclear reactions as discussed in Section 6 can take place in the high temperatures and densities which occur in stellar interiors, and it remains to be seen whether these reactions can release enough nuclear potential energy to satisfy the solar energy needs.

Hydrogen and helium are by far the most abundant elements, and it seems reasonable to look for them to provide the necessary energy. The discussion in Section 6 suggests that reactions which build heavy particles out of light ones are probably needed, and so one might see whether hydrogen can somehow be converted into helium. Normal hydrogen consists of a proton with an electron orbiting around it, while normal helium has two protons and two neutrons in the nucleus and two outer electrons. This H atom has a mass of about 1.008 amu (atomic mass units), while the He atom is about 4.004 amu. The He atom has slightly less than four times the mass of the H atom; the mass-energy equivalence indicates that if four H atoms can be converted into an He atom, energy will be released. Since 1 amu = 1.66×10^{-24} g, equation (6.3) shows that the nuclear potential energy change when four H atoms combine to produce one He atom is

$$E = c^2(\mathcal{M}_{\mathrm{He}} - 4\mathcal{M}_{\mathrm{H}})$$

$$= 9 \times 10^{20}(4.004 - 4.032) \times 1.66 \times 10^{-24}$$

$$= -4.2 \times 10^{-5} \text{ erg}$$

The minus sign indicates that the potential energy is lost. Since four H atoms have a mass of 6.69×10^{-24} g, the energy released when one gram of hydrogen is converted to helium is about 6.3×10^{18} erg.

The mass of the Sun is about 2.0×10^{33} g; if this were all hydrogen and if it could all be converted into helium, then the nuclear energy supply due to this conversion would be about 10^{52} ergs. There is reason to believe that only about 10% of this hydrogen is available to be converted into helium, since only the very central regions are sufficiently hot for nuclear reactions to take place at an appreciable rate; therefore, the nuclear energy supply in the Sun is

$$E_n \approx 10^{51} \text{ erg} \tag{21.23}$$

A comparison of this number with (21.19) shows that the nuclear energy is some $2\frac{1}{2}$ orders of magnitude greater than the thermal energy in the Sun, so it is sufficiently large for the known needs of the Sun.

It might be expected that (21.23) is a lower limit, since more nuclear energy could be released by the conversion of helium into heavier elements yet. It is easily seen from the data in Appendix C that if one gram of He^4 were somehow converted to Fe^{56}, the most abundant isotope of iron, the energy release would be about 1.6×10^{18} erg. This corresponds to the most efficient use possible of the nuclear fuels, so the conversion of hydrogen to helium is seen to release about 80% of the available nuclear energy.

Saying that hydrogen can be converted into helium is rather different from giving the details of just how this is to be done. Plausible mechanisms for this were suggested by C. F. von Weizsäker, H. Bethe, and C. L. Critchfield in the late 1930s, and it is now believed that most stars are deriving their luminosities from one or both of two competing processes, both of which convert hydrogen into helium. These are known as the proton-proton (PP) reaction and the carbon-nitrogen (CN) cycle. Each of these has more than one variation but the main reactions are as follows:

$$\left.\begin{aligned} 2(H^1 + H^1) &\rightarrow 2(H^2 + \beta^+ + \nu) \\ 2(H^2 + H^1) &\rightarrow 2(He^3 + \gamma) \\ He^3 + He^3 &\rightarrow He^4 + 2H^1 \end{aligned}\right\} \tag{21.24}$$

$$\left.\begin{aligned} C^{12} + H^1 &\rightarrow N^{13} + \gamma \\ N^{13} &\rightarrow C^{13} + \beta^+ + \nu \\ C^{13} + H^1 &\rightarrow N^{14} + \gamma \\ N^{14} + H^1 &\rightarrow O^{15} + \gamma \\ O^{15} &\rightarrow N^{15} + \beta^+ + \nu \\ N^{15} + H^1 &\rightarrow C^{12} + He^4 \end{aligned}\right\} \tag{21.25}$$

In the above reactions, β^+ stands for the positron, which will very quickly find an electron, with the result that the pair will be annihilated with the emission of radiation. The symbol γ stands for a γ-ray photon emission, and ν is the neutrino. The neutrino has very little interaction with matter, so little that it will probably pass through the entire star and escape without further interactions. Since the energy carried off by the neutrinos is not included in the observed luminosities of the stars, the reaction energies must be corrected for this loss.

The PP reaction is represented by (21.24), while (21.25) is the CN cycle. The PP reaction is the more direct one, since it starts by combining two protons to form a deuteron or heavy hydrogen nucleus. The latter then builds up to He^3 and then He^4, although there are other reactions through which the PP reaction sometimes ends. The PP reaction is favored by two points: it involves only the most abundant elements, and only a relatively small coulomb barrier need be overcome for the reactions in (21.24). This is balanced by one unfavorable point, namely that the β^+ emission that must occur in the first reaction is extremely unlikely. This means that when two protons collide, they usually just scatter off with no nuclear reaction. When the β^+ emission to form H^2 does occur, the next two reactions take place relatively quickly to finish the PP reaction. The first two reactions must take place twice before the last one can occur, and this is indicated in (21.24).

The CN cycle is of a different sort. Both carbon and nitrogen appear as catalysts; i.e., they are not used up in any way, yet they are essential to the cycle. The cyclic property of (21.25) with respect to C and N is apparent: one could have started with any of the reactions and worked around to the previous one without any difference. It happens that the capture of a proton by the N^{14} nucleus is the least likely reaction to occur, so when it does take place, the others follow relatively quickly. This means that in a region in which the CN cycle has been going on for long enough, most of the carbon and nitrogen will have been converted to N^{14}, and the other reactions are waiting for this bottleneck.

The coulomb barrier which a proton must overcome if it is to be captured by carbon or nitrogen is relatively large. At low temperatures, very few protons can be captured, and so the PP reaction will dominate. The CN cycle is much more sensitive to temperature, however, so as T increases, the relative contribution of the CN cycle to the total energy generation rapidly goes up. At very high temperatures, the CN cycle dominates, and the PP reaction is of no importance.

In order for the energy generation rate to be determined, the rate of each reaction must be known. This means that for each reaction one must be able to calculate the number of collisions which take place and the probability that a collision will result in the desired reaction. The collision rate can be calculated as a function of temperature, density, and abundances from the properties of gases, as discussed in Section 5. The probability that a collision of given energy results in the relevant nuclear reaction is analogous to the transition probabilities involved in the absorption and emission of photons by atoms. This probability is expressed in terms of what is known as the cross section of the nuclear reaction, and the cross section must be measured in the laboratory or be calculated by means of quantum mechanics.

Experimental data on the nuclear cross sections of interest in astrophysics are of somewhat limited value. The reason for this is that nuclear reactions under conditions found in stars take place extremely slowly, too slowly for laboratory data to be obtained under similar conditions. The laboratory data must be obtained at much higher energies for which a measurable yield can be produced. The extrapolation of the high-energy laboratory data to the low-energy stellar conditions involves some uncertainty, although there is much confidence in many of the important cross sections known today. In the case of the PP reaction, the source of information is completely theoretical.

There are other nuclear reactions which are of importance under certain circumstances, and some of these will be referred to later.

If the relevant reaction rates are known, one can then calculate ε, the energy released by nuclear reactions per gram of material and per second. This will be a function of the density, the temperature, and the composition of the material, so one can express the energy generation in the form

$$\varepsilon = \varepsilon(\rho, T, \text{composition}) \tag{21.26}$$

in the same way that the absorption coefficient is expressed. The equations for stellar structure are now complete.

Obtaining the Interior Structure. It is worth bringing together all of the important relations for stellar structure:

$$\frac{d\mathcal{M}(r)}{dr} = 4\pi r^2 \rho \tag{21.27}$$

$$\frac{dP}{dr} = -G\frac{\mathcal{M}(r)\rho}{r^2} \tag{21.28}$$

$$\frac{dL(r)}{dr} = 4\pi r^2 \rho \varepsilon \tag{21.29}$$

$$P = \frac{k}{m}\rho T \tag{21.30}$$

and

$$L(r) = -\frac{16\pi\sigma}{3}\frac{r^2}{k_R\rho}\frac{dT^4}{dr} \tag{21.31}$$

or

$$P = \text{constant} \times \rho^{\gamma} \tag{21.32}$$

Equation (21.31) holds in radiative regions, and equation (21.32) takes its place when convection occurs. It is understood that ε, κ_R, m, and γ can be calculated as functions of composition and the local conditions. There are then five equations for the five unknown functions $\mathcal{M}(r)$, $L(r)$, T, P, and ρ, and mathematical theory indicates that it should be possible to solve them.

It has already been pointed out that one needs to know the chemical composition in order to determine the energy-generation rate, the absorption coefficient, the quantity γ appearing in equation (21.32), and the average mass per particle m. This means that the composition must be known before the structure equations can be solved, so composition is an input parameter for model interiors just as it is for model atmospheres.

Suppose that the density were known at all points in the star. In order for this to be known, however, the radius R of the star must also be known, since R is the value of r for which the density goes to zero. The quantity R is then another input parameter for the model interior.

With R and $\rho(r)$ known, equation (21.27) can be integrated to give $\mathcal{M}(r)$, and (21.28) can then be integrated to yield the pressure at all points. The equation of state (21.20) can then be solved for the temperature (actually the equation of state and the ionization equation are solved together for T and m), and equation (21.29) can finally be integrated to give $L(r)$. In this fashion the entire structure of the star can be obtained, including the total mass of the star $\mathcal{M}(R) = \mathcal{M}$ and the emitted luminosity $L(R) = L$. It is worth pointing out that both the surface gravity and the effective temperature follow from the values of $\mathcal{M}$, L, and R, and so even the detailed structure of the atmosphere can be computed for the model star by using the method discussed in Section 19.

How does one know the correct $\rho(r)$ relation in the beginning? One does not know it, so it must be guessed. If the guess is correct, equation (21.31) will automatically be satisfied for all points in radiative zones, and equation (21.32) will be satisfied for all points in convective regions in the star. If these equations are not satisfied, then the $\rho(r)$ relation will have to be revised and the process will have to be repeated. It can be shown that the solution obtained is unique.

It would appear that stars which are completely convective are easy to handle, since equation (21.32) is very simple; however, this equation cannot be used unless the value of the constant has been determined, and this can be found for such stars only from a detailed atmospheric model. The solution in this case is not at all simple.

The above method of solution of the structure equations, which is essentially the same as that for solving the model atmosphere equations, was only chosen for illustration. This is not the actual method of solution used in practice, but the result is the same: chemical composition plus one other input parameter uniquely determines the entire structure of the star.

The other input parameter besides composition need not be the radius, but it could be the mass, the luminosity, the central temperature, or the like. Mass is usually chosen as the independent quantity, since it appears in many respects to be more fundamental than the others. The statement that the mass and composition of a star uniquely determine the entire structure of the star is known as the Russell-Vogt theorem, after H. N. Russell and H. Vogt.

Although the Russell-Vogt theorem mentions only mass and composition, it is apparent that another quantity must also be involved in order to make the structure completely unique. A star with uniform composition is burning hydrogen and converting it to helium, and this does not take place at the same rate at all points in the star. The rate at which the composition changes with time at a point depends on how fast the nuclear reactions are taking place there and on whether or not convection is present at that point to help keep the material well mixed. Since these can be calculated, the change of composition can be found as a function of time for any point in the star. This leads to the same conclusion found in previous sections: the structure of a star depends on its original composition plus two more quantities, the latter usually being taken as the mass and age of the star. Since the nuclear reactions take place very deep in the star, the composition found from the atmosphere should be the same as the original composition, unless the star is well mixed throughout. For most stars this complete mixing does not occur.

Homology Transformations. Two stars are said to be homologous if they are built alike in the sense that all physical variables have the same relative variation within the two stars. Thus, if the temperature falls to one-half of its central value at a certain fraction of the distance out of the first star, it will fall off to one-half of its central value at the same fraction of the distance out of the second star. Let the following dimensionless quantities be defined:

$$\left.\begin{aligned} x &= \frac{r}{R} & \mathcal{M}_* &= \frac{\mathcal{M}(r)}{\mathcal{M}} & L_* &= \frac{L(r)}{L} \\ \rho_* &= \frac{\rho(r)}{\rho_c} & P_* &= \frac{P(r)}{P_c} & T_* &= \frac{T(r)}{T_c} \end{aligned}\right\} \tag{21.33}$$

The subscript c indicates the center of the star. For stars which are homologous to each other, the functions $\mathcal{M}_*(x)$, $L_*(x)$, and the like are the same. If the structure of a star is known, then the structure of any star homologous to the first can be found by a simple change of scale, i.e., by the appropriate homology transformation.

Consider equation (21.27) for the mass distribution. If $\mathcal{M}(r)$ and r are eliminated by means of (21.33), one finds

$$\frac{\mathcal{M}}{R}\frac{d\mathcal{M}_*}{dx} = 4\pi R^2 x^2 \rho$$

or

$$\frac{1}{4\pi x^2}\frac{d\mathcal{M}_*}{dx} = \frac{R^3 \rho(x)}{\mathcal{M}}$$

The left side of this equation depends only on the function $\mathcal{M}_*$ and the value of x; therefore, if one stays at the same relative position in the stars (so that x is constant) and if one stays with stars which are homologous to each other (so that $\mathcal{M}_*$ is the same function), then the left side of the above expression is a constant. Obviously, the right side must also be constant, so

$$\rho(x) = \text{constant} \times \frac{\mathcal{M}}{R^3} \tag{21.34}$$

Equation (21.34) is the homology transformation for the density. For stars that are built alike, the density at any point in the interior scales as the mass divided by the cube of the radius.

Consider now the pressure equation (21.28). If P, $\mathcal{M}(r)$, and r are eliminated by means of (21.33), one has

$$\frac{P_c}{R}\frac{dP_*}{dx} = \frac{P}{P_* R}\frac{dP_*}{dx} = -\frac{G\mathcal{M}\mathcal{M}_*\rho}{R^2 x^2}$$

or

$$-\frac{x^2}{GP_*\mathcal{M}_*}\frac{dP_*}{dx} = \frac{\mathcal{M}\rho}{PR}$$

Once again the left side of this equation is constant for constant x and homologous stars, so one has

$$P(x) = \text{constant} \times \frac{\mathcal{M}\rho(x)}{R}$$

Using (21.34), one finally has

$$P(x) = \text{constant} \times \frac{\mathcal{M}^2}{R^4} \tag{21.35}$$

The equation of state, (21.30), can be used to fix the temperature transformation. If the average mass per free particle m is a constant, then it is apparent that

$$T(x) = \text{constant} \times \frac{P(x)}{\rho(x)} = \text{constant} \times \frac{\mathcal{M}}{R} \tag{21.36}$$

It should be emphasized that equations (21.34), (21.35), and (21.36) do not indicate how the physical quantities vary throughout a star; they indicate how the quantities vary from one star to another if x is held constant. As an example, suppose that star 2 has twice the mass and twice the radius of star 1. If these two stars are homologous with each other, then (21.36) indicates that the temperature at any point in star 2 is the same as that at the corresponding point in star 1. Equation (21.35) shows that the pressure at any point in star 2 is only $\frac{1}{4}$ of the pressure at the corresponding point in star 1, and the densities from (21.34) are also $\frac{1}{4}$ those in star 1. If the structure of either star is known, these homology transformations provide an easy way to find the structure of the other star.

Homology transformations for the luminosity can also be obtained, but they introduce additional uncertainties. For example, one can easily find from equation (21.29) that

$$L = \text{constant} \times \mathcal{M}\varepsilon \tag{21.37}$$

One thus needs to know how the energy generation depends on conditions before this can be made a function of mass and radius alone. Likewise, equation (21.31) lead to an expression which depends on the transformation properties of the mean absorption coefficient.

The energy-generation formula, equation (21.26), is often expressed in the following form:

$$\varepsilon = \varepsilon_0 \rho T^n \tag{21.38}$$

This form does not follow from the physics of the situation, but is only a mathematical convenience; therefore, the exponent n is not actually a strict constant. If n is considered a constant, then the above two equations yield

$$L = \text{constant} \times \frac{\mathcal{M}^{2+n}}{R^{3+n}} \tag{21.39}$$

For the proton-proton reaction, the exponent n has a value of about 4 in the relevant temperature range, while the carbon-nitrogen cycle has a temperature exponent of about 15–20. Equation (21.39) then leads to quite different results, depending on whether a star derives its energy primarily from the PP reaction or from the CN cycle.

An alternative expression for the luminosity can be found as follows: according to equation (8.5), luminosity is proportional to the square of the radius times the fourth power of the effective temperature. Now, the effective temperature is only slightly greater than the surface temperature of a star ($T_e = 1.189T_0$ in the Eddington approximation of equation (19.20)), so T_e should be very nearly a temperature at a fixed value of x. Thus it should transform as in equation (21.36), and one has

$$L = \text{constant} \times \frac{\mathcal{M}^4}{R^2} \tag{21.40}$$

Equations (21.39) and (21.40) together give

$$R = \text{constant} \times \mathcal{M}^{(n-2)/(n+1)} \qquad L = \text{constant} \times \mathcal{M}^{(2n+8)/(n+1)} \tag{21.41}$$

As an example of an application of the above equations, main sequence stars of mass less than or equal to that of the Sun derive their energy primarily from the proton-proton reaction, for which the index n is about 4. Equations (21.41) then indicate that $R \sim \mathcal{M}^{2/5}, L \sim \mathcal{M}^{16/5}$. Considering the approximate nature of these relations, this is in very good agreement with the observed mass-luminosity relation for main-sequence stars. This also indicates that the surface gravity, which is proportional to $\mathcal{M}/R^2$, should increase only very slowly as one proceeds up the main sequence.

These homology transformations are of limited use, for the assumption that two stars would be built alike is very restrictive. When should one expect this assumption to be valid? Two stars will be simple scale models of each other if all of the major physical processes have the same relative importance in the two stars. For example, both radiation and gas pressures compete in the equilibrium of a star; therefore, a star in which the gas pressure is dominant cannot be expected to be homologous with a star in which radiation pressure is dominant. Radiation, convection, and conduction all compete with each other in bringing energy out of a star. Stars in which the relative importance of these is significantly different cannot be expected to be homologous.

A very large number of very complicated processes take place in all stars, and one cannot expect to change any stellar parameter by much without upsetting some of the important balances that exist between competing processes; therefore, the homology transformations should be expected to be valid only among stars of similar characteristics. One should have some confidence in such a transformation between a G0 V star and one of spectral type G5 V; the confidence should drop considerably if the range is extended from F5 V to G5 V, although it may still give useful results; one may not obtain anything useful if the range goes from B0 V to G5 V.

The present section is concerned with the physical basis for interior investigations. The results of these investigations and their interpretation are discussed in Section 22.

22. Stellar Evolution

The theoretical approach to stellar evolution is based on model stars, and theory and observation are brought together in H-R diagrams. The reason for this is that an H-R diagram represents the most important observations which are fairly easily obtained.

Both observational and theoretical evidence have been given that the structure of a star is determined by its original composition, its mass, and its age. Magnetic fields, rapid rotation, and membership in a close binary system can cause important exceptions to this, but they will generally be assumed to be absent. Uncertainties in the physical theory can cause a model star to differ from a real star of the same parameters, so the comparison between theory and observations can help improve the theory as well as give a physical interpretation to the observations.

Interpretation of the Main Sequence. One should be able to reproduce the characteristics of most stars by calculating models of all reasonable values of the input parameters of composition, mass, and age. Stars in the solar neighborhood undoubtedly represent a wide range of these parameters, but it is more instructive to proceed more slowly by considering the parameters one at a time.

Suppose that one constructs a series of stellar models having different masses but the same age and composition. In particular, let the ages be essentially zero so that the composition is still uniform throughout the stars. If one plots such a series of models in an H-R diagram, one finds that they fall along the main sequence. The models with high mass are on the upper left, and the lower masses fall toward the lower right. The main sequence appears to be populated by stars which are not old enough to have changed appreciably from their original properties.

Now let the composition be changed and then the above calculations be repeated. The new series will define its own main sequence, and it will not be much different from the main sequence of the first group of models if the composition change is not unreasonably large.

Evolution Past the Main Sequence. Aging effects can be studied by constructing a series of models of constant mass but varying age. It is true that the star loses mass as the nuclear burning takes place, but this is such a small amount that one can consider the total mass of a star as remaining constant. One is less certain about mass loss through ejection of material into space. It is believed that normal stars do this only at certain critical periods, if at all, so that one can consider the evolution to take place largely at constant mass; however, this is a potential source of error, as the subject is not at all well understood.

A star begins its career on the main sequence with a homogeneous composition. As hydrogen is being converted to helium in the central parts, the features of the star change, very slowly at first, more rapidly as the hydrogen becomes more nearly depleted at the center. The star has become somewhat brighter but is still near its original position on the main sequence when the hydrogen content in the center is too low for nuclear reactions to supply enough energy to support the star against gravity. Gravitational contraction then becomes an important energy source as the inner parts of the star contract, while the outer parts expand. The star rapidly becomes brighter and cooler and moves upward and to the right in the H-R diagram, toward the red-giant region.

The contraction of the core of the star causes the central region to become more hot and dense, and hydrogen burning in shells further out from the center can take place. Eventually conditions in the center will become extreme enough for the nuclear burning of helium to form carbon in the so-called triple alpha process. The star will then settle down to a second period of nuclear burning, although the helium-burning giant stage will not last nearly as long as the hydrogen-burning main-sequence stage.

The details of the change from the main-sequence to the giant phase depend on the mass. For moderate and low masses, degeneracy of the electron gas in the helium core slows down the core contraction, so that the helium burning is delayed. The more massive stars which burn hydrogen by means of the carbon-nitrogen cycle have convective cores. This keeps the material well mixed and provides a larger supply of hydrogen for the central region than would otherwise be available. Low-mass stars which use the proton-proton reaction do not have convective cores, and in them the composition varies smoothly between the partially depleted center and the unaffected regions farther out.

The further evolved the models become, the less reliable they are. It is likely that the most massive stars become supergiants, while less massive ones become only giants. It is certainly true that the more massive stars evolve off the main sequence more rapidly. It was pointed out in Section 21 that the Sun has enough hydrogen in its central regions to last for about 10^{10} years under present conditions. This means that its main-sequence lifetime is about this long, and for any other star,

$$t_{ms} \approx 10^{10} \frac{\mathcal{M}}{L} \tag{22.1}$$

where $\mathcal{M}$ and L are the mass and luminosity in solar units. The mass-luminosity relation along the main sequence is about $L \sim \mathcal{M}^3$, to a good approximation in the general region of the Sun, so it follows that the more massive stars burn up their nuclear fuel more rapidly than the less massive ones. Main-sequence lifetimes as short as 10^7 years and as long as 10^{13} years are to be expected.

To summarize, stars begin the nuclear burning part of their lives on the main sequence. They evolve off the main sequence after almost all of the hydrogen in their centers has been converted to helium, and any star far from the main sequence must be either highly evolved or too young to have yet reached it. Differences in the original composition cause small but important differences in the position of the main sequence. The main sequence of the stars in the solar neighborhood is rather broad, owing partly to the range of chemical compositions and partly to the range in ages of the stars. The main sequence and the giant branch of the H-R diagram are heavily populated because stars spend a large part of their lives in these regions. If stars pass very quickly through a certain stage, there is small probability of finding many stars in that stage, and the corresponding region of the H-R diagram will be poorly populated.

Star Clusters. The above points are extremely important and are based on model star calculations. While they are consistent with the observations, in general outline, one would like to have a stronger observational confirmation. Fortunately, this is possible. Star clusters make it possible to check many details of evolution theory directly.

A star cluster is a group of stars all of which are bound together by their mututal gravitation. Many clusters are known, and they range in size from a few stars to hundreds of thousands of them. The largest clusters are often nearly spherical in shape and are called globular clusters. A second type of cluster which is usually much less populous than the globulars is known as the galactic or open cluster. The most important point about clusters, relevant to the present subject, is that there is good evidence that all members of a given cluster were formed at about the same time and out of the same kind of material. The H-R diagram of a cluster should therefore indicate what a group of stars of constant age and composition is like. By comparing different clusters, one can obtain direct observational evidence of aging and compositional effects.

Another advantage of clusters is that the members of one are all at essentially the same distance from the Earth. If the distance of a cluster is not well known, as is often the case, the individual luminosities cannot be accurately found; however, differences in apparent magnitude are the same as differences in absolute magnitude, so a plot of color vs. apparent magnitude will be essentially the same as an H-R diagram. In such a diagram the only thing not known is the zero point of the vertical scale, but this does not affect the shape of the curves.

The H-R diagrams of galactic clusters are usually simpler than those of globular clusters, and they will be considered first. Figure 22.1 is a schematic H-R diagram for three galactic clusters which nearly cover the range shown by these objects. All three main sequences come together, more or less, at the lower right. Cluster 1 has a main sequence which extends up to very hot, luminous stars plus a small number of cool supergiants. Cluster 2 has its main sequence veer above that of 1 and end at stars which are less hot and luminous than those at the bright end of cluster 1. Cluster 2 has red giants but no supergiants, and the gap between the giant branch and the main sequence (the so-called Hertzsprung gap) is smaller than in cluster 1. Other clusters are known which have a main sequence that veers upward at still lower temperatures and luminosities and that ends sooner. Their giant branches come closer to the main sequence and have lower luminosities. The cluster 3 in the figure has no Hertzsprung gap, and a more or less continuous sequence connects the giant branch to the main sequence.

One further point should be made here: abundance analyses for galactic clusters indicate that, while composition differences among clusters are not absent, neither are they very great. It follows, therefore, that the differences in the H-R diagrams illustrated in Figure 22.1 are due primarily to aging effects.

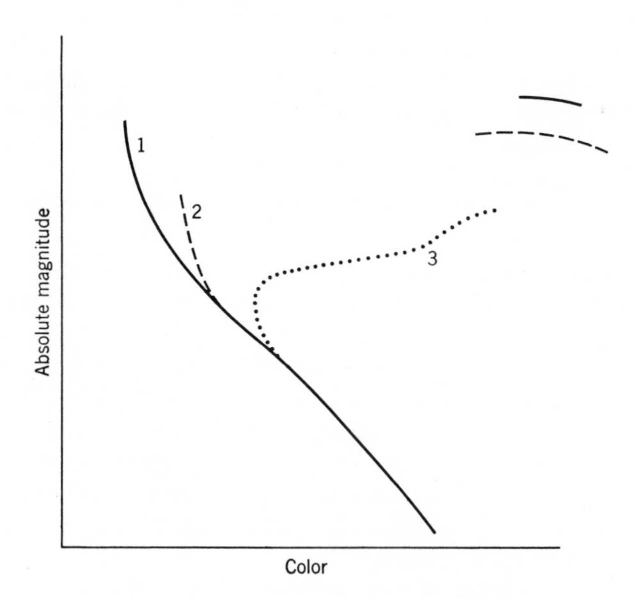

Figure 22.1 H-R diagrams of galactic clusters.

The model calculations indicate that a cluster of age zero should have only a main sequence. As the cluster ages, the most massive stars will evolve off the main sequence toward the red giant or supergiant region. Those slightly less massive will have evolved slightly up from their age-zero position, but they will still be in the main-sequence region. Since the evolution from main sequence to giant takes place very rapidly, the chances are small of finding stars in the Hertzsprung gap. Stars of much smaller mass are still in their age-zero positions. This represents cluster 1, a very young galactic cluster in which only the most massive stars have evolved away from the main sequence.

Cluster 2 is somewhat older. It is old enough for still smaller mass stars to have evolved off the main sequence, and these have produced the red giants shown. The supergiants have burned themselves out completely by this time, and the turn-off point, at which the main sequence leaves the age-zero main sequence, has come down to smaller masses. Stars below this turn-off point are not sufficiently massive to have been appreciably affected by aging.

Cluster 3 is still older. All of its most massive stars have long ago burned themselves out, and the turn-off point is still further down the main sequence. The fact that the Hertzsprung gap has closed must mean that stars do not evolve as rapidly here as did their more massive counterparts. This may be due to the degeneracy effects mentioned earlier which delay the onset of the helium burning, and it may also be due in part to the fact that more stars are formed with these masses than with the much higher masses.

Figure 22.2 shows a schematic picture of a typical globular-cluster H-R diagram. It is similar in many ways to that of an old galactic cluster, except that there is a horizontal branch extending to the left of the giants. This branch has a small region known as the RR Lyrae gap, in which nothing but RR Lyrae variables appear. The suggestion is that any star which evolves into this region becomes a variable star.

It is not easy to compare the H-R diagrams of galactic and globular clusters. The latter are so far away that their distances are poorly known, and usually most or all of the main sequence is too faint to be observed accurately. It is also difficult to distinguish age and composition effects, since there is a significant difference in composition between galactic and globular clusters. The globular clusters have a much smaller amount of the heavy elements than the galactic clusters, and this does have some important effects on the details of the H-R diagrams. For example, the glubular-cluster giants appear to be more luminous than those in the very old galactic clusters, and the globular-cluster main sequences are somewhat bluer. These are probably due to the composition differences, and globular cluster stars appear to be similar to the subdwarfs mentioned in Section 19.

Galactic clusters occur at all ages from very young to very old, while the globular clusters are all very old. The oldest galactic clusters have turn-off

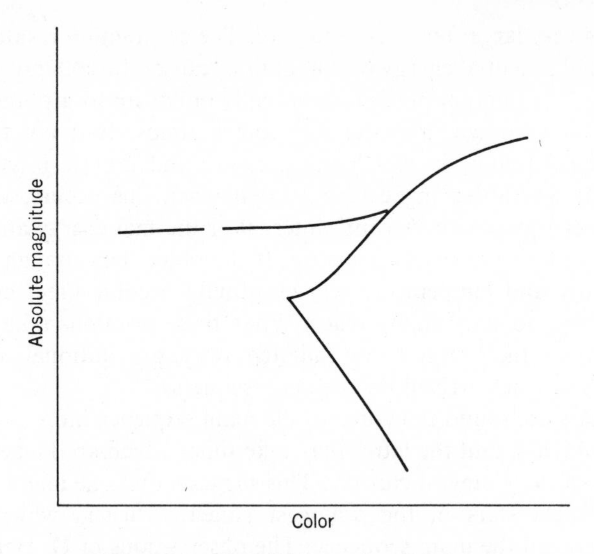

Figure 22.2 H-R diagram of a globular cluster.

points which occur at about the position of the Sun, and equation (22.1) then indicates that these clusters are around 10^{10} years old. Globular clusters are about this age also, but the youngest galactic clusters are perhaps under 10^7 years old.

The agreement between the theoretical models and the observations is quite remarkable, and there is little doubt that the main features of the main sequence to giant evolution as sketched above are correct. It should be emphasized, however, that the quantitative results are often subject to a rather large uncertainty.

Pre-Main-Sequence Evolution. It is suggested above that stars begin their lives on the main sequence, but it is apparent that they must evolve onto the main sequence from some previous state. This previous state must be the interstellar medium, the tenuous gas and dust that exist between the stars, for where else could the stars come from?

The interstellar medium is not exactly uniform, there being large irregularities in the density. The gravitational attraction of the material will try to draw it together, but the thermal motions of the atoms are usually great enough to resist. Occasionally the density at a given region will become great enough for gravitational contraction to win out, and the material in this region will condense out of the general medium. The specific conditions under which this will occur are not at all well understood, and this has been the subject of many investigations in recent years.

When an object of stellar mass has first condensed out of the interstellar medium, it is very large, tenuous, and cool. The contraction results in a loss of gravitational potential energy which, at first, causes the collapse to proceed more rapidly. The internal pressure eventually builds up to a point at which the collapse slows down, and the star comes almost (but not quite) into hydrostatic equilibrium, in which the pressure and gravity forces balance each other. It was noted in Section 21 that when this occurs, the loss of gravitational energy goes partly into internal or thermal energy and is partly radiated away; thus the object heats up. If the object has enough mass, the internal density and temperature will eventually become great enough for nuclear reactions to start taking place. When these reactions take place fast enough to supply the energy being radiated away, gravitational contraction stops and the star has arrived on the main sequence.

Massive stars are found to evolve to the main sequence more rapidly than those of small mass, and the latter may take times which are longer than the ages of some of the youngest clusters. This suggests that one might be able to observe low-mass stars in the youngest galactic clusters which are still contracting toward the main sequence. The observations of G. Herbig, P. P. Parenago, M. Walker, and others seem to confirm this. Irregular variables known as T Tauri stars are observed in this situation, and they are thought to be in an evolutionary phase just prior to arriving on the main sequence. They occur where the interstellar matter is unusually dense, and they lie above the zero-age main sequence where theory says such objects should be. The clusters that they are members of are not old enough for stars of this small mass to have evolved off the main sequence, so they are not in a post-main-sequence phase.

Large numbers of small, dark globules are observed in regions in which the interstellar density is large. It is tempting to interpret these as stars in the making, but this is as yet uncertain.

It is mentioned above that a star will arrive on the main sequence if it has sufficient mass. If it is too small, the internal density and temperature will never become great enough for the main-sequence hydrogen-burning reactions to take place, and the object will continue to contract until degeneracy sets in, after which it will slowly radiate away what excess internal energy it has. According to S. S. Kumar, the lower limit to the main sequence is somewhat under 0.1 solar mass, the precise value depending on the composition.

Post-Giant Evolution. The post-giant evolutionary phase is much more speculative than the phases which have been discussed so far. Both the physical theory and the observations are more uncertain. There is reason to expect that the same general processes which bring a star to the red-giant region also take it away again; i.e., gravitational contraction brings on more extreme conditions which ignite new nuclear reactions; the new fuels are soon depleted, contraction begins once more, and so on.

This cycle of processes cannot go on indefinitely, for at least two reasons. In the first place, the most stable elements are those around iron in the periodic table. If one starts with hydrogen and builds up heavier elements by nuclear fusion, the reactions will give off energy only until the iron region is reached. Any further reactions will absorb energy from the surroundings instead of giving it off to the surroundings, and such reactions are called endothermic instead of the more usual exothermic ones. If endothermic reactions become important in a star, then the thermal energy in the star is being used up all the more rapidly instead of being replenished, and the above cycle of processes will come to an abrupt halt.

The second point is that, if the matter in a star is allowed to become degenerate, it will resist further contraction, and again, the above cycle of processes will come to an end. Since matter becomes degenerate if the density becomes too large at a given temperature, it may appear that all stars should simply end as degenerate objects; however, there is a complication: it was shown by S. Chandrasekhar that a star cannot become completely degenerate (actually it is only the electrons that become degenerate) unless its mass is less than a certain critical limit. This limit depends on the composition of the star, but it is about 1.4 solar masses. In effect, a star more massive than the Chandrasekhar limit will increase its temperature as it contracts in such a way that it cannot ever become completely degenerate, no matter how large the density becomes.

The white-dwarf stars are evidence for the existence of such degenerate stars. White dwarfs get their name from the fact that many of them are rather hot, thus having white colors, yet they fall far below the main sequence. This means that they must have very small radii (more planetary than stellar in size). Several of them have well-determined masses, and they are all well below the Chandrasekhar limit. The very high densities that follow, 10^4 to 10^6 times that of water, confirm that the electron gas in the main part of white dwarfs is very highly degenerate.

There is little doubt that white dwarfs are in the final stage of stellar evolution prior to being completely burned out. They would be expected to be largely if not completely depleted of their hydrogen. Some white dwarfs do show no hydrogen in their spectra, but many have very strong hydrogen lines. This is not surprising, since E. Schatzman has shown that the gravitational field in a white dwarf is so strong that the elements will be separated according to their masses; the lighter hydrogen, if any is left, will be buoyed up to the surface where its spectrum can be observed.

It is somewhat marginal whether the Galaxy is old enough for a one-solar-mass star to have evolved to the white dwarf stage, but it certainly is not old enough for a star of one-half solar mass to have done so; yet white dwarfs with mass this small are known. This is evidence that some stars lose a large fraction of their mass at some stage of their evolution. Also, although white dwarfs are very faint and cannot be observed at great distances, the number of them known is very large; this gives support to the idea that many if not all stars go through the white-dwarf stage, and again suggests that mass loss must be important in the evolution of many stars.

An interesting idea for getting rid of mass involves the endothermic nuclear reactions. If such a reaction were very sensitive to temperature and density, it could be turned on quite suddenly during one of the contraction stages. This could rapidly drain a large amount of thermal energy from the interior of the star, causing a catastrophic collapse of the star. The subsequent release of energy might blow off a large part of the mass of the star, and this type of mechanism has been advanced in several forms to account for supernovae.

The mass loss need not be this violent. A. J. Deutsch has found evidence that red giants are undergoing a mild form of mass loss, so many massive stars may get below the degenerate mass limit during their helium burning phase. Also, any star which is rotating must rotate faster as it contracts, as a result of the conservation of angular momentum. If too much contraction takes place, the outer layers become unstable and material is ejected. D. N. Limber suggests this mechanism to explain Wolf-Rayet stars, which are early-type stars of peculiar characteristics.

Another type of object that should be mentioned here is the planetary nebula. This consists of a very hot central star that is surrounded by a mass of gas which has obviously been ejected by the star. The name comes from the planetary appearance of some of them in the telescope. The precise place of planetaries in evolution is not known, but D. E. Osterbrock shows that, by making plausible assumptions, one can conclude that the number of planetaries which have existed in the Galaxy is a significant fraction of the total number of white dwarfs. Planetary nebulae may represent a brief stage in the evolution of many or most stars.

There is little doubt that mass loss is an important part of the late evolution of many stars. L. V. Kuhi finds that it is important for pre-main-sequence stars, and V. G. Fessenkov and A. G. Massevitch and others have even suggested that it is important for main-sequence stars. Certainly the lack of understanding of the precise mechanism of mass loss is a major defect in evolution studies.

The differences in understanding of the pre- and post-giant phases of stellar evolution should now be rather clear. Both theoretical star models and observations have led to a fairly good understanding of many details of early evolution, but for late evolution not much more than some general principles are known. Many peculiar types of stars are observed which may represent special stages in the lives of common stars, but their precise role in evolution is not known. It is only a question of time until enough of the relevant physics is understood so that one will be able to evolve a model star through its entire life.

Successive Generations of Stars. The Galaxy consists of stars of all ages from zero to one or two times 10^{10} years. These stars were formed from condensations in the interstellar medium, but it is pointed out above that much of this stellar material is later returned to the interstellar space. Much of this ejected matter has been at least partly exposed to nuclear reactions in the stars. Since the general effect of these reactions is to form heavy elements out of light ones, the matter ejected by an old star will have a higher abundance of heavy elements than that which originally formed the same star. It follows that the interstellar medium is slowly being enriched with heavy elements, and very young stars should have a significantly different composition from the very old stars. In this picture, each succeeding generation of stars has a slightly higher heavy-element content.

This is at least partly confirmed by the observations. Globular clusters are certainly all very old, and it is known that they are deficient in the heavy elements. On the other hand, some of the oldest galactic clusters appear to have little composition differences from the youngest ones even though they approach the globular clusters in age. There certainly is a correlation of composition with age, but it does not appear good enough for composition alone to be a very reliable age indicator. It is possible that the very earliest generations of stars, which included the globular-cluster members and the subdwarfs, contained an unusually large number of high-mass stars that produced a large enrichment of the interstellar medium very quickly. Subsequent generations have then produced only a small further enrichment, so that age and composition have become less strongly correlated. Whether the Galaxy started out as pure hydrogen is not known, but it is possible that some helium and heavier elements were present at the beginning of star formation.

Although this concludes the section on stellar evolution, the subject is very closely connected with the material yet to be covered; therefore, it will come up again many times in the remaining sections.

23. Interstellar Matter

It is obvious from many photographs that there is material between the stars. The word nebula was given to the cloud-like structures often seen illuminated by starlight. A group of stars which is so far away that the individual stars cannot be seen may appear much like a nebula, and this has led to the word nebula being applied also to galaxies; however, this is no longer very common, and this meaning will not be used here.

Although nebulae were known many years ago, they were usually thought to be rather special objects, and it was generally believed that the space between the stars is both empty and transparent to starlight. This belief has slowly changed until today it is known that the interstellar medium is hardly less important than the stars themselves in modern astronomy.

Absorption and Reddening of Starlight. A number of early investigations gave evidence that starlight is dimmed by the material between the stars, but they were not given much credence until R. J. Trumpler in 1930 showed very strong evidence that distant star clusters are very appreciably dimmed by obscuring matter between the Earth and the clusters.

The treatment of this interstellar absorption is very much like that of atmospheric extinction in Section 10. Let σ_λ be the absorption coefficient in cm^{-1} at wavelength λ. Then the optical thickness τ_λ between Earth and a star at distance r is

$$\tau_\lambda = \int_0^r \sigma_\lambda \, ds \tag{23.1}$$

The observed flux F_λ is then related to $F_{0\lambda}$, the flux that would be observed if space were transparent, by the equation

$$F_\lambda = F_{0\lambda} e^{-\tau_\lambda} \tag{23.2}$$

In magnitudes the star has been dimmed by the amount Δm_λ, where [see equation (10.5)]

$$\Delta m_\lambda = 1.086 \tau_\lambda \tag{23.3}$$

Equation (9.8) gives the relation between apparent magnitude, absolute magnitude, and distance, but it is written with the assumption that space is transparent. Suppose that an absorption of Δm_λ magnitudes actually occurs in the starlight. Neither distance nor the absolute magnitude is affected, but the apparent magnitude is increased by Δm_λ. Then it is apparent that the more general relation is

$$m_\lambda = M_\lambda - 5 + 5 \log r + \Delta m_\lambda \tag{23.4}$$

The amount of the absorption Δm_λ must be known before spectroscopic parallax methods can be used. If one were unaware of interstellar absorption and found a distance based on apparent magnitudes, the distance found, r', would be related to the true distance, r, by

$$5 \log r' = 5 \log r + \Delta m_\lambda$$

or

$$r' = r \, 10^{0.20 \Delta m_\lambda} \tag{23.5}$$

Thus the calculated distance would always be greater than the true distance, and this leads to a tendency to overestimate the distances of stars.

The determination of Δm_λ at any wavelength or for any magnitude system can be rather involved. If one could find the distance to a star by a method which is independent of the apparent magnitude of the star, then knowledge of absolute magnitude (from the spectrum, for example) and the apparent magnitude would immediately yield the value of the interstellar absorption term. The trouble is that there are few methods of finding the distance of a star which do not involve the apparent magnitude. Trigonometric parallaxes could be used in this way, but interstellar absorption is appreciable only for stars which are too far away to allow accurate trigonometric parallax measurements.

A common method of finding the absorption is to consider two stars, 1 and 2, of the same spectral type. Let $\delta m_\lambda = m_\lambda(2) - m_\lambda(1)$, the difference in apparent magnitude between the two stars. Since the stars are of the same spectral type, they should have very nearly the same absolute magnitudes. Then equation (23.4) leads to

$$\delta m_\lambda = 5 \log \frac{r(2)}{r(1)} + \Delta m_\lambda(2) - \Delta m_\lambda(1) \tag{23.6}$$

The difference in apparent magnitude is due in part to different distances and in part to different interstellar absorptions.

Figure 23.1 shows what a plot of δm_λ versus $(1/\lambda)$ might look like. (It is usual to make such plots as a function of the reciprocal of the wavelength rather than of the wavelength itself.) The shape of this curve depends only on the Δm_λ's, but the numerical values of δm_λ depend also on the relative distances to the two stars. If there were the same amount of absorption for both stars, then the curve would be perfectly flat. The fact that the curve rises

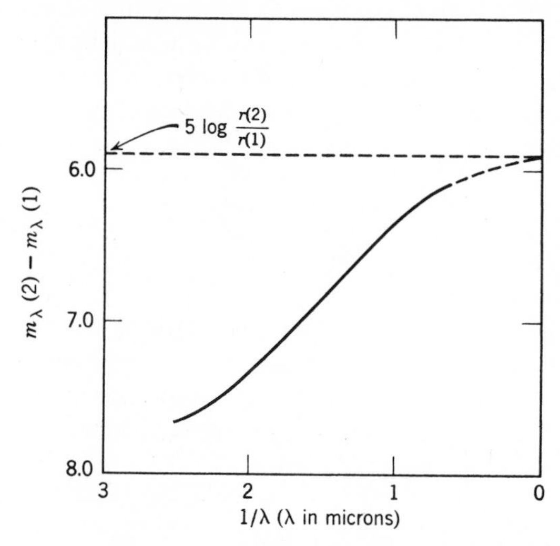

Figure 23.1 A hypothetical reddening curve.

toward the right (toward longer wavelengths) means that star 2 is getting brighter relative to star 1; that is, the absorption of the light from star 2 relative to that from star 1 is decreasing.

Theory shows that the absorption should vanish for infinitely long wavelengths, i.e., as $(1/\lambda) \to 0$. The dotted part of the figure is an extrapolation of the curve to infinite wavelength, and the corresponding value of δm_λ should then equal $5 \times \log [r(2)/r(1)]$ as equation (23.6) indicates. With the relative distances of the two stars known, the curve immediately yields $\Delta m_\lambda(2) - \Delta m_\lambda(1)$ for any wavelength. Now, if one chooses for star 1 a star close enough to the Earth for its dimming by interstellar matter to be negligible, then $\Delta m_\lambda(1) = 0$, and the absorption of star 2 for any wavelength is known.

The role of star 1 in the above example is only to show what the energy distribution in an unobscured star is. A different approach is to consider equation (23.4) applied to two different wavelengths in the same star. Then

$$m_{\lambda_1} - m_{\lambda_2} = M_{\lambda_1} - M_{\lambda_2} + \Delta m_{\lambda_1} - \Delta m_{\lambda_2} \tag{23.7}$$

The distances terms go out, but now the absolute magnitudes must be considered. Since absolute magnitudes are not affected by interstellar absorption, $M_{\lambda_1} - M_{\lambda_2}$ is the same as the intrinsic color of the star at these wavelengths, i.e., the same as the observed color would be in the absence of this absorption. The difference between the observed color and the intrinsic color of a star is known as the color excess of the star,

$$E_{12} = (m_{\lambda_1} - m_{\lambda_2}) - (M_{\lambda_1} - M_{\lambda_2}) = \Delta m_{\lambda_1} - \Delta m_{\lambda_2} \tag{23.8}$$

If the intrinsic color of a star is known, as from an unobscured star of the same spectral type, then the color excess can be found from the observed color. By extrapolating the observations to a wavelength $\lambda_2 \to \infty$, one obtains from the color excess the absorption at any wavelength λ_1.

If the absorbing properties of the interstellar medium were uniform, a simpler procedure would be possible. Consider the ratio

$$R = \frac{\Delta m_{\lambda_2}}{E_{12}} = \frac{\Delta m_{\lambda_2}}{\Delta m_{\lambda_1} - \Delta m_{\lambda_2}} \tag{23.9}$$

where R is known as the ratio of total to selective absorption for the magnitude systems specified by λ_1 and λ_2. Now suppose that the absorbing properties of the medium are uniform in the sense that the absorption coefficient at one wavelength depends on physical conditions in the same manner as that at any other wavelength. In other words, the ratio

$$\frac{\sigma_{\lambda_1}}{\sigma_{\lambda_2}} = f(\lambda_1, \lambda_2) \tag{23.10}$$

depends on the wavelengths but not on temperature or any of the other properties of the interstellar medium. Then one has, from equations (23.1) and (23.3),

$$\frac{\Delta m_{\lambda_1}}{\Delta m_{\lambda_2}} = \frac{\int_0^r \sigma_{\lambda_1}\, ds}{\int_0^r \sigma_{\lambda_2}\, ds} = \frac{\int_0^r f(\lambda_1,\lambda_2)\sigma_{\lambda_2}\, ds}{\int_0^r \sigma_{\lambda_2}\, ds}$$

where the integrations are over the distance from the Earth to the star. By assumption $f(\lambda_1,\lambda_2)$ does not depend on position, so it can be taken out of the above integral. The result is

$$\frac{\Delta m_{\lambda_1}}{\Delta m_{\lambda_2}} = f(\lambda_1,\lambda_2) \tag{23.11}$$

and equation (23.9) then gives

$$R = \frac{1}{f(\lambda_1,\lambda_2) - 1} \tag{23.12}$$

Thus the ratio of total to selective absorption is a constant for any two magnitude systems. It can be determined once and for all, and thus a measurement of the color excess of a star immediately yields Δm_{λ_2} through equation (23.9).

In the (U, B, V) system it is usually assumed that a good approximation is given by

$$R_{BV} = \frac{\Delta m_V}{E_{BV}} = 3.0 \tag{23.13}$$

Suppose that an A0 V star is observed to have an apparent visual magnitude of $V = 10.0$ and an apparent blue magnitude of $B = 10.5$. The observed color is $(B - V) = 0.5$, while the intrinsic color of this kind of star is $(B_0 - V_0) = 0.0$. (See Appendix B.) The star then has a color excess of 0.5 mag, and equation (23.13) indicates that the star has been dimmed by 1.5 mag in the visual. The absorption in the blue is seen to be 2.0 mag.

Equation (23.13) is based on a large number of observations, but large deviations from this have been found for the general region of the Orion nebula. H. L. Johnson believes that large variations in R are common, and if he is correct, then the absorbing properties of the interstellar medium are not as uniform as is generally supposed.

The fact that Figure 23.1 is nearly a straight line means that the interstellar absorption is approximately proportional to $(1/\lambda)$. If this wavelength dependence is assumed, then it is apparent from (23.10) that $f(\lambda_1, \lambda_2) = \lambda_2/\lambda_1$, and equation (23.12) then yields

$$R = \frac{\lambda_1}{\lambda_2 - \lambda_1} \tag{23.14}$$

Since the mean wavelengths of the B and V systems are about $\lambda\lambda$ 4450 and 5500, respectively, this gives $R_{BV} \approx 4.2$, not far from the value given in (23.13).

The amount of interstellar absorption is greater at shorter wavelengths. This results in the blue light from a star being dimmed by a greater amount than the red light, and stars appear redder than they would in the absence of interstellar absorption. This interstellar reddening makes it necessary for one to be very careful about using observed colors to infer temperatures. Without further information, one cannot be certain whether a star with $(B - V) = +1.0$ is cool or whether it is heavily reddened.

Correcting apparent magnitudes for absorption requires a knowledge of the intrinsic colors of the stars. This is a major problem for distant stars for which there are no nearby stars of the same spectral type. Cluster membership sometimes helps, but one is not always justified in assuming that the obscuration is the same for all members of a cluster.

The variation of the amount of absorption with distance and with direction supplies information about the distribution of the interstellar medium. The absorbing material has a rather irregular distribution, but one can very roughly write

$$\Delta m_\lambda = A_\lambda r \tag{23.15}$$

where A_λ is the average absorption in magnitudes per unit distance. The value of A_λ in the visual region is in the neighborhood of 0.5 mag per 10^3 pc, but the form of equation (23.15) should be assumed only when very rough results are desired.

Interstellar Dust. So far, the nuisance value of the interstellar medium has been emphasized, i.e., how it complicates other measurements of interest. Now a more direct interest will be taken in the material responsible for this absorption.

The stars are composed mainly of hydrogen and helium, and so one might expect the interstellar medium also to be mainly composed of these gases. This is correct, but it is not the gas that is responsible for the interstellar absorption. The absorption and the reddening of starlight are caused by tiny grains or dust particles which exist in space. The absorption coefficients of gases do not fit either the amount or the wavelength dependence observed for this absorption.

Small particles of atomic or molecular size scatter visible light according to Rayleigh scattering, which has a λ^{-4} wavelength dependence. (The word absorption is used to include both true absorption, in which a photon is removed from the radiation field, and scattering, in which a photon merely has its direction changed.) This is the type of scattering that takes places in the Earth's atmosphere and is responsible for red sunsets and the blue sky. On the other hand, particles which are very much larger than the wavelength of light will scatter in a way that is independent of wavelength. Both theory and experiment indicate that the interstellar absorption (or scattering), which has a λ^{-1} wavelength dependence, must be due to particles which are about the same size as the wavelength of visible light. The dust particles, therefore, must be about 10^{-5} to 10^{-4} cm in size.

Suppose that the grains are of a typical size a in radius. Then the cross section area which one grain presents is πa^2 cm^2, approximately, and this will also be approximately the effective area from which the grain can absorb or scatter light. In other words, πa^2 will be roughly the absorption coefficient in cm^2/particle. (The absorption coefficient of particles is usually written $\pi a^2 Q$ where Q is known as the efficiency of absorption. The quantity Q can be quite large, but it is usually of the order of unity.) Then if there are N_g grains per cm^3, the absorption coefficient is

$$\sigma \approx \pi a^2 N_g \qquad \text{cm}^{-1} \tag{23.16}$$

This does not have the proper wavelength dependence, but this is only an order-of-magnitude calculation. From equations (23.1), (23.3), and (23.15) it is seen that $\sigma \approx A$, where A is the absorption in magnitudes per unit length. Since a typical value of A is about 0.5 mag per 10^3 pc $= 1.6 \times 10^{-22}$ mag per cm, it follows that

$$N_g \approx \frac{1.6 \times 10^{-22}}{\pi a^2} = \frac{0.5 \times 10^{-22}}{a^2} \tag{23.17}$$

With grain sizes in the 10^{-5} to 10^{-4} cm range, there need be only some 10^{-12} to 10^{-14} grains per cm^3 to account for the observed absorption. It is only over extremely large distances that such small particle densities can add up to an appreciable extinction.

If each grain has a mass density about the same as water, then the mass of a grain is about $4\pi a^3/3 \approx 4a^3$. Then the contribution of the dust to the mass density of interstellar space is

$$\rho_g \approx 4a^3 N_g \approx 2a \times 10^{-22} \qquad \text{g cm}^{-3} \tag{23.18}$$

Typical densities of the dust are then of the order of 10^{-26} g cm^{-3}.

Frequently a region of high density of the interstellar dust will almost completely block off the light of the stars beyond it. Such a region may be easily visible because of the bright regions around it, and it is known as a dark nebula. Dark nebulae can be studied by means of star counts. A star count is simply a count of the number of stars of given apparent magnitudes in a small area of the sky. If a dark nebula occurs at a certain distance, this will be reflected in a sudden decrease in the observed number of stars at an apparent magnitude corresponding, on the average, to the distance of the nebula. The distance of the nebula and the amount of obscuration it presents can be determined in this fashion. Star counts are discussed in greater detail in Section 24.

If a bright star is near a region of high density of the interstellar medium, it will light up the region and produce what is known as a bright nebula. There are two types of bright nebula. If the source star has a sufficiently high temperature, early B or O in spectrum, then the light from the nebula will be primarily from emission lines of gas atoms in the nebula. Such emission nebulae are discussed later in this section. If the nearby stars are not hot enough for an emission nebula, the light is found to be reflected starlight. The dust grains reflect the starlight in all directions, and if the source star is bright enough (as opposed to hot enough), this will be visible in a reflection nebula. Reflection nebulae apparently differ from dark nebulae only in that the latter do not have any bright stars situated so as to light them up. Examples of bright and dark nebulae are shown in Figure 23.2.

Since the transmitted starlight is reddened by the dust, the part that is scattered should be more blue than the original starlight. It is observed that reflection nebulae are generally more blue than the source stars. Also, scattering produces polarization of the scattered light, i.e., the electric fields of the scattered light have a preferred direction of vibration. Observations show that the reflection nebulae are highly polarized, as expected.

Analyses of the amount of light coming from a reflection nebula seem to indicate that the dust grains scatter nearly all of the light incident upon them, and that there is thus very little true absorption. The grains have a very high albedo, like snow.

The dust grains block out the light from distant stars, but they compensate by scattering into the line of sight light from nearby stars. It has been estimated that up to one-third of the light from the Milky Way is diffuse radiation that has been scattered by the interstellar dust. This causes a small amount of light to reach the Earth from all directions, not just from the directions in which stars happen to occur.

The problem of the formation of the dust has been investigated by many persons. J. H. Oort and H. C. van de Hulst have suggested that an equilibrium has been set up between the dust and the gas. Large molecules will occasionally form which then grow by accretion of gas atoms. The composition of the grains will be determined both by the composition of the gas and by the relative probabilities that the different kinds of gas atoms will stick to a grain with which they collide. The relative velocities of atoms and grains in a cloud of dust and gas are small, so the collisions tend to increase the grain sizes as the atoms stick. But relative velocities between different clouds are rather large, so collisions between different clouds are sufficiently energetic that they tend to evaporate material from the grains, decreasing their sizes. Perhaps steady-state conditions have been set up between these competing processes

Figure 23.2 Examples of bright and dark nebulae in the Galaxy. The Horsehead nebula is at the top, the Orion nebula in the middle, and the Cone nebula at the bottom. (Mount Wilson and Palomar Observatories.)

in many regions, with the result that a nearly constant distribution of grain sizes has been reached. This would explain the fairly uniform reddening properties which have been found, while exceptional regions, such as those near very hot stars, could also be understood.

It has been generally believed that the grains are primarily ices of water, methane, and ammonia, plus small amounts of other impurities. Recent balloon observations were analyzed by R. E. Danielson, N. J. Woolf, and J. E. Gaustad in an attempt to detect absorption by the interstellar ice grains. They could not find the expected absorption, and so the grains may have a composition rather different from what had been supposed.

Alternate theories on the origin and chemistry of the grains have been proposed by J. R. Platt and by F. Hoyle and N. C. Wickramasinghe. The latter suggest that the grains are graphite particles formed in the atmospheres of certain cool carbon-rich stars and ejected into the interstellar medium.

The Polarization of Starlight. In 1949 J. S. Hall and W. A. Hiltner discovered that the light coming from many stars is polarized to a small degree. This differs from the polarization of the reflection nebulae mentioned above in that the latter is a polarization of light which has been scattered, while this is a polarization of light which has been transmitted, i.e., light which has avoided being scattered. The important point is that polarized light shows a preferred direction of vibration, and in trying to find out why this one direction is singled out among all of the others, one can often discover many characteristics of the material which emitted, absorbed, or scattered the light.

The polarization is strongly correlated with the reddening of starlight, so the dust grains are the suspected source. Rayleigh scattering by the gas atoms will polarize the scattered light, but the transmitted light is not polarized. This means that the interstellar gas is not the source of the observed polarization of starlight.

How can dust grains produce polarization of the transmitted starlight? Light with the electric field vibrations in a certain direction must be preferentially scattered away, so that the transmitted light vibrates more in the direction at a right angle to this direction. This means that the grains must have a preferred direction of their own, and therefore, the grains must tend to be elongated. This conclusion is not surprising, but it does not go far enough. Each grain would tend to polarize the light, but if the long axes of the grains were pointing at random, then the effect of a large number of them would be unpolarized light. There must also be some mechanism to cause a partial alignment of the grains, so that the preferred direction can be maintained over large regions of space. The only reasonable mechanism for this that has been found is a magnetic field. Thus the discovery of the polarization of starlight led directly to the inference of the existence of interstellar magnetic fields.

An alignment mechanism suggested by L. Davis and J. L. Greenstein appears to be the most likely. Collisions with gas atoms cause the grains to spin, and the magnetic field tends to cause this spin to occur around the short axis, and it tends also to orient this short axis parallel to the field. Further collisions would tend to destroy this orientation, so the amount of polarization should be a complicated function of the properties of the grains, the density of the interstellar gas, and the strength and the uniformity of the magnetic field. Reasonable estimates of the relevant quantities lead to an interstellar magnetic field in the solar neighborhood of around 10^{-5} to 10^{-6} gauss. For comparison, the field at the surface of the Earth is about one-half gauss. While the interstellar field is quite small, it is probably important, or even dominant, in any large-scale structure in the Galaxy.

Excitation and Ionization in Space. The interstellar gas is less obvious than the dust because it does not produce strong absorbing effects at all wavelengths. The gas can be detected in three ways: by emission lines and continua near very hot stars, by absorption lines superposed on the spectra of distant stars, and by emission at radio wavelengths. Before these processes are studied in any detail, however, the special excitation and ionization conditions which prevail in interstellar space must be considered.

In Section 4 the excitation and ionization conditions of a gas in thermodynamic equilibrium were studied. In the interiors and in much of the atmospheres of stars, conditions are sufficiently close to TE that the Boltzmann and Saha equations are valid, although there are exceptions to this in the extreme outer parts of the atmospheres. In interstellar space conditions are so far from TE that the use of the Boltzmann and Saha equations without modification can lead to errors of many orders of magnitude.

The very large departures from TE in space are due not to the low material densities, since TE can occur at any density of matter, but to the very low density of radiation. It was pointed out in Section 2 that in TE the radiation field is uniquely specified by the value of one quantity, the temperature. In space the radiation comes from stars of all temperatures, and an average photon has an energy which corresponds to a black-body temperature of perhaps 7000°K; however, all photons in space are extremely rare, and the total energy density of radiation corresponds to a black body of only about 3°K. This tremendous difference between the quality and the quantity of the radiation is responsible for the large deviations from TE in interstellar space.

Let L and U represent the lower and upper levels of a bound-bound transition that can take place in an atom. Then transitions from L to U can take place through collisions and the absorption of radiation, while downward transitions $U \rightarrow L$ can occur through collisions, induced emission, and spontaneous emission, as mentioned in Section 4. The notation $P\,(\text{col } L \rightarrow U)$ will be used to indicate the probability that an atom in level L will have a collisional transition to level U per unit time, and abs, IE, and SE will stand for absorption, induced emission, and spontaneous emission, respectively. Then if $N\,(L \rightarrow U)$ is the total number of transitions from L to U per unit volume and per unit time,

$$\frac{N\,(L \rightarrow U)}{N} = \frac{N_L}{N}\,[P\,(\text{col}\ L \rightarrow U) + P\,(\text{abs}\ L \rightarrow U)] \tag{23.19}$$

and

$$\frac{N\,(U \rightarrow L)}{N} = \frac{N_U}{N}\,[P\,(\text{col}\ U \rightarrow L) + P\,(\text{IE}\ U \rightarrow L) + P\,(\text{SE}\ U \rightarrow L)] \tag{23.20}$$

In these equations N_L and N_U are the number of atoms per unit volume in the given levels, and N is the total number of atoms in the given ionization stage per unit volume.

In TE the number of transitions $L \rightarrow U$ is equal to the number of opposite transitions, so the above two equations can be set equal to each other. One can show that in TE it is possible to go even further and to equate the collisional terms of (23.19) and (23.20) and then separately equate the radiative terms of these equations. In the general case, however, this cannot be done.

The probabilities of the collisional transitions in the above equations are proportional to the matter density, since the collisional rate increases with the number of particles per unit volume. For the same reason the probabilities of absorption and induced emission are proportional to the density of radiation. In contrast, the probability of spontaneous emission is an atomic constant and does not depend on the physical conditions.

In space the densities of both matter and radiation are many orders of magnitude less than their values in stars, and both terms on the right side of equation (23.19) are very small. Accordingly, the number of upward transitions per neutral atom is drastically reduced in space as compared with that in a star. Since TE does not prevail, the number of transitions $U \rightarrow L$ does not necessarily equal the number $L \rightarrow U$; however, energy still must be conserved, so the energy involved in all upward transitions (from all lower levels to all possible upper levels) must equal that in all possible downward transitions. The fact that all upward transitions in interstellar space are rare, therefore, requires that all downward transitions also be rare. (It cannot come down unless it has first gone up.)

Now consider equation (23.20). According to the above arguments, all three terms on the right side must be very small. The first two terms are automatically small in interstellar space, since they are proportional to the density of matter and that of radiation, respectively. The third term is different. The probability of spontaneous emission is independent of the physical conditions, so it is just as large in space as it is in a star. The only way to make this third term small is to make the ratio N_U/N small. In other words, in interstellar space virtually all of the atoms in a given ionization stage are in the ground state; very few are excited to higher bound levels. As a result, the Boltzmann excitation equation is not valid.

The above arguments would not hold for an excited level if the probability of spontaneous emission from it to all lower levels were very small. Levels exist for which this is true, and they are known as metastable levels. In TE an atom in a metstable level will soon absorb a photon and leave the level (or be knocked out of it by a collision if the material density is not too low), but in space this cannot occur very often. Metastable levels are considerably overpopulated in interstellar space as compared with other excited levels.

Ionization conditions can be studied in much the same fashion as excitation conditions. An atom can be ionized by collision or by absorption of a photon (photo-ionization) if the photon has enough energy. De-ionization can occur by collision, by induced emission, or by spontaneous emission. In all three types of de-ionization, the ion must find a free electron to recombine with. In collisional de-ionizations the energy of recombination is given to some nearby third particle, while for induced de-ionizations the recombination is caused by the nearby passage of a photon similar to the one which is emitted. The important difference from excitation is in the spontaneous transitions. Spontaneous de-ionizations depend on the availability of the free electrons and, unlike the spontaneous de-excitations, have a probability that depends on physical conditions.

Let N_a and N_i be the number of neutral atoms and ions, respectively, per unit volume. As is noted above, almost all of these will be in their respective ground states in interstellar space. Also let $N\,(a \rightarrow i)$ be the number of ionizations per unit volume per second, and $N\,(i \rightarrow a)$ be the number of de-ionizations per volume per second. Then, using a notation similar to that of equations (23.19) and (23.20), one can write

$$N\,(a \rightarrow i) = N_a[P\,(\text{col}\ a \rightarrow i) + P\,(\text{abs}\ a \rightarrow i)] \tag{23.21}$$

To emphasize that the de-ionization probabilities are all proportional to the number of free electrons per unit volume N_e, these probabilities will be written as the product of N_e and the corresponding P. Then

$$N\,(i \rightarrow a) = N_i N_e[P\,(\text{col}\ i \rightarrow a) + P\,(\text{IE}\ i \rightarrow a) + P\,(\text{SE}\ i \rightarrow a)] \tag{23.22}$$

With these definitions, the three P's in equation (23.22) are proportional to the mass density, the radiation density, and a constant, respectively.

As long as ionization conditions do not change with time, equations (23.21) and (23.22) must be equal, whether TE holds or not. Then one has

$$\frac{N_i N_e}{N_a} = \frac{P\,(\text{col}\ a \rightarrow i) + P\,(\text{abs}\ a \rightarrow i)}{P\,(\text{col}\ i \rightarrow a) + P\,(\text{IE}\ i \rightarrow a) + P\,(\text{SE}\ i \rightarrow a)} \tag{23.23}$$

In TE the collision terms in equations (23.21) and (23.22) exactly balance, and the remaining radiation terms in (23.23) can be shown to reduce to the usual Saha relation, equation (4.8). The low densities of interstellar space generally make the third term of the denominator much larger than the first two, so the above can be reduced to

$$\frac{N_i}{N_a} = \frac{P\,(\text{col } a \to i) + P\,(\text{abs } a \to i)}{N_e P\,(\text{SE } i \to a)} \tag{23.24}$$

If the degree of ionization of an element is desired, then equation (23.24) must be solved for successive ionization stages of that element. The numerator is a function of the density and the velocity distribution of the particles and of the intensity of the radiation field. Since the denominator contains the electron density N_e, this equation must be solved simultaneously with similar equations for the elements which are providing the free electrons. This illustrates the complexity of the problem, but in practice one can often make further simplifying assumptions so that the problem can be readily solved.

Usually the collision term in (23.24) can be neglected, and so this equation is just the ratio of the absorption term to N_e times the spontaneous emission term. The numerator is thus proportional to the intensity of the radiation field, while N_e is proportional to the density of the matter. If both the density of matter and of radiation are decreased by the same factor in going from a star to interstellar space, as is sometimes the case, then the ionization conditions in space will be similar to those in the star. This does not mean that the Saha ionization equation is valid; it only means that ions which are common in the star can be expected to be common in space.

H II *Regions.* Hydrogen is the most abundant element in the interstellar gas, and its ionization will be considered in some detail. Hydrogen has only two ionization stages, neutral or H I, and ionized or H II. Regions in interstellar space in which the hydrogen is nearly all neutral are called H I regions, and H II regions are those in which it is essentially all ionized.

Since most of the neutral hydrogen in space is in the ground state, the only photons it can absorb are those which correspond to transitions from the ground state to the higher bound levels and to the continuum. These are the Lyman lines and the Lyman continuum of hydrogen. The first excited state is about 10.2 eV above the ground level, and a photon of this energy has a wavelength of 1215 Å. A photon of longer wavelength than this does not have enough energy to be absorbed by hydrogen in the ground state. A photon of wavelength shorter than 1215 Å cannot be absorbed either, unless it happens to correspond to one of the other Lyman lines or unless it has enough energy to ionize the hydrogen. It takes about 13.6 eV to ionize hydrogen from the ground state, and to do this a photon must have a wavelength of less than about 912 Å. Such a photon is known as a Lyman continuum or Lc photon.

Consider a region very close to a hot star. The density of Lc photons will be fairly high, and the second term in the numerator of equation (23.24) will be very much greater than the denominator. Under these conditions almost all of the hydrogen is ionized. Any neutral atoms formed by recombination will soon be re-ionized by another Lc photon.

As one moves away from the star the radiation becomes diluted by the inverse-square law, and the flux of the Lc photons falls off accordingly. (Actually the flux of Lc photons falls off somewhat more rapidly than this, since some of them are used up in ionizing the few neutral atoms.) The collision term can be neglected, so the numerator of (23.24) falls off quite rapidly. The denominator, on the other hand, remains nearly constant. Hydrogen is so abundant that, as long as it is nearly all ionized, it supplies nearly all of the free electrons and N_e is essentially the same as the density of hydrogen ions.

Eventually a point will be reached for which the numerator of (23.24) is no longer very much larger than the denominator, and at this point the number of neutral atoms begins to produce an appreciable absorption of the Lc photons. This causes a rather rapid drop in the flux of Lc photons, and the hydrogen quickly becomes essentially all neutral. B. Strömgren was the first to show theoretically that the boundary between H I and H II regions is usually quite sharp. The H II regions are often called Strömgren spheres.

Figure 23.3 is based on calculations made by R. E. Williams, and includes the ionization of both hydrogen and helium. The stars are assumed to radiate like black bodies, which is probably a very poor approximation for these high temperatures. Helium has three ionization stages, and the ionization edges correspond to photon wavelengths of 504 Å and 228 Å. Very close to the star there are enough photons below 228 Å to keep the helium twice-ionized, but soon these photons are used up and the helium becomes once-ionized. The hotter the star, the more photons below 228 Å it radiates and the larger is the He III region. The helium becomes neutral at a distance for which the photons below 504 Å are used up, and it is interesting that the He II region can be either larger or smaller than the H II region, depending on the temperature of the star (and on the relative abundances of hydrogen and helium).

When an Lc photon of frequency ν ionizes a hydrogen atom, the freed electron escapes with a kinetic energy given by

$$KE = \tfrac{1}{2}mv^2 = h\nu - 13.6 \quad \text{eV} \tag{23.25}$$

The kinetic energy of the electron is the excess of the photon energy over the ionization potential of hydrogen. This free electron will suffer many elastic collisions with other free particles, sharing its energy with them. Although conditions are far from TE, these elastic collisions between the particles tend to make them move with very much the same velocity distribution they would

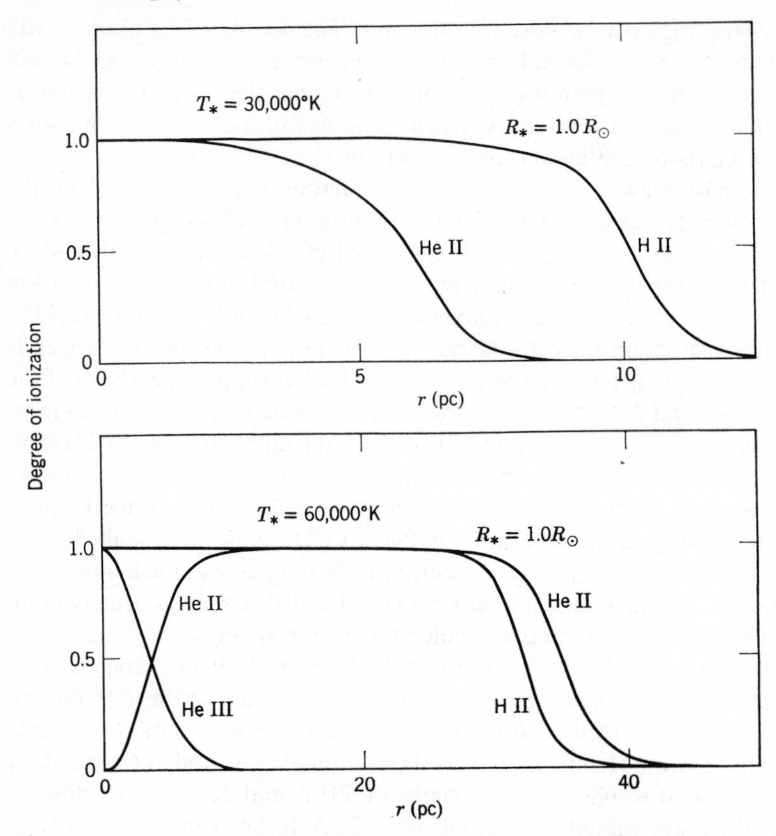

Figure 23.3 Ionization of interstellar hydrogen and helium near stars of effective temperatures of 30,000°K (top) and 60,000°K (bottom). The assumed density is one atom per cubic centimeter and $N_{\mathrm{He}}/N_{\mathrm{H}} = 0.15$.

have in TE at some temperature. This temperature is called the kinetic temperature T_k, or sometimes the electron temperature T_e (not to be confused with the effective temperature). The kinetic temperature of an H II region obviously must be less than the temperature of the exciting star, unless there is some other mechanism for putting energy into it.

The free electron may suffer an inelastic collision with some particle and give up part of its energy in exciting or ionizing the other particle, or it may gain kinetic energy in de-exciting it. Eventually it will recombine with a proton to form a neutral atom plus a photon. If the recombination is to the ground state, then this will be another Lc photon and it will not go far before it is again absorbed in a photo-ionization process. This will continue until the recombination is to one of the excited levels, and then the atom will drop to the ground state by emitting the appropriate photons through spontaneous bound-bound emission. The original Lc photon is thus broken up into two or more photons of smaller energy, one of which is a Lyman line photon. The other photons will immediately escape from the H II region, but the Lyman line photon will be absorbed and emitted until it in turn is broken up into two or more photons, one of which is a Balmer photon formed by a transition down to the second level, and another is an Lα photon, formed by a transition from the second level to the first one. The Balmer photon will also readily escape, but the Lα photon will continue to be absorbed and emitted a tremendous number of times by the neutral hydrogen until it disappears in a more complicated process.

The result of the above processes is that each Lc photon is broken up into several photons of longer wavelengths, one of which is a Balmer photon (line or continuum). Thus H II regions should have strong Balmer line emission. Further, if one could measure the rate of Balmer emission in all lines and continuum, one would be able to fine the rate at which the exciting star emits Lc photons. This then gives a measure of the temperature of the star. This process was first studied by H. Zanstra.

The kinetic temperature is quite important in ionization theory, as both collisional ionizations and recombinations occur at rates which are very temperature-sensitive. The material will come to a kinetic temperature such that it loses energy at the same rate it receives it. An H II region receives its energy from the exciting star or stars, and it loses its energy by the radiation it emits which can escape without being further absorbed. A very important cooling mechanism is provided by the metastable levels of certain elements, notably oxygen and nitrogen.

Ionized nitrogen and once and twice-ionized oxygen have metastable states a few electron volts above the ground states. Radiative transitions between these levels and the ground states are unlikely to occur because of the small transition probabilities; otherwise, the states would not be metastable. Collisional transitions with the ground states, however, can occur if the kinetic temperature is not too low, and in H II regions electron collisions keep these metastable states fairly well populated.

Although spontaneous transitions back to the ground level from a metastable state are very unlikely per unit time, they will occur eventually if nothing else happens. Interstellar space with its low densities is precisely the place where nothing else is likely to happen, so radiative transitions from the metastable levels back to the ground state do occur, and they compete with the collisional de-excitations. They are known as forbidden transitions because they do violate certain approximate rules governing transition probabilities. Forbidden lines are often among the strongest emissions in an H II region, and they provide an important way to get rid of excess kinetic energy of the electrons. The lines act as a thermostat to keep the kinetic temperature from getting much above 10^4 °K, even though the exciting star may be many times hotter than this.

It is understandable why emission nebulae occur only near very hot stars. The number of Lc photons radiated by a star is very sensitive to the star's temperature, and only very hot stars produce enough to have much of an H II region. The various recombinations that occur in an H II region lead to the emissions of an emission nebula. In H I regions there are very few free electrons, so there are very few recombinations and there is very little emission. A relatively cool star cannot have a large H II region, so any large bright nebula around such a star must be of the reflection type.

The brightness of an interstellar emission line depends on the number of emitting atoms in the line of sight, if the absorption is small, as is usually the case. Then the observed intensity of an emission line can be expressed as a function of the abundance of the relevant element, the electron density N_e, the kinetic temperature T_k, and the geometry of the emitting region. By measuring a number of lines for which the transition probabilities are known, one is able to get an idea of the conditions in the emitting gas. The procedure, however, is not simple.

One complicating feature was discovered by I. S. Bowen. Certain lines of O III and N III (twice-ionized oxygen and nitrogen) are sometimes far stronger than the abundances or the conditions should allow, and Bowen explained this in terms of an emission line of He II. By chance this helium line at 304 Å coincides almost exactly with a line connecting the ground state of O III. This excites a series of O III lines, one of which happens to coincide with an N III line that also arises from the ground state. Thus a number of lines of O III and N III are excited by this helium line, and if one were not aware of this pair of coincidences, one would be led to erroneous conclusions from the analysis of these lines.

Abundances found for the interstellar gas do not appear to differ significantly from those found in stars. Densities of the gas in the solar neighborhood average around 10^{-24} g cm^{-3}, or about one hydrogen atom per cubic centimeter. In the more intense emission nebulae, densities 10^3 or more times this sometimes occur. The gas is about 10^2 times more abundant than the dust.

The H II regions are not likely to be static configurations. The kinetic temperature in an H II region is much higher than that in an H I region, and so if the densities of the two regions are equal, the pressures will not be. The high-pressure H II region should expand into the surrounding H I region. Eventually pressure equality will be obtained, but the time this takes is not very short compared with the main-sequence lifetimes of the very early-type stars. Thus H II regions are likely to be expanding and in a highly turbulent state. Shock waves set up at the boundaries of the hot, expanding regions possibly account for some of the light of observed nebulosities.

The H II regions are similar in many respects to planetary nebulae, and methods of investigating their properties are much the same. The main differences are due to the higher densities and the greater uniformity of the planetaries. In view of the likely position of planetaries as near the end of their lives as stars, it is interesting that they apparently do not have major composition differences from the normal H II regions.

Absorption Lines. In 1904 J. Hartmann first observed and gave the correct interpretation for an interstellar absorption line, although his explanation was not generally accepted for many years. Many such lines have been observed by now, and it is not surprising that most of them are resonance lines; i.e., the lower levels are usually the ground states of the relevant atom or ion.

Two lines of Ca II at $\lambda\lambda$ 3968 and 3934, which are known as the H and K lines, are among the strongest of the observed interstellar absorption lines. The strengths of these lines in the spectra of different stars have been found to be well correlated with the distances of the stars, and this suggests a fairly uniform distribution of the gas and provides a rough method of estimating the distance of certain stars. The interstellar H and K lines are easily observed only in the hotter stars, as cooler stars have strong H and K absorption of their own which masks the interstellar absorption lines.

The interstellar absorption lines can be analyzed in much the same manner used for stellar lines. Curve-of-growth methods can give abundances and excitation and ionization conditions. In some respects the analysis is much simpler than in the case of stellar atmospheres. One great simplification is that the emitting region is separated from the absorbing region. Also, the low densities in space cause some of the complicated broadening mechanisms to be of negligible importance. Most of the absorption lines observed from the Earth's surface arise in the H I regions, so they are formed under very different conditions from the interstellar emission lines.

As an example, consider two lines of the same stage of ionization of the same element. If they are both weak lines on the linear part of the curve of growth, then equation (20.13) will be valid for both. Both lines arise from the ground state, and so both will have the same number of absorbing atoms, and both will obviously have the same path length L. The two equivalent widths will differ from each other only because of the different f values in equation (20.10), so the equivalent widths will be in the same ratio as the f values if weak line assumption is correct. In this case equation (20.13) will give the the number of absorbers in the line of sight, and knowledge of the distance of the star will then yield the average number of absorbers per unit volume N.

If the observed equivalent widths are not in the same ratio as the f values, then at least one of the lines must be strong enough for saturation to have set in. In this case the equivalent width is a function of the kinetic temperature and the turbulent velocities as well as the abundance of the absorbers. By measuring the lines in stars of different distances, an observed curve of growth can be constructed. Comparing this with the theoretical curve, as in the stellar atmosphere case, then yields abundances, kinetic temperatures, and turbulent velocities.

If lines from successive ionization stages of a single element can be measured, then an application of equation (23.24) can fix the mean value of the electron density N_e, so the degrees of ionization of the different elements can be found.

There are, naturally, some complicating features. For one thing, quantities obtained refer to an average over the entire path length between the star and the Earth. Conditions can be quite nonuniform over this path, and this is particularly important when data obtained from one line are applied to another. If the two lines are formed in different regions, then large errors can result from not taking this into account.

The nonuniformity of the gas can be directly observed. When the stronger lines are observed under very high resolution, they often show several separate components. These are caused by several gas clouds being in the line of sight, and the lines from the different clouds are separated by their doppler shifts. A particular danger in the curve-of-growth procedure is that several clouds do not combine their lines into the same total equivalent width as one cloud with the same total number of absorbing atoms.

The abundance ratio of calcium to sodium as determined by the interstellar absorption lines is considerably lower than that found from stellar atmospheres. Some of this may be due to the above complications. It has also been suggested that calcium tends to condense on the grains more readily than sodium, and this could significantly deplete its supply in the gaseous state. There are some unusual differences in the way that the calcium and the sodium line strengths correlate with interstellar reddening which may support the latter conclusion. A real difference in the abundances of these elements between stars and interstellar space seems unlikely in view of the many other agreements.

The 21-cm Line. The state of atomic hydrogen which is usually called the ground state is actually composed of four separate quantum mechanical states. One of these is the true ground level, and the other three form a degenerate energy level slightly above this. This splitting is called hyperfine structure and is caused by the quantum mechanical way that the spin of the electron combines with the spin of the nucleus. Hyperfine splitting for hydrogen amounts to only 6×10^{-6} eV, and a photon with this energy has a wavelength of 21.2 cm. In the mid-1940s H. C. van de Hulst suggested that the 21-cm radiation from atoms in interstellar space might be observable, and in 1951 it was discovered by H. I. Ewen and E. M. Purcell.

It is difficult to overestimate the importance of the discovery of this one line on the subsequent development of astronomy. For one thing the great mass of neutral hydrogen in space was suddenly rendered visible. Also, the absorption by the interstellar grains is essentially absent at the radio wavelengths, so the 21-cm observations can be used to obtain information about regions that are not visible at optical wavelengths. This was the first line observed at radio wavelengths, and it provided the first (and until recently the only) means of finding radial velocities of radio sources which have not been identified in visible light. The importance of these points in galactic structure is discussed in Section 25.

Consider a region which emits 21-cm radiation. If the region is uniform and emits like a black body at temperature T, then the intensity coming from the region, from equations (19.5), (19.7), and (19.9), is

$$I_\lambda = B_\lambda(T)(1 - e^{-\tau_\lambda}) \tag{23.26}$$

where τ_λ is the optical thickness of the region for wavelength λ. Although temperatures in H I regions are very low, they are still high enough for the Rayleigh-Jeans relation (2.9) to be valid at radio wavelengths. Then equation (23.26) becomes

$$I_\lambda = \frac{2ckT}{\lambda^4}(1 - e^{-\tau_\lambda}) \tag{23.27}$$

It is important to realize that, for an extended object like an H I region, it is possible to determine intensities; for stars, which are essentially point sources, only fluxes can be determined.

It is apparent from equation (23.26) that the observed intensity is less than or equal to the Planck function. If one measures a region of high optical thickness, $\tau_\lambda \gg 1$ and $I_\lambda = B_\lambda(T)$. Such regions do exist, and they allow the temperature T to be directly determined. This temperature is a measure of brightness of a black body having the same intensity, and so it is called a brightness temperature.

Once the brightness temperature of H I regions has been found from the regions of large optical thickness, equation (23.27) can be applied to other regions in order to obtain the value of τ_λ. Since optical thickness is a measure of the number of absorbing atoms along the line of sight [see equation (20.9)], knowledge of τ_λ plus an idea of the geometry of the emitting region can yield the density of the neutral hydrogen.

The 21-cm observations show that the neutral hydrogen has a rather irregular distribution, with an average density near the Sun of about one atom per cubic centimeter. This would indicate that about 80% of the matter near the Sun is in stars, the rest comprising the interstellar medium. The region near the Sun appears to have more gas and dust than usual, however, and in the Galaxy as a whole is probably only a few per cent interstellar matter by mass.

It has been suggested that much of the interstellar hydrogen is in the molecular form instead of atomic, and molecular hydrogen is extremely difficult to detect. If this is correct, then interstellar matter may be more abundant than indicated above.

The temperatures of H I regions are measured at about 125°K. This is a higher temperature than the persons who calculate the heat balance of the medium would like, and large amounts of molecular hydrogen would make the discrepancy worse. Perhaps there is some heating mechanism which has not been properly taken into account.

A number of other radio lines have also been observed. In 1959 N. S. Kardashev predicted that the recombination of electrons and protons in H II regions should produce emission due to transitions in the very high quantum levels, and a number of these lines have been observed. One example is the 6-cm line which corresponds to the transition between the levels $n = 110$ and $n = 109$. These very large electron orbits do not exist except under the low-density conditions of interstellar space.

Radio lines due to the hydroxyl radical OH have also been observed, both in absorption and in emission. The lines are produced in H II regions, and they often show many puzzling features. For example, the relative intensities are often quite different from what one would expect, and the excitation mechanism is not known. One of the more interesting properties of these lines is that they occasionally show evidence of variability.

The Radio Continuum. A continuous background of radio emission at all wavelengths and coming from all directions is observed. This emission has sharp maxima in certain directions, and these maxima were at one time known as radio stars. The low resolution in position of radio telescopes makes accurate direction-finding difficult, but many of the radio stars have since been identified with special objects within our Galaxy or as external galaxies. It is now thought that radio emission from individual stars is a negligible part of that observed (neglecting the Sun), so the idea of radio stars has been dropped.

The radio continuum consists of a large number of discrete sources superposed on a general background which comes from all directions. This emission was discovered by K. G. Jansky in the early 1930s. The general background has a maximum in the direction of the galactic center, and it is undoubtedly associated with the distribution of interstellar matter.

A thermal source of radiation obtains its energy from the kinetic motions of the atoms, and part of the radio emission must come from thermal sources. For example, free-free transitions of hydrogen in H II regions certainly occur, and these are an important part of the emission from some of the discrete sources; however, the radio continuum as a whole has a wavelength dependence which looks nothing like a thermal source at any reasonable temperature, so much of the radio emission must be nonthermal in nature. It has been suggested by I. S. Shklovsky and others that synchrotron radiation is responsible for the nonthermal component of the radio emission, and this is generally accepted today.

Synchrotron radiation arises when high-energy charged particles encounter a magnetic field. Because of their charges, the particles cannot pass freely through the magnetic field but are accelerated into paths which spiral around the field. On being accelerated by the field, they emit radiation. In certain cases, such as the Crab nebula, synchrotron radiation may make an appreciable contribution to the visible as well as to the radio emission. This provides another reason for suspecting the existence of large-scale magnetic fields in the Galaxy and in other galaxies. The strengths of the fields are not well determined, but fields of 10^{-6} to 10^{-5} gauss similar to those indicated by the polarization of starlight are consistent with the observations.

The source of the high-energy charged particles is not known. The H II regions do not have sufficiently high temperatures. It has been suggested that the particles may come from supernova explosions, and this is supported by the strong synchrotron radiation of the Crab nebula, an old supernova. Whether this is correct and whether these high-energy particles are connected with cosmic rays is also not known. Observations in recent years have detected a general expansion of the interstellar medium out of the galactic nucleus, both for our Galaxy and for others. The source of this expansion may be connected with the synchrotron particles.

Synchrotron radiation and radio emission from galaxies are discussed in much greater detail in Section 26.

PROBLEMS

1. Star 1 has an effective temperature $T_e = 6000°K$, and star 2 has $T_e = 6100°K$. The relative prominence of lines of Fe I and Fe II is the same in the two stars. What can one expect about the relative prominence of lines due to Ca I and Ca II in the two stars? Note problem 6 of the Introduction.

2. A star has a gray atmosphere in which the Eddington approximation of equation (19.20) is valid. Determine the fraction of the outward ($\theta = 0$) intensity which originates at various optical depths in the star, if it radiates like a black body.

3. A uniform slab of thickness L and temperature T radiates like a black body. The absorption coefficient σ_ν is small everywhere except for a strong line at frequency ν_0. Determine the intensity at ν_0 relative to neighboring frequencies for various limiting cases of the optical thickness of the slab.

4. Lines D_1 and D_2 are two lines of neutral sodium at wavelengths of λ 5896 and λ 5890, respectively. Both have the ground state as the lower level, and D_2 has twice

as large an f value as D_1. Determine the ratio of the equivalent widths of the two lines under various conditions of formation.

5. A neutral sodium line at λ 4748 has an f value of 1.8×10^{-3}. The lower level has a statistical weight of 2 and an excitation potential of 2.11 eV. A neutral aluminum line at λ 5557 has an f value of 4.0×10^{-3}, and its lower level has a statistical weight of 2 and an excitation potential of 3.15 eV. These two absorption lines have about the same equivalent width in the Sun. If they are both on the linear part of the curve of growth, determine the relative abundance of sodium to aluminum in the Sun. Assume that the lines are formed under conditions of $T = 5040°K$, $\log P_e = 0.48$.

6. In the model star for which the density is constant, find the position at which the energy generation ε has fallen off to 10% of its central value. Use equation (21.38) for $n = 4$ and for $n = 20$.

7. A star has measured apparent magnitudes of $V = 8.0$, $B = 8.3$. If its spectrum is found to be A0 V, is its spectroscopic parallax likely to be more or less accurate than its trigonometric parallax?

REFERENCES

There are a number of general works on astrophysics, including References 2 and 3 of Chapter II plus the following:

1. Ambartsumian, V. A. (Ed.). *Theoretical Astrophysics*, Pergamon, London 1958.
2. Dufay, J. *Introduction to Astrophysics: The Stars*, Dover, New York, 1964.
*3. Goldberg, L., and L. H. Aller. *Atoms, Stars, and Nebulae*, Blakiston, Philadelphia, 1943.
4. Pecker, J.-C., and E. Schatzman. *Astrophysique Général*, Masson, Paris, 1959.
5. Rosseland, S. *Theoretical Astrophysics*, Clarendon Press, Oxford, England, 1936.

Section 17. The (U,B,V) systems were presented in the following:

6. Johnson, H. L., and W. W. Morgan. *Astrophys. J.*, **117**, 313, 1953.

Black-body colors on the (U,B,V) systems are given in the following:

7. Arp, H. C. *Astrophys. J.*, **133**, 874, 1961.

This reference also gives the (U,B,V) response functions as affected by atmospheric extinction according to W. G. Melbourne. Effective temperatures and bolometric corrections are discussed by D. L. Harris III on p. 263 of Reference 4 of Chapter I. The measurement of stellar energy distributions is the subject of the following:

8. Code, A. D., in *Stellar Atmospheres* (J. L. Greenstein, Ed.), Univ. of Chicago Press, Chicago, p. 50, 1960.
9. Oke, J. B. *Ann. Rev. Astron. Astrophys.*, **3**, 23, 1965.

* There are on the popular or semi-popular level.

The interferometer and the measurement of stellar diameters are discussed in Reference 1 of Chapter I, in Reference 2 above, and in the following:

*10. Brown, H. R. *Sky and Telescope*, **28,** 64, 1964.

Section 18. A very good history of early stellar spectroscopy is given by the following:

11. Curtiss, R. H., in *Handbuch der Astrophysik*, **V,** Part 1, 1, 1932.

Modern classification methods and their calibration are the topics of several articles in Reference 4 of Chapter I. The H-R diagram is discussed in the following:

*12. Hack, M. *Sky and Telescope*, **31,** 260, 1966.

Section 19. A good discussion of the development of the theory of stellar atmospheres is contained in the article by B. Strömgren on p. 172 of Reference 2, Chapter II. There are many works on the theory of radiative transfer and stellar atmospheres, and some of the more appropriate are the following:

13. Pecker, J.-C. *Ann. Rev. Astron. Astrophys.*, **3,** 135, 1965.
14. Münch, G., in *Stellar Atmospheres* (J. L. Greenstein, Ed.), Univ. of Chicago Press, Chicago, p. 1, 1960.
15. Aller, L. H. *The Atmospheres of the Sun and Stars*, Second Edition, Ronald Press, New York, 1963.
16. Unsöld, A. *Physik der Sternatmosphären*, Second Edition, Springer-Verlag, Berlin, 1955.

Various aspects of the outer solar atmosphere and solar physics are contained in the following references:

17. Brandt, J. C., and P. Hodge. *Solar System Astrophysics*, McGraw-Hill, New York, 1964.
18. Pagel, B. E. J. *Ann. Rev. Astron. Astrophys.*, **2,** 267, 1964.
19. Newkirk, G. A. *Ann. Rev. Astron. Astrophys.*, **5,** 213, 1967.
*20. Athay, R. G. *Science*, **143,** 1129, 1964.
*21. Wilson, O. C. *Science*, **151,** 1487, 1966.
*22. Parker, E. N. *Scientific American*, **210,** 4, 66, 1964.

Section 20. For references pertinent to line formation and abundance determinations, see Reference 2 of Chapter II, References 1–5, 15, and 16 above, and the following:

23. Böhm, K.-H., in *Stellar Atmospheres* (J. L. Greenstein, Ed.), Univ. of Chicago Press, Chicago, p. 88, 1960.
24. Aller, L. H., in *Stellar Atmospheres* (J. L. Greenstein, Ed.), Univ. of Chicago Press, Chicago, pp. 156 and 232, 1960.
25. Aller, L. H. *The Abundances of the Elements*, Interscience, New York, 1961.
26. Cayrel, R., and Cayrel de Strobel, G. *Ann. Rev. Astron. Astrophys.*, **4,** 1, 1966.

A determination of the abundances of a large number of elements in the Sun was made in the following:

27. Goldberg, L., E. A. Müller, and L. H. Aller. *Astrophys. J. Suppl.*, **V,** 45, 1, 1960.

Section 21. References to stellar structure include:

28. Aller, L. H., and D. B. McLaughlin (Eds.). *Stellar Structure*, Univ. of Chicago Press, Chicago, 1965.
29. Chandrasekhar, S. *An Introduction to the Study of Stellar Structure*, Dover, New York, 1957.
30. Schwarzschild, M. *Structure and Evolution of the Stars*, Princeton Univ. Press, Princeton, 1958.
31. Wrubel, M. *Handbuch der Physik*, **LI,** 1, 1958.

Section 22. Among the many possible references to stellar evolution are the article by R. L. Sears and R. R. Brownlee, in Reference 28 above, and the following:

32. Arp, H. C. *Handbuch der Physik*, **LI,** 75, 1958.
33. Burbidge, E. M., and G. R. Burbidge. *Handbuch der Physik*, **LI,** 134, 1958.
34. Eggen, O. J. *Ann. Rev. Astron. Astrophys.*, **3,** 235, 1965.
35. Stein, R. F., and A. G. W. Cameron (Eds.), *Stellar Evolution*, Plenum Press, New York, 1966.
*36. Brownlee, R. R., and A. N. Cox. *Sky and Telescope*, **21,** 252, 1961.
*37. Ezer, D., and A. G. W. Cameron. *Sky and Telescope*, **24,** 328, 1962.
*38. Hack, M. *Sky and Telescope*, **31,** 333, 1966.
*39. Struve, O. *Sky and Telescope*, **24,** 261, 1962.

Reference 35 consists of the proceedings of a conference held in 1963.

Section 23. Interstellar matter is covered in Reference 3 of Chapter II, and in References 1–5 above. A good review of recent work is the following:

40. Dieter, N. H., and W. M. Goss. *Rev. Modern Physics*, **38,** 256, 1966.

Other references include:

41. Dufay, J. *Galactic Nebulae and Interstellar Matter*, Philosophical Library, New York, 1957.
42. Greenberg, J. M. *Ann. Rev. Astron. Astrophys.*, **1,** 267, 1963.
43. Greenstein, J. L., in *Astrophysics* (J. A. Hynek, Ed.), McGraw-Hill, New York, p. 526, 1951.
44. Kahn, F. D., and J. E. Dyson. *Ann. Rev. Astron. Astrophys.*, **3,** 47, 1965.
*45. Struve, O. *Sky and Telescope*, **21,** 269, 1961.
*46. Vandervoort, P. O. *Scientific American*, **212,** 2, 90, 1965.

IV

Galaxies and Cosmology

Stars appear to occur only in huge aggregates known as galaxies, and this final chapter is concerned with the properties of galaxies. Section 24 describes the most direct way of trying to find the true distribution of stars around us. The characteristics of our own galactic system are discussed in Section 25, while galaxies in general are the subject of Section 26. Finally, some of the modern thoughts on, and methods of approach to, the problem of the nature of the Universe as a whole are discussed in Section 27.

24. Star Counts

Many attempts have been made to determine the arrangement and distribution of the stars. Galileo noted in 1609 in his telescope that the Milky Way is actually composed of a very large number of faint stars. Since the Milky Way is a narrow band which extends around the sky in a great circle, this must mean that most stars are concentrated toward a plane and that the Sun is in this plane. This simple observation shows that the stars are not uniformly distributed in all directions.

In more recent times star counts have been made in an effort to find their distribution. A simple example will illustrate how this can be carried out. Suppose that all stars have the same absolute magnitude M, that the number of stars per cubic parsec D is everywhere a constant, and that there is no interstellar absorption. Let $N(m)$ be the total number of stars observed of apparent magnitude m or brighter. If $r(m)$ is the distance of stars of apparent magnitude m, then $N(m)$ is simply the number of stars closer than $r(m)$. By the above assumptions, this is

$$N(m) = \tfrac{4}{3}\pi r^3(m) D \tag{24.1}$$

Equation (9.8) relates apparent and absolute magnitudes and distance. If this is solved for the distance, one finds

$$r(m) = 10 \times 10^{0.2(m-M)} \tag{24.2}$$

If this is substituted into equation (24.1) one obtains

$$N(m) = \tfrac{4}{3}\pi \times 10^3 D \times 10^{0.6(m-M)} \tag{24.3}$$

With knowledge of M and observations of $N(m)$ for any apparent magnitude, one could find the space density D of the stars. An interesting feature of this equation is that the relative increase of the number of stars with increasing apparent magnitude is independent of D, m, and M:

$$\frac{N(m+1)}{N(m)} = 10^{0.6} = 3.98 \tag{24.4}$$

The simple relations of equations (24.3) and (24.4) are very easy to check, and it is found that they are not satisfied by the star counts. The difference is in the sense that the number of faint stars observed is much less than that predicted by the above equations. At least one of the three basic assumptions which were made is not satisfied.

It is important to note that the observation that stars are concentrated toward the Milky Way, or the galactic plane, does not affect the validty of equation (24.4), although it does require a modification of equation (24.3). Equation (24.4) holds as long as the density of stars remains constant for all points within the same small solid angle at all distances; however, D can be different in one direction from its value in another without affecting this equation.

The first assumption is that all stars have the same absolute magnitude M. This is known to be incorrect, as there are extremely large variations in M from one star to another. The quantity $\phi(M)\,\Delta M$ is defined as the function which gives the number of stars per unit volume which have absolute magnitude between $(M - \frac{1}{2}\,\Delta M)$ and $(M + \frac{1}{2}\Delta M)$, and $\phi(M)$ is known as the luminosity function. If the luminosity function were the same everywhere, then equation (24.4) would still be satisfied, even though the stars were not all of the same M. This is easily seen from the fact that it would be valid for stars of any given absolute magnitude, and if all of these were added together, it would be valid for all stars.

It is known that the luminosity function is not the same everywhere. For example, regions of high density of the interstellar matter have a much greater than average likelihood to have large numbers of young, very luminous stars. Also, it is well known that the luminosity function of a globular cluster is not the same as that of a galactic cluster. Nevertheless, known variations in the luminosity function would produce rather small deviations from equation (24.4), much smaller than those observed.

The effects of interstellar absorption will now be examined, with the assumption that D is independent of distance being retained. If $A(r)$ is the absorption in magnitudes for a star at distance r, then equation (23.4) indicates how this affects the relation between m, M, and r. It is seen that equation (24.2) is multiplied by the factor $10^{-0.2A(r)}$ and equation (24.3) by the factor $10^{-0.6A(r)}$. Thus if $N_0(m)$ is what the count would be in the absence of absorption,

$$N(m) = N_0(m)10^{-0.6A(r)} \tag{24.5}$$

It was pointed out in Chapter III that interstellar absorption is important, and it is not surprising that the effects are in the same direction as the observation show; i.e., absorption increases with distance and, therefore, cuts down on the number of faint stars observed. This same effect, of course, could be obtained by an actual decrease in the density of stars at large distances. It is not generally possible to completely separate the density and the absorption effects, and this is why star counts have only a limited use in determining the distributions of stars in space.

Let $D(r)$ now be the actual density of stars per cubic parsec at distance r and let r' and $D'(r')$ be the distance and the density which one would obtain if absorption were not taken into account. Then r' and r are related by

$$r' = r10^{0.2A(r)} \tag{24.6}$$

that is, the absorption always makes the star appear fainter and therefore farther away than it actually is. All stars which lie between r and $r + dr$ appear to lie between r' and $r' + dr'$, and so one has

$$4\pi r'^2 D'(r')\, dr' = 4\pi r^2 D(r)\, dr \tag{24.7}$$

The above relations then yield

$$\frac{D'(r')}{D(r)} = \frac{10^{-0.6A(r)}}{1 + 0.46r\dfrac{dA(r)}{dr}} \tag{24.8}$$

A uniform absorption of $\frac{3}{4}$mag per 10^3 pc is not unreasonable, and this gives $A(r) = 0.00075r$. Equation (24.8) then yields the values given in Table 24.1. J. L. Greenstein (see Reference 43 of Chapter III) gives a similar table

TABLE 24.1. THE EFFECTS OF INTERSTELLAR ABSORPTION

r (pc)	r' (pc)	$D'(r')/D(r)$
100	104	0.87
500	595	0.51
1000	1410	0.26
2000	4000	0.075
4000	16000	0.007
5000	28000	0.002

for larger absorption of one magnitude per 10^3 pc. It is apparent that tremendous distortions occur in the distances and star densities found for very distant objects. The uncertainties in the absorption will completely mask any information star counts might have on density variations at great distances. Star counts are still useful for relatively nearby regions and for detecting the properties of dark nebulae, as mentioned in Section 23.

Because of the neglect of the absorption, early attempts to determine the structure of the Galaxy from star counts always resulted in a star density which fell off with distance. This caused the Sun to be erroneously placed at or near the center of the Galaxy. Early in this century J. C. Kapteyn found such a model from star counts. The "Kapteyn Universe," as his model was called, had most stars within about 1000 pc. At greater distances the stars rapidly thinned out, and the edge of the system was some 10,000 pc from the Sun.

From Table 24.1 one finds that stars which Kapteyn thought were at the edge of the system may actually only be some 3000 pc from the the Sun. It is now known that the Sun is not at the center of the Galaxy, and 3000 pc is only about $\frac{1}{3}$ of the distance of the Sun from the center. This is not a criticism of Kapteyn's work; it is only an illustration of how reasonable-appearing assumptions can lead to completely erroneous conclusions.

25. Galactic Structure

The Galaxy is the name given to the large collection of stars and interstellar matter of which the Sun is a member. The position of the Sun and of the absorbing material in space are such as to make it difficult to obtain a good overall view of the Galaxy. Its general properties have been understood only since the first few decades of this century, and important details are still being filled in.

Early Models of the Galaxy. Early ideas of the properties of the Galaxy were based on star counts. These ideas were in error because of the effects of interstellar absorption. It is important to realize that the neglect of this absorption causes a star to appear more distant than it actually is, but it also caused the Galaxy to seem much smaller than it is. As is explained in Section 24, this is due to the fact that the distortion in distance gave evidence of a large drop in star density that actually does not exist, and this density decrease was identified with the edge of the Galaxy.

About 1920 Harlow Shapley proposed a new model of the Galaxy which was based on the distribution of the globular clusters. Globular clusters are not spread uniformly over the sky, but most of them appear in the general direction of the constellation Sagittarius. Also, while most stars are concentrated toward a plane and must, therefore, be part of a flat system, the clusters are not so concentrated; still, the clusters are symmetrically placed about a point that is in this plane. Shapley used this as evidence that the system of the globular clusters is closely related to the system of stars that defines the galactic plane. If the clusters are more or less centered about a point in the direction of Sagittarius, the system of stars might also be, and the Sun is not at the center.

Shapley found the distance of the nearer globular clusters by means of their RR Lyrae variables. He assumed that these variables satisfy the period-luminosity relation which had been found for classical Cepheids only a short time earlier. The nearer clusters were then used to calibrate the distances for clusters which do not contain RR Lyrae stars and for those which are too far away for these variables to be observed. These methods included getting the mean absolute magnitudes of the most luminous stars in a cluster, the the mean absolute magnitude of a cluster as a whole, and the mean diameter of a cluster. In this fashion Shapley was able to determine the size of the globular cluster system. By his assumption that the center of this system is the same as the center of the Galaxy, he was able to find the distance of the Sun from this center. Although his distances are now known to have been too large by factors of up to three, his model of the Galaxy was correct in general outline. There was still considerable support for the star-count models, which supported a much smaller Galaxy with the Sun near the center, and debate on this continued for a number of years.

The position of the Sun in the Galaxy makes it difficult to determine its overall structure. The view of other galaxies is a tremendous help in this, but it took a long time for all astronomers to be convinced that external galaxies are just that and not a part of our Galaxy. Galaxies have been known for a long time, but a rather curious set of circumstances lent evidence for their being relatively small, nearby objects. Their spectra indicated a stellar content, but reflection nebulae also show the spectra of the illuminating stars, so this was not conclusive. A. van Maanen published proper motions for some galaxies indicating that they were rotating with rather short periods, much too short for objects the size of the Galaxy. These observations were much later shown to be due to systematic errors, but they were very effective in keeping many astronomers from the correct interpretation of the nature of galaxies.

In many galaxies novae were discovered. This would have been conclusive, for the absolute magnitudes of novae in our Galaxy were approximately known; however, an object which is now known to have been a supernova also appeared in M 31, the Andromeda galaxy. It was then uncertain whether the one bright object or the many fainter ones were true novae, and this made the difference between M 31 being an external galaxy and its being a relatively small, nearby object within the Galaxy.

The argument on the nature of the galaxies raged for the first quarter of this century, but in the mid 1920s Edwin P. Hubble was able to resolve and identify Cepheids in M 31. This finally put an end to the controversy and allowed astronomers to infer properties of the Galaxy from the appearance of the external galaxies. There still remained a large systematic error in the sizes. Shapley's model overestimated the size of the Galaxy, and the zero-point error in the period-luminosity relation for the Cepheids caused the distances and the sizes of external galaxies to be underestimated. For a while it appeared that our Galaxy is a monster far larger than any other, but is now believed to be larger than average but not an unusual spiral galaxy.

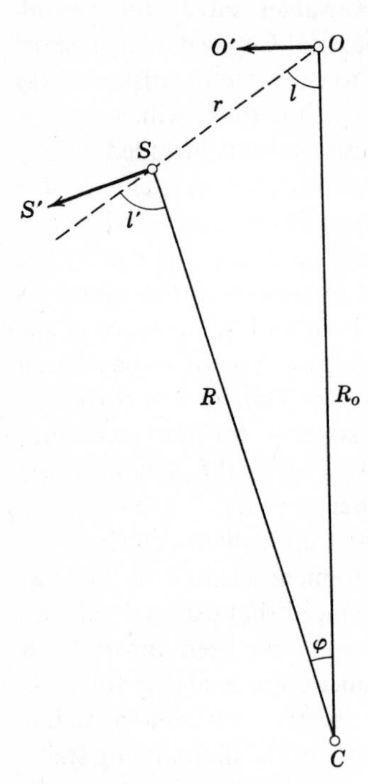

Figure 25.1 Differential galactic rotation.

Galactic Rotation. The fact that the Galaxy is flat suggests rotation. Bertil Lindblad and Jan H. Oort pioneered work on galactic rotation as far back as the 1920s. If the stars were essentially at rest, the gravitational attractions would cause the system to collapse. Since the collapse is not observed, the stars must be moving in orbits of some kind about the galactic center. These orbits need not be the elliptical paths of the simple two-body problem. If most of the mass of the Galaxy is spread over a large volume, the orbits can be quite different. If the orbits are directed at random, then the stars should exist in a roughly spherical distribution about the center. The fact that the stars are concentrated in a plane indicates that the galactic orbits are not directed at random, but they must tend to be directed in this plane. The order or near uniformity in these orbits of the stars gives the effect of rotation to the Galaxy as a whole.

The following analysis follows a similar one by Oort. The purpose is to illustrate the effects resulting from the assumption of a very simple form of galactic rotation, namely, that the stars all move in circular orbits in the galactic plane. In Figure 25.1, point O is the position of the Sun and S is another star. Both move in the plane of the paper in circular orbits around the point C, the galactic center. The orbital velocities are V_o and V, and these are directed along OO' and SS', respectively, these directions being normal to the radii R_o and R. The distance between the stars is r, and l is the galactic longitude of the star at S.

The observed radial velocity of the star at S is simply the difference between the radial components of the two orbital velocities. From the figure it is seen that this is

$$V_r = V \sin l' - V_o \sin l \tag{25.1}$$

The sine formula applied to triangle OSC gives

$$\sin(\pi - l') = \sin l' = \frac{R_o}{R} \sin l$$

so the radial-velocity relation can be written

$$V_r = \left(V \frac{R_o}{R} - V_0 \right) \sin l \tag{25.2}$$

The various physical variables will in general depend on position. It will now be assumed that the Galaxy has axial symmetry. This means that the axial angle ϕ shown in Figure 25.1 does not enter into the values of the variables, and so these variables can depend only on the radius R. One can then write a Taylor expansion for V:

$$V(R) = V_o + \left.\frac{dV}{dR}\right)_o (R - R_o) + \cdots \tag{25.3}$$

Another assumption will now be made: the distance to the galactic center is very large compared to the distance to the star; that is, (r/R) is very much smaller than one. If two quantities are each small, the product of them will be much smaller still. The common approximation will now be made that any term that contains the product of two or more such small quantities can be neglected. This is not necessary, and it means that this analysis will be valid only for stars which are relatively near the Sun; however, it makes the algebra much simpler, and it still demonstrates all of the important physical ideas.

The above assumption means that ϕ is a very small angle. Drop a perpendicular from S onto OC. Then the top segment of OC is $(r \cos l)$, and the bottom segment is approximately R, so

$$R_o - R = r \cos l$$

The higher-order terms in equation (25.3) which are indicated by $(\cdots)$ all contain the product of at least two small quantities, so they can be neglected.

Then this relation is

$$V(R) = V_o - \left.\frac{dV}{dR}\right)_o r \cos l \tag{25.4}$$

The symbol $(dV/dR)_o$ means that the indicated derivative is to be evaluated at the position of the Sun.

The approximation being used allows one to write

$$\frac{R_0}{R} \cong R_o(R_o - r \cos l)^{-1} \cong 1 + \frac{r}{R_o} \cos l$$

When these approximations are used, equation (25.2) for the radial velocity becomes

$$V_r = \left[\left(V_o - \left(\frac{dV}{dR}\right)_o r \cos l\right)\left(1 + \frac{r}{R_o}\cos l\right) - V_o\right]\sin l$$

If one carries out the indicated multiplication and drops the terms containing two or more small factors, one finds

$$V_r = \frac{1}{2}\left[\frac{V_o}{R} - \left(\frac{dV}{dR}\right)_o\right] r \sin 2l \tag{25.5}$$

The tangential velocity observed for the star is the difference between the tangential components of the orbital velocities. Figure 25.1 indicates this to be

$$V_t = V \cos l' - V_o \cos l \tag{25.6}$$

But the figure shows that $l' = l + \phi$, so

$$\cos l' = \cos l \cos \phi - \sin l \sin \phi$$

But ϕ is a small angle, so $\cos \phi \simeq 1$. The sine formula applied to triangle *OSC* gives

$$\begin{aligned} \sin \phi &= \frac{r}{R}\sin l \\ &\simeq r(R_o - r\cos l)^{-1}\sin l \\ &\simeq \frac{r}{R_o}\sin l \end{aligned}$$

If the above approximations and equation (25.4) are used in (25.6), one finds

$$\begin{aligned} V_t &= \left(V_o - \left(\frac{dV}{dR}\right)_o r \cos l\right)\left(\cos l - \frac{r}{R_o}\sin^2 l\right) - V_o \cos l \\ &= -\left[\frac{V_o}{R_o}\sin^2 l + \left(\frac{dV}{dR}\right)_o \cos^2 l\right] r \end{aligned}$$

But $\qquad \sin^2 l = \frac{1}{2}(1 - \cos 2l) \qquad \cos^2 l = \frac{1}{2}(1 + \cos 2l)$

and so the tangential velocity expression becomes

$$V_t = \left\{-\frac{1}{2}\left[\frac{V_o}{R_o} + \left(\frac{dV}{dR}\right)_o\right] + \frac{1}{2}\left[\frac{V_o}{R_o} - \left(\frac{dV}{dR}\right)_o\right]\cos 2l\right\} r \tag{25.7}$$

Equations (25.5) and (25.7) are the desired relations for the observed radial and tangential velocities. They can be simplified by defining the quantities A and B as follows:

$$A = \frac{1}{2}\left[\frac{V_o}{R_o} - \left(\frac{dV}{dR}\right)_o\right] \qquad B = -\frac{1}{2}\left[\frac{V_o}{R_o} + \left(\frac{dV}{dR}\right)_o\right] \tag{25.8}$$

One then has
$$V_r = Ar \sin 2l \tag{25.9}$$
$$V_t = Br + Ar \cos 2l \tag{25.10}$$

The tangential velocity is not directly measurable, and so equation (25.10) is usually replaced by the expression for the proper motion. Using equation (11.4), one obtains, from (25.10),

$$\mu = 0.211(B + A \cos 2l) \tag{25.11}$$

The numerical factor in (25.11) is valid for proper motions measured in ″/yr, velocities in km/sec, and distances in pc. The units of A and B are km/(sec pc), although their physical dimensions are simply $(\text{time})^{-1}$.

Equations (25.9)–(25.11) are the famous Oort equations of differential galactic rotation, and A and B are the Oort constants. Stars which have different galactic orbits have different orbital velocities, and the Oort equations express these differences as functions of the quantities which determine them.

The form of the Oort equations can be visualized without much difficulty. For example, a star with $l = 0$ (or $l = \pi$) will be moving in the same direction as the Sun but faster (or more slowly) because of its smaller (or larger) orbit; therefore the observed radial velocity should vanish for directions toward or away from the galactic center. Also, a star with $l = \pm\pi/2$ has the same orbital speed and keeps the same distance from the Sun, so the radial velocity vanishes for these directions too. The result is that there are four directions along which the observed radial velocity should vanish, and the double sine relation of equation (25.9) is qualitatively explained. The proper motion of a star with $l \pm \pi/2$ is such that it would appear to go once around the celestial sphere during the orbital period. This is a proper motion of (V_o/R_o) radians per second, which is what equations (25.8) and (25.11) predict if the proper units are used.

The assumptions that both stars are in the galactic plane and that (r/R) is a small quantity were made for simplicity, but they are not necessary. By considering only stars that are relatively nearby and close to the galactic plane, one can be sure that the equations are applicable; however, the analysis is easily generalized to include any part of the Galaxy.

The assumption that the orbits are circular can also be modified. It is known that the peculiar velocities of the stars in the solar neighborhood, i.e., the velocities with respect to the average of the group, are nearly (but not exactly) random, and so all of these stars cannot be moving in exactly circular orbits. Perhaps the average velocity in the solar neighborhood corresponds to the circular velocity. Each star would then have a velocity equal to the circular value plus its own peculiar velocity, the latter being nearly random. With this interpretation the relative motion between two stars is composed of the peculiar motions plus a term due to the differential galactic rotation (the difference in the circular velocities at the positions of the two stars), and the Oort equations apply only to the latter.

The peculiar velocities are nearly independent of position, while the differential rotation velocities increase with distance. In terms of proper motion, the peculiar motions decrease with distance, while the rotation contribution remains the same for all distances. In both cases the relative contribution of the rotation effect to the observed motions must increase with distance. Within the solar neighborhood the rotational effect is negligible. (This is essentially how the solar neighborhood is defined in Section 11.) At very great distances from the Sun, the galactic-rotation effect becomes dominant, although if r is larger than about 2000 pc or so the small-distance approximation is no longer valid. The peculiar motion of the Sun is reflected in all of the observed motions, and a correction should be made for it. It is usually assumed that the local standard of rest has the circular velocity around the galactic center.

Numerous investigations have shown that the equations of differential galactic rotation are satisfied. The general procedure is to divide stars into groups according to their distances. A plot of radial velocity vs. galactic longitude should be a double sine wave as in equation (25.9). The peculiar velocities will produce a scatter on top of this curve, but the amplitude of it (Ar) increases with distance. If this amplitude can be measured, Ar is known for the group in question. If the relative distances of two such groups is known, the value of A follows. Both A and B can be found from the proper motions. Equation (25.11) shows that the proper-motion effect does not depend on distance, but, as is mentioned above, the scatter introduced by the peculiar motions will decrease with distance. The observed motions agree with other observations, including the globular cluster work of Shapley, in determining the direction of the galactic center. There is little doubt that the assumption of nearly circular orbits about a distant center is basically correct.

Unfortunately there is a rather large scatter in the measured values of the Oort constants. Most of this is probably due to poor distance determinations caused by uncertainties in absolute magnitudes and in the corrections for interstellar absorption. It may also happen that the general orbits deviate more from circular than is usually assumed. Numerous "best" values of A and B have been published, and B. J. Bok in his draft report for IAU Commission 33, 1964, gives the following consensus values:

$$\begin{aligned} A &= 15 \text{ km/(sec kpc)} = 0.49 \times 10^{-15} \text{ sec}^{-1} \\ B &= -10 \text{ km/(sec kpc)} = -0.32 \times 10^{-15} \text{ sec}^{-1} \end{aligned} \qquad (25.12)$$

In these equations kpc stands for kiloparsec = 1000 parsecs, the standard unit of distance in galactic structure.

From equations (25.8), one finds

$$A + B = -\left.\frac{dV}{dR}\right)_o \qquad A - B = \frac{V_o}{R_o} \tag{25.13}$$

A circular orbit of radius R_o has a circumference $2\pi R_o$. With a velocity of V_o, the period of one revolution is

$$P = 2\pi\frac{R_o}{V_o} = \frac{2\pi}{A - B} \tag{25.14}$$

The above values of A and B give a period of 7.8×10^{15} sec $= 2.5 \times 10^8$ yr. At the Sun's position, the Galaxy rotates once in about a quarter of a billion years. Note that

$$\left.\frac{dV}{dR}\right)_o = -5 \text{ km/(sec kpc)} \tag{25.15}$$

This means that the circular velocity is decreasing as one moves out from the center, at least in the neighborhood of the Sun, so the rotation period of the Galaxy increases further out from the center. The Galaxy is not rotating like a solid body in which V is proportional to R.

The quantities V_o and R_o can be found from observations. The circular velocity in the solar neighborhood may be roughly measured from the radial velocities of the nearby galaxies. The accuracy is not great, but a velocity of 200–300 km/sec is indicated. The quantity R_o may be found more accurately. The distributions of many different types of objects indicate a center of symmetry located between 8–10 kpc from the Sun. Bok's report mentioned above gives

$$R_o = 10 \text{ kpc} \qquad V_o = 250 \text{ km/sec} \tag{25.16}$$

These values are consistent with equation (25.12).

The 21-cm Observations. When the Oort constant A is known, equation (25.9) can be used to find the distance of an object. (This is true if its direction is not along one of the four critical directions that cause the sine term to vanish and if the object is far enough away that its peculiar velocity can be ignored.) This is an important method for finding distances, and it is not restricted to stars. This is the basis for finding the distribution of neutral hydrogen from 21-cm observations as mentioned in Section 23. The interstellar matter must move in galactic orbits just as the stars do. The 21-cm observations are particularly important because the radio waves are not appreciably affected by the interstellar absorption, so the neutral hydrogen can often be observed at much greater distances in the galactic plane than stars can. Again, for large distances equation (25.9) must be appropriately modified, but the principle is the same.

In Fig. 25.2, point O represents the position of the local standard of rest, and C is the galactic center. A 21-cm observation is made in the galactic plane along galactic longitude l, and radiation is received from any point along this line of sight that has a high enough concentration of neutral hydrogen. Points A, B, D, and E represent such points in the figure, and the arrows indicate the rotation velocities at these points. If there were little or no radial velocity spread among these points, the radiation from all of them would be received at the same wavelength. The absorptions and the emissions from all of these points would add together, and one would be looking into an optically thick region (if there were enough neutral hydrogen), as described in Section 23. This is what occurs toward $l = 0$ or $l = \pi$. Toward other longitudes, however, the differential galactic rotation causes the points at different distances to have different radial velocities. The resulting doppler shifts bring about a small separation in wavelength between the radiations from the different points. If hydrogen were uniformly spread over the line of sight, the signal would be a smooth function of wavelength over a range corresponding to the extreme radial velocities. The observed signal is actually irregular, indicating that the hydrogen has a patchy distribution.

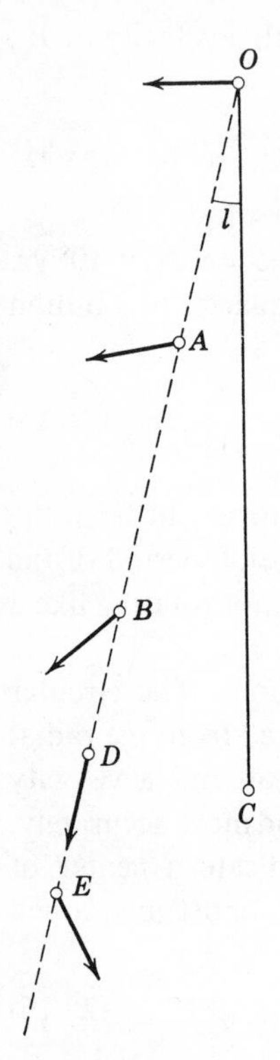

Figure 25.2 The 21-cm observations.

Each maximum in the signal comes from a region of high density. If the rotation curve of the Galaxy is known, i.e., if the circular velocity V is known as a function of R, then equation (25.2) can immediately give the V and R of each high-density region having an observed radial velocity. Note that this relation assumes circular orbits, but it is not restricted by the small-distance approximation. By making these measurements at different latitudes and longitudes, one can find the positions of the high-density regions, and the spiral structure of the Galaxy can be traced out.

Point D in the figure has a circular orbit that is tangent to the line of sight. Note that points B and E, which are equal distances on either side of D, will have equal radial velocities. This means that their signals will coincide in wavelength; however, region B is closer to the observer and should subtend a larger solid angle, so measurements made in latitude can possibly separate the two regions.

An interesting thing about point D is that, of the points along the line of sight, it has extremum values for both R and V. Obviously, R is a minimum value, and V has a local maximum or minimum, depending on what the rotation curve is doing at that point. In any case, equation (25.2) shows that the radial velocity is also an extremum for point D. Thus the greatest (or least, depending on longitude) doppler shift measured at a given longitude must belong to point D. But for this point, $R = R_o \sin l$, and equation (25.2) is

$$V_r \text{ (point } D) = V - V_o \sin l \tag{25.17}$$

The observed extreme radial velocity for any longitude then determines V. Since the R of this point is known, the rotation curve $V(R)$ is found for all points within R_o. This method obviously does not work for regions having radii greater than R_o. Use of this 21-cm data also is based on the assumption that the hydrogen moves in circular orbits, and this may not always be justified.

The Distribution of Mass. An idea of the mass of the Galaxy can be obtained from the above data. Any object moving in a circle of radius R_o with a velocity V_o suffers an acceleration toward the center of amount V_0^2/R_o. This acceleration in the galactic orbits must be due to the gravitational attraction of the matter in the Galaxy. If all the matter were concentrated at the center, then it would produce the acceleration $(G\mathcal{M}/R_o^2)$, where G is the gravitational constant and $\mathcal{M}$ is the mass of the Galaxy. If these two expressions are equated, one obtains

$$\mathcal{M} = \frac{R_o V_o^2}{G} \tag{25.18}$$

The above numerical values give $\mathcal{M} = 2.9 \times 10^{44}$ g $= 1.5 \times 10^{11} \mathcal{M}_o$ where $\mathcal{M}_o$ is the mass of the Sun. Since not all of the galactic matter is concentrated at the center, this value is only approximate; however, it is probably accurate within better than a factor of two.

More accurate mass estimates of the Galaxy can be made. The 21-cm data can give the orbital velocity of the neutral hydrogen at different distances from the center, i.e., the rotation curve of the Galaxy, as described earlier in this section. The rotation curve is a function not only of the mass of the Galaxy but also of how this mass is distributed. An example is given in Section 26 of how the rotation curve can be analyzed to yield the mass distribution.

The component of a star's velocity perpendicular to the galactic plane is similar to that of a one-dimensional harmonic oscillator. The star is accelerated back toward the plane by an amount that depends on the density of matter in the plane. Analyses of these z-component motions, as they are called, can yield information about how matter is distributed in the Galaxy.

These are very complicated dynamical problems, and the data are not very complete, so models of the mass distribution of the Galaxy are still being improved. The total mass seems to be close to the $1.5 \times 10^{11}\mathcal{M}_o$ derived above from much simpler considerations. Much of this mass is in the central nuclear bulge, most of the rest is highly concentrated toward the galactic plane. There is some matter of extremely low density which, like the globular clusters, has little or no concentration toward this plane, although it is symmetric about the galactic center. This is the galactic halo.

These dynamical investigations give information on the total density of matter. When the observed matter in stars is subtracted from this, the remainder must be due to stars too faint to be seen and to the interstellar matter. As long ago as the early 1930s Oort determined an upper limit for the density of interstellar matter in this fashion. The early estimates indicated that about as much mass should be in the interstellar matter as in the stars, but the 21-cm data give much smaller amounts than the Oort limit. As is mentioned in Section 23, the Galaxy as a whole appears to contain only a small percentage of interstellar matter, although molecular hydrogen may bring this up considerably if it is as abundant as is sometimes suggested. The numbers involved are not accurately enough determined for a serious discrepancy to exist.

High-Velocity Stars and Elliptical Orbits. Most stars in the solar neighborhood have peculiar velocities of no more than a few tens of kilometers per second, and these are nearly random. There is a small but definite tendency for the peculiar velocities to be directed toward or away from a certain direction. This phenomenon is known as star streaming. Kapteyn suggested that two groups of stars are intermingling, the relative motion of the groups giving rise to the observed preferred direction. This preferred direction happens to be toward the galactic center, and this indicates a possible connection with galactic rotation. Lindblad showed that star streaming results from small deviations from circular orbits that the stars have, and this explanation fits well into the picture of differential galactic rotation.

Since the circular velocity in the solar neighborhood is about 250 km/sec, and since most stars have peculiar velocities less than 10% of this value, it follows that most stars near the Sun have galactic orbits that are very nearly circular. The number of stars in the solar neighborhood having a given peculiar velocity decreases as the velocity becomes larger. Two special things happen when the velocities become greater than about 60–65 km/sec. First, the number of stars of very high peculiar velocity, while being a very small fraction of the total number, is still much greater than the number one would expect from an extrapolation of the observed number with low velocities. Second, while the motions of the low-velocity stars are nearly random, the high-velocity stars have a very striking asymmetry in their peculiar motions. No high-velocity star is moving toward galactic longitude $l = 90°$, and almost none is directed anywhere toward the hemisphere centered on this longitude. This direction of avoidance is the direction of galactic rotation of the solar neighborhood.

These points suggest that the stars in the solar neighborhood consist of at least two separate groups, one of which is predominant at low velocities and the other (or another) at high velocities. Differences in peculiar velocity are also differences in orbital velocity, so the two groups must have significantly different galactic orbits. Lindblad showed that the asymmetries in the motions of the high-velocity stars could be understood on the basis of galactic rotation. It is worth considering this in some detail for the simplified assumption of two-body elliptical orbits. The two-body approximation is adequate for illustration purposes, but in practice greater accuracy is sometimes needed.

Two masses $\mathcal{M}_1$ and $\mathcal{M}_2$ move in elliptical paths about each other with velocities V_1 and V_2. The total energy of the system is constant and is the sum of the kinetic and the potential energies. The energy equation looks very much like relation (3.7), except that the forces here are gravitational instead of electrostatic:

$$E = \tfrac{1}{2}\mathcal{M}_1 V_1^2 + \tfrac{1}{2}\mathcal{M}_2 V_2^2 - \frac{G\mathcal{M}_1\mathcal{M}_2}{R} \tag{25.19}$$

where R is the separation of the masses. The two bodies have equal momenta, so

$$\mathcal{M}_1 V_1 = \mathcal{M}_2 V_2$$

and the relative velocity V of one mass with respect to the other is

$$V = V_1 + V_2 = V_1\left(1 + \frac{\mathcal{M}_1}{\mathcal{M}_2}\right)$$

The kinetic energy can then be expressed in terms of the relative velocity as

$$\begin{aligned} KE &= \tfrac{1}{2}\mathcal{M}_1 V_1^2 + \tfrac{1}{2}\mathcal{M}_2 V_2^2 \\ &= \tfrac{1}{2}\mathcal{M}_1 V_1^2\left(1 + \frac{\mathcal{M}_1}{\mathcal{M}_2}\right) \\ &= \frac{1}{2}\frac{\mathcal{M}_1\mathcal{M}_2}{\mathcal{M}_1 + \mathcal{M}_2} V^2 \end{aligned} \tag{25.20}$$

The product of the masses divided by their sum is known as the reduced mass of the system. Equation (25.19) for the total energy now becomes

$$E = \frac{1}{2}\frac{\mathcal{M}_1\mathcal{M}_2}{\mathcal{M}_1 + \mathcal{M}_2} V^2 - \frac{G\mathcal{M}_1\mathcal{M}_2}{R} \tag{25.21}$$

It is desired to express the energy E in terms of the constants of the orbit, and this can be done easily from Kepler's three laws. When the two masses are at their closest distance, the separation, according to equation (12.4) is

$$R(\theta = 0) = a(1 - e)$$

where a is the semi-major axis and e is the eccentricity. Equations (12.5) and (12.7) for the areal velocity and the period-orbital size relations combine to give

$$R^2 \frac{d\theta}{dt} = [G(\mathcal{M}_1 + \mathcal{M}_2)a(1 - e^2)]^{1/2}$$

At the point of closest approach ($\theta = 0$), the relative velocity has no component in the radial direction, so it must be entirely in the θ-direction; therefore,

$$V(\theta = 0) = R \frac{d\theta}{dt}$$

When the above three equations are used to eliminate R and $(d\theta/dt)$, the relative velocity is found to be

$$V^2(\theta = 0) = \frac{G(\mathcal{M}_1 + \mathcal{M}_2)}{a} \frac{1 + e}{1 - e} \tag{25.22}$$

This gives the maximum relative velocity. Since the energy is constant, E should have the same value when $\theta = 0$ as at any time. If equation (25.22) is substituted into (25.21), and if the above value for $R(\theta = 0)$ is also used, then one finds

$$E = -\frac{G\mathcal{M}_1\mathcal{M}_2}{2a} \tag{25.23}$$

The total energy is negative, just as in the electrostatic case discussed in Section 3. Equations (25.21) and (25.22) together give, for the relative velocity at arbitrary time,

$$V^2 = G(\mathcal{M}_1 + \mathcal{M}_2)\left(\frac{2}{R} - \frac{1}{a}\right) \tag{25.24}$$

Equation (25.24) gives the relative velocity at any point in the orbit as a function of the distance between the bodies and the relevant orbital parameters. In the present context, this will be applied to a star moving in the gravitational field of the Galaxy. With $\mathcal{M}_1$ being the mass of a star and $\mathcal{M}_2 = \mathcal{M}$ being the mass of the Galaxy, it is apparent that the former can be neglected in the velocity relation. Also, if equation (25.18) is used to eliminate the mass of the Galaxy, then (25.24) can be written

$$V^2 = V_o^2 R_o\left(\frac{2}{R} - \frac{1}{a}\right) \tag{25.25}$$

Finally, this relation will be used primarily for stars which are observed in the solar neighborhood, for which R is essentially equal to R_o. For these objects, the velocity is

$$V^2 = V_o^2\left(2 - \frac{R_o}{a}\right) \tag{25.26}$$

The quantity V is the velocity of a star relative to the center of the Galaxy; i.e., it is the orbital velocity of the star. Since V_o and R_o are both fixed quantities, equation (25.26) indicates that the orbital velocity of any star observed in the solar neighborhood depends only on the semi-major axis of its orbit. Measurements of V then yield the size of the orbit.

Equation (25.26) indicates that V increases with the orbit size. This may seem strange at first, but it must be remembered that equation (25.26) gives the orbital velocity at a fixed point in space. The largest velocities for stars which come through the solar neighborhood must occur for those which have large orbits which extend far beyond the solar neighborhood, with $a \gg R_o$. Such orbits must have high eccentricities, or they would not extend as far in as the solar neighborhood. The maximum velocity is $\sqrt{2}\, V_o$, which is the escape velocity at the position of the Sun. The smallest velocity is zero, and it belongs to a star in the outer part of a very long, narrow orbit with $a = \frac{1}{2}R_o$. Any orbit smaller than this does not extend out from the center as far as the solar neighborhood. Figure 25.3 illustrates some of the orbits of stars observed in the solar neighborhood. Point O is the position of the Sun and C is the center of the Galaxy. The dashed curve is the orbit of the local standard of rest, which is assumed to be circular.

The peculiar velocity of a star is the difference between its orbital velocity and that of the local standard of rest. The local standard of rest is moving with speed V_o in the direction of longitude $l = 90°$. If all stars moved in the galactic plane and had the same direction of revolution, then the largest peculiar velocity in the direction of $l = 90°$ would be $V\,(\max) - V_o = 0.4\, V_o$. This would be for the stars of very large orbital velocity overtaking the Sun. The greatest peculiar velocity in the opposite direction, toward galactic longitude $l = 270°$, would be V_o. This would belong to stars of essentially zero orbital

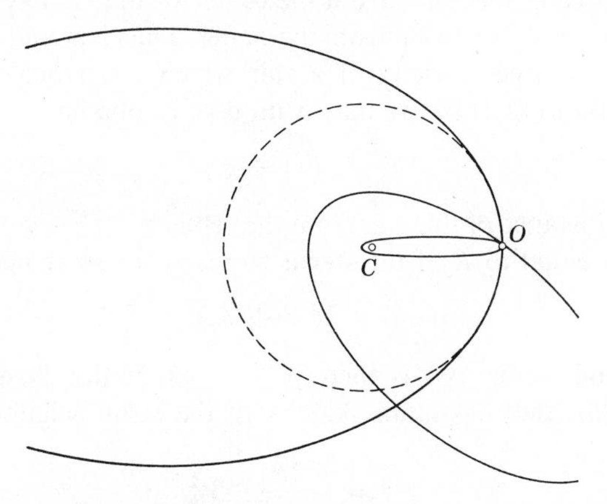

Figure 25.3 Different orbits of stars observed in the solar neighborhood

velocity being overtaken by the Sun. It is seen that a natural asymmetry in the motions occurs. All stars with peculiar velocities greater than $0.4V_o$ must be moving in directions other than toward $l = 90°$; otherwise, they would have orbital velocities greater than the velocity of escape of the Galaxy. Such stars would have long ago escaped from the Galaxy, unless some mechanism can be discovered for accelerating new stars up to these high speeds.

It is apparent that the galactic orbit of a star can be found from the direction and the magnitude of its peculiar velocity. All high-velocity stars, of course, have rather highly elliptical orbits. Observations show that these orbits are not confined to the galactic plane, but appreciable motions perpendicular to this plane occur. If the orbits can actually be retrograde, i.e., directed toward $l = 270°$ in the galactic plane, then peculiar velocities as large as $(1 + \sqrt{2})\, V_o$ could be observed in that direction. This would be the escape velocity in a retrograde orbit, and the asymmetry in the motions of the high-velocity stars would be even more pronounced.

The above description gives a good qualitative understanding of the high-velocity stars; however, the interpretation of the 60–65 km/sec limit in the peculiar velocities toward $l = 90°$ as a velocity of escape cutoff is probably incorrect. If this a velocity-of-escape effect, then $0.4V_0 = 60–65$, and $V_o = 150–160$ km/sec. This is probably too low for the circular velocity in the solar neighborhood, which, as equation (25.16) indicates, is probably closer to 250 km/sec. This discrepancy exists even when corrections for deviations from two-body orbits are made. The observed cutoff is more likely due to the finite size of the Galaxy.

Equation (25.26) shows that the largest orbital velocities observed in the solar neighborhood are those of stars with very large orbits, i.e., stars which are only temporarily in the solar neighborhood and which spend most of the time at much greater distances from the center of the Galaxy. Let R' be the maximum distance stars can get from the center. Then if a' and e' are the semi-major axis and the eccentricity of a star which can reach this maximum distance, equation (12.4) shows that, with $\theta = \pi$, one has

$$R' = a'(1 + e')$$

The minimum distance of this star from the center is $a'(1 - e')$, and this must be less than or equal to R_o if the star is to reach the solar neighborhood. It follows that

$$a' \leqslant \tfrac{1}{2}(R' + R_o)$$

The largest semi-major axis is then $\frac{1}{2}(R' + R_o)$. If this is substituted into equation (25.26), the maximum velocity in the solar neighborhood, V', is given by

$$V'^2 = \frac{2R'}{R' + R_o} V_o^2 \tag{25.27}$$

If V' can be observed, then this relation gives R', the maximum distance any star in the solar neighborhood can get from the galactic center.

The cutoff of peculiar velocities at about 63 km/sec toward $l = 90°$ indicates that no star in the solar neighborhood has an orbital velocity in that direction greater than about $(V_o + 63)$. Perhaps this is the maximum velocity for any direction. If so, then it is to be identified with V' of equation (25.27). With $V_o = 250$, one has $V' = 313$ km/sec, and (25.27) gives $R' = 3.6 \times R_0 = 36$ kpc.

An effective radius of the Galaxy is probably considerably less than 36,000 pc, but a few stars may reach this far out. This shows how the size of the Galaxy can be estimated from the motions of stars near the Sun, but the estimation is based only on stars with highly elliptical orbits. Stars with large circular orbits never come into the region of the solar neighborhood.

There are complications to the above analysis. For example, one is not certain that the local standard of rest does have exactly the circular velocity. Also, this standard is not a perfectly well-defined quantity, since the value of the solar motion depends on the kind of stars used to define it, even when the obvious high-velocity stars are excluded. There is evidence that many low-velocity stars may be in groups which have systematic motions that differ from the circular velocity, and the same may be true of the interstellar matter. The latter is particularly important because much of the knowledge of galactic structure is based on 21-cm data and the assumption that the neutral hydrogen moves in circular orbits.

Stellar Populations. The fact that many high-velocity stars have large motions normal to the galactic plane means that such stars are not strongly concentrated toward it. Their motions carry them far above and below the galactic plane. In this respect they are similar to the globular clusters. The globular clusters also have very large peculiar velocities, and these are directed toward the hemisphere of longitude 270°. The globular clusters move in highly elliptical orbits, and most of their peculiar motions are a reflection of the orbital velocity of the solar neighborhood.

The distributions of positions and orbits of high-velocity stars are similar to those of globular clusters, and the similarity does not end with this. The high-velocity stars are also found to be deficient in their content of heavy elements as compared with most of the nearby stars. This characteristic was interpreted in Section 22 as an indication of great age. The subdwarfs mentioned in Section 19 are distinguished by their low abundance of the heavy elements, and, as a group, they have large peculier velocities and a rather small concentration toward the galactic plane. It would appear that globular-cluster members, high-velocity stars, and subdwarfs are all members of a general group of stars that has significant differences from most of the stars in the solar neighborhood.

In the mid 1940s Walter Baade was able for the first time to resolve individual stars in the nuclear region of a spiral galaxy. He found that the stars in the nucleus were quite different as a group from those which populate the spiral arms. The latter type of stars he called Population I and the former Population II. By studying objects in the Galaxy and in other galaxies, astronomers found that the differences between the stellar populations included the points mentioned above, namely, type of galactic orbit, distribution about the galactic plane, and chemical composition. An additional point is that interstellar matter is strongly associated with Population I but not with Population II.

The assumption that there are only two populations is obviously artificial. There is probably a near continuum of characteristics covering the whole range. The finite resolving power of the observations causes objects with a range of properties to be lumped together into one category, as in the case of spectral types. Astronomers often use five populations instead of only two: intermediate and extreme for both Populations I and II, plus a Disc Population in the middle. All of the parameters described above vary rather smoothly from Extreme Population I through Extreme Population II. (The latter is sometimes called the Halo Population because such objects are common in the galactic halo.) It is often difficult to determine the classification of an individual star, for some Population II stars have nearly circular orbits, and Population I stars sometimes have large peculiar velocities.

It has long been realized that the Galaxy is a spiral system, but our position in it is such that the spiral arms cannot be directly observed. When Baade found that supergiants, O and B stars, and large concentrations of interstellar matter, all Extreme Population I objects, occur only in the arms of other spirals, W. W. Morgan and associates examined the distribution of such objects near the Sun. In this way part of the spiral structure of the Galaxy was first observed. The point is that the total density of stars does not appear to be much greater in the arms than between them, so the arms do not stand out when the total stellar population is examined. But the large groups of very early-type stars called O associations by V. A. Ambartsumian, and other Extreme Population I objects do delineate the arms. Shortly after Morgan's work, the 21-cm data became available, and this is the source of most of the present information on the spiral structure of the Galaxy. The Sun is near the inner edge of a spiral arm, and this is why most of the stars in the solar neighborhood are of Population I.

Evolution of the Galaxy. Stars are formed in the regions in which the interstellar matter has a large density. The halo stars of Extreme Population II form a system that is almost spherically symmetric about the center of the Galaxy. This suggests that at the time of their formation, perhaps some 1.0–1.5×10^{10} years ago, the interstellar gas was also in a spherical distribution and that it was considerably deficient in the heavy elements. The gas and dust apparently were subjected to a rapid collapse to the galactic plane. This is evidenced by the fact that the Disc Population is still very old, but it is well concentrated toward the plane. Today star formation appears to be confined to the spiral arms, and it is not known why the interstellar matter is now concentrated in the arms. Both the rotation of the Galaxy and the galactic magnetic field are probably important in determining the distribution of the interstellar matter.

It is pointed out in Section 22 that some of the material in the stars is later returned to the interstellar medium. The enrichment of the latter in heavy elements as time goes on is expected, and the variation of composition with stellar population is qualitatively understood. All of the stellar material cannot be returned, and so the gas and dust are slowly but surely being used up. Gas densities must have been much greater in the past than they are now, and so star formation and heavy-element enrichment must have proceeded at a greater rate long ago. The observed luminosity function, which gives the number of star per unit volume with given absolute magnitude, depends on the star formation and evolution rates. These rates are closely related to the density and composition of the interstellar matter, so the luminosity function contains clues concerning the evolutionary history of the Galaxy. This field has been investigated by many persons, but the uncertainties are still too great for any reliable conclusions other than those of a rather general nature.

An example similar to one given by E. E. Salpeter makes a good illustration. Let $\phi(L,t)$ be the luminosity function, so that $\phi(L,t)\,dL$ is the number of stars per unit volume having luminosities between L and $L + dL$ at any time t. The time $t = 0$ is taken as the beginning of star formation, so all stars have ages less than or equal to t. Let $\psi(L,t)\,dL\,dt$ be the number of stars per unit volume with luminosity in the range dL which are formed out of the interstellar medium between t and $t + dt$. This is the star-formation rate, and it plus the evolution rate determine the luminosity function. It will be assumed for simplicity that a star of luminosity L spends a time interval $T(L)$ on the main sequence, and then it stops shining. Then it is apparent that

$$\begin{aligned} \phi(L,t) &= \int_0^t \psi(L,t')\,dt' \qquad &\text{if } T(L) > t \\ \phi(L,t) &= \int_{t-T}^t \psi(L,t')\,dt' \qquad &\text{if } T(L) < t \end{aligned} \tag{25.28}$$

If the star formation rate ψ is assumed to be independent of the time (probably a poor approximation), then one has

$$\begin{aligned} \phi(L,t) &= t\psi(L) \qquad & T(L) > t \\ \phi(L,t) &= T(L)\psi(L) \qquad & T(L) < t \end{aligned} \tag{25.29}$$

Theories of stellar evolution allow one to estimate $T(L)$, and it is apparent from equations (25.29) that $(d\phi/dL)$ should be discontinuous at the luminosity for which $T(L) = t$. Thus the observed luminosity function yields the star-formation rate and the age of the Galaxy, t.

Let ρ be the density of interstellar matter. Then ρ is being decreased by the formation of new stars, and it is being increased by the shedding of material by old stars which have essentially completed their evolution. If L_o is the luminosity of a star whose lifetime is just equal to t, then only stars of luminosity greater than L_o have had enough time to complete their evolution and help to increase ρ. If α is the mass which a star ejects back to the interstellar medium, then the conservation of mass in the interstellar medium indicates that

$$\frac{d\rho}{dt} = -\int_0^\infty \mathcal{M}(L)\psi(L)\, dL + \int_{L_o}^\infty \alpha(L)\psi(L)\, dL \tag{25.30}$$

where $\mathcal{M}(L)$ is the mass of a star of luminosity L. Knowledge of the star-formation rate and of mass loss by stars permits one to determine the rate at which the density of interstellar matter is changing. One could write equation (25.30) separately for hydrogen, for helium, and for the heavy elements, so one could study the rate of enrichment in heavy elements of the interstellar matter and the composition of stars as a function of their age of formation.

Although the present example is grossly oversimplified, it does illustrate how the observed luminosity function can be combined with theories of stellar evolution to obtain information about the evolution of the stellar and the interstellar content of the Galaxy.

26. Galaxies

Classification of Galaxies. There are three general types of galaxies: spirals, ellipticals, and irregulars. Spirals have a cluster of stars in the center known as the nucleus, and spiral arms are prominent in the outer regions. The arms are usually attached to opposite sides of the nucleus, but sometimes they are attached to opposite ends of a bar that passes through the nucleus. The latter are known as barred spirals. All spirals are very flat systems, and large amounts of absorbing dust are often apparent in them.

Elliptical galaxies show little or no structure of any kind. As the name suggests, the shape is elliptical. They occur with varying degrees of flatness, and little interstellar matter is apparent. Irregular galaxies include those which do not fit into the other classes. They have no obvious symmetry in their appearance, and they have large amounts of interstellar matter.

When one is studying a new type of object, it is desirable to classify the objects according to some easily observed property. It may be advantageous to modify the classification at a later time in order to give it a better physical

Figure 26.1 NGC 205, a small elliptical galaxy which is a companion of M 31. (Mount Wilson and Palomar Observatories)

basis, but a classification that is based primarily on observations, rather than interpretations of observations, will not need very much changing as new physical theories develop. Such a classification for galaxies was devised by E. P. Hubble many years ago. A tremendous amount of information has been collected since then, but Hubble's classification is still probably the most common one in use.

Hubble classified the ellipticals according to their observed degree of flatness. They are designated by the letter E followed by a number from 0–7. The E0 galaxies are essentially circular in appearance, while E7 galaxies are the most flat of the ellipticals.

Appearing at an intermediate point between the ellipticals and the spirals are the S0 galaxies. These are flatter than E7 galaxies and show little or no evidence of spiral structure. The spirals are placed in a sequence along which the nuclear regions become less conspicuous and the arms become more prominent. (Actually there are two parallel sequences, one for normal spirals and one for barred spirals.) The designations Sa, Sb, and Sc (or SBa, SBb, and SBc for barred spirals) represent spirals at different points along this sequence. Our Galaxy is a normal Sb spiral.

The irregular galaxies are usually placed beyond the Sc spirals, although they are sometimes considered as separate from the rather smooth sequence running from E0 through Sc. Hubble emphasized that this classification is based only on appearance, and a scheme of evolution is not implied.

Figure 26.2 Galaxy M 31, an Sb spiral. (Mount Wilson and Palomar Observatories)

True Shape of the Ellipticals. The apparent shape of an elliptical galaxy depends on its true shape and its orientation in the plane of the sky. Since the orientation is not known, the true shape of any galaxy is not known; however, by assuming that the orientations are at random, one can determine something about the relative probabilities of various true shapes.

Figure 26.3 shows an elliptical galaxy which is assumed to have the true shape of an oblate spheroid, a figure obtained by rotating an ellipse about its small axis. The figure is in the plane defined by the minor axis and the direction to the Earth. Let a, b be the true semi-major and semi-minor axes of the galaxy, while a, b' are the apparent axes. (The major axis of an oblate spheroid is seen to be independent of its orientation, so $a' = a$.) The line PQ is directed toward Earth and is tangent to the galaxy, so the perpendicular distance from this line to the center is the apparent semi-minor axis b'. The true and apparent ellipticities are defined by

$$q = \frac{b}{a} \qquad q' = \frac{b'}{a} \tag{26.1}$$

If θ is the angle between the direction to Earth and the minor axis, one can show from the figure that

$$\cos^2 \theta = \frac{q'^2 - q^2}{1 - q^2} \tag{26.2}$$

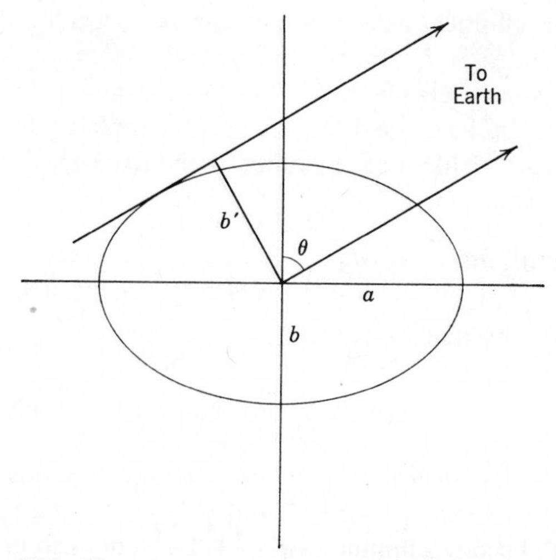

Figure 26.3 The true and apparent shapes of elliptical galaxies.

(See problem 3 at the end of this chapter.) Equation (26.2) gives the relation between the orientation of the galaxy and the true and apparent ellipticities. Not all values of θ are equally probable. If the orientations are random, then the probability that the angle lies in the range $d\theta$ is proportional to the solid angle extended by that range. According to equation (1.2), this probability, normalized to unity for the range in θ from 0 to $\pi/2$, is simply $\sin\theta\, d\theta$.

Let $P(x)\,dx$ represent the probability that some quantity x lies in the range x to $x + dx$. Further, let $P(q,q')\,dq\,dq'$ be the probability that q lies in the range dq and that q' lies in the range dq'. Then from above $P(\theta)\,d\theta = \sin\theta\,d\theta$. Now, the probability that both q and q' are in a given pair of ranges is the same as the probability that q is in the desired range and that the angle θ is such that true ellipticity q looks like q', that is, that angle θ satisfies (26.2). Then

$$P(q,q')\,dq\,dq' = P(q)\,dq\,\sin\theta\,d\theta$$

If (26.2) is used to eliminate θ, the result is

$$P(q,q')\,dq\,dq' = \frac{q'P(q)\,dq\,dq'}{\sqrt{(1-q^2)(q'^2-q^2)}} \tag{26.3}$$

A galaxy cannot appear more flat than it actually is, so q' is never less than q. The probability of any q' is then obtained by integrating (26.3) over all q values which can lead to that q':

$$P(q')\,dq' = q'\,dq'\int_0^{q'} \frac{P(q)\,dq}{\sqrt{(1-q^2)(q'^2-q^2)}} \tag{26.4}$$

The index of an elliptical galaxy is the nearest integer to $10(1 - q')$. The highest observed index is 7, and for this q' is the in range 0.25–0.35. Thus $q' \geqslant 0.25$, and this must also be the lower limit for q.

If true ellipticities in the range 0.25–1.00 occur completely in a random way, i.e., if one value is just as likely as any other, then $P(q) = \frac{4}{3}$. With this assumption equation (26.4) becomes

$$P(q')\,dq' = \tfrac{4}{3}q'\,dq' \int_{0.25}^{q'} \frac{dq}{\sqrt{(1 - q^2)(q'^2 - q^2)}} \tag{26.5}$$

The probability of any index n is then

$$P(n) = \int_{q_L}^{q_U} P(q')\,dq' \tag{26.6}$$

where the above limits cover the range in q' that corresponds to the given n. For an E3 galaxy, for example, $n = 3$ and q' is within the range 0.65–0.75. The quantity $P(q')$ becomes infinite as $q' \to 1$, but it does so in such a manner that the integral of $P(q')$ over any finite range remains finite, as it should.

Figure 26.4 is a plot of the probability of any index vs. the index. The x's are the calculations from equations (26.5) and (26.6), while the o's are the original observations of Hubble. The drop in the computed curve at E0 is due to the fact that the range in q' values is only one-half of that leading to the other index values, i.e., for an E0 galaxy q' is between 0.95 and 1.00. The differences

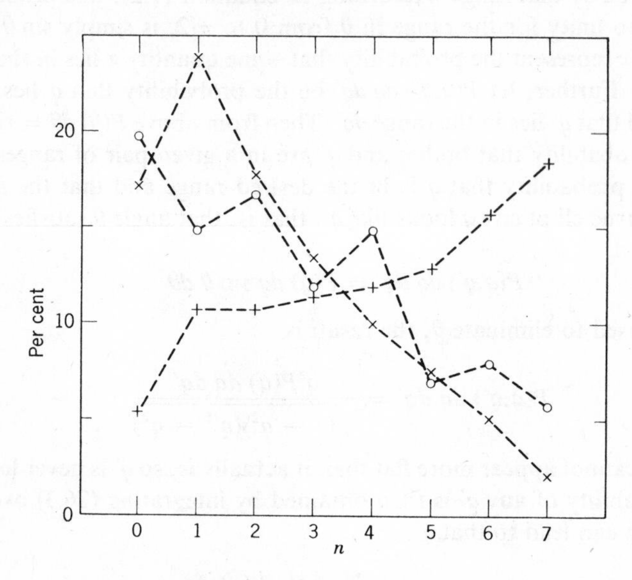

Figure 26.4 Shapes of ellipticals. o-observed; ×-calculated from random distribution +-calculated from all true shapes being E7.

between the two curves are due to the true ellipticities being distributed in a nonrandom fashion, to observational error, and to fluctuations in the statistics. Highly flattened systems appear more abundant than predicted, and more recent data tend to confirm this; however, the differences are probably not great enough for the assumption of random true shapes to be a very bad one. Thus, although the true shape of any galaxy is uncertain because of the unknown orientation, the probability that it lies in any given range is fairly well known.

For purposes of illustration, the +'s in Figure 26.4 show what the observed distribution would be if all elliptical galaxies had the true shape of an E7. This is so different from the observed curve that it is apparent that most ellipticals must have true shapes which are much more nearly spherical than is true for E7's.

Distribution in Direction. The apparent distribution of galaxies with direction is far from uniform. Very few are seen in or near the galactic plane. This is due to the fact that the obscuring dust in the Galaxy is heavily concentrated toward the galactic plane, so the Galaxy is optically thick in this plane in visible light. One must observe away from this plane in order to find clear enough regions for external galaxies to be readily seen. This indicates that one must show care in the interpretation of observed irregularities in the distribution of galaxies: these irregularities may be due to nonuniformity in the true distribution of galaxies, or they may be due to the patchy distribution of dust in the Galaxy. There are a number of ways in which the two cases can be distinguished, including examining the color excesses of nearby stars in the same direction. To be safe, one should observe only at relatively high galactic latitudes.

The true distribution of galaxies with direction is uniform or isotropic on a large scale, but it is irregular on the small scale. This is due to the clustering tendency of galaxies. It appears that most if not all galaxies are members of clusters of galaxies.

G. O. Abell divides clusters of galaxies into two general types, regular and irregular. Regular clusters are almost spherically symmetric and contain possibly many thousands of members. The irregular clusters show little symmetry, and they can be large or small. Our Galaxy is a member of a very small irregular cluster known as the Local Group. There are 17 known members of the Local Group plus a few other possible members. The irregular clusters contain all types of galaxies, but the regular clusters seem to contain only E and S0 galaxies. The galaxies in a given cluster are not distributed at random, but they have a strong tendency to occur together in subgroups within the cluster.

There is evidence that the clusters themselves are not uniformly distributed, but that they tend to lump together into clusters of clusters of galaxies,

Figure 26.5 Galaxy NGC 1300, a barred spiral of type SBb. (Mount Wilson and Palomar Observatories)

although this evidence is not accepted by all astronomers. Evidence is lacking about yet higher orders of clusterings.

Distances. In order to determine much about galaxies one must be able to find their distances. As may be expected, distances of galaxies are found by comparing the observed and the intrinsic properties of objects they contain. The two main problems are finding such objects to observe and knowing their intrinsic properties. The tremendously large distances between galaxies make all except the most luminous objects useless as distance indicators. The Cepheid variable stars are quite valuable in this respect.

Cepheids are known to have absolute magnitudes which are correlated with their periods. If the period-luminosity relation has been properly calibrated, one can find the absolute magnitude of any Cepheid by measuring its period. The usual comparison of apparent and absolute magnitudes then yields the distance, at least if the correction for interstellar absorption is known. In addition to interstellar absorption, there are three important sources of uncertainty in the application of this method: errors in the measured calibration, intrinsic scatter in the period-luminosity relation, and possible systematic differences between Cepheids in different galaxies.

The first point, mentioned in Section 16, is still of some concern. The second point has been largely alleviated by means of color measurements. There is a scatter of perhaps one magnitude for Cepheids of the same period; however, the brighter ones of a given period are bluer in color than the fainter ones, so observations of both period and color reduce the scatter considerably. The third point can be caused by differences in chemical composition between galaxies. It is suspected of being important in certain cases, and there is evidence that Cepheids in different regions of our Galaxy can have systematically different intrinsic properties, presumably for the same reason.

Unfortunately, Cepheids are observed in only a few galaxies, partly because not all galaxies contain them and partly because most galaxies are too far away. It was pointed out in Section 16 that classical Cepheids are Population I objects, but elliptical galaxies appear to contain mostly or entirely Population II objects. Also, while Cepheids are very luminous, they are not luminous enough to be visible in galaxies beyond the Local Group. The Cepheids are still quite valuable, however, for they are used to calibrate other methods of estimating the distances of the more remote galaxies.

The brightest stars in spiral or irregular galaxies have absolute visual magnitudes in the neighborhood of $M_v = -9$. This is much brighter than the Cepheids, so they can be seen in much more distant galaxies. Bright novae appear to have about the same absolute magnitude at maximum; supernovae are very much brighter even than this, but not enough is known about their luminosities for them to be very important distance indicators, and their occurrence is quite rare and unpredictable. The apparent sizes of H II regions and the apparent sizes and integrated magnitudes of globular clusters are also used to estimate the distances of galaxies which contain them.

As one considers galaxies which are more and more remote, one is eventually limited to distance estimates which are based on the apparent size and brightness of whole galaxies. The accuracy of this is not as bad as the wide range of sizes and luminosities of galaxies might suggest. By considering only the several brightest galaxies in the large clusters, one probably has objects within a rather narrow range of intrinsic properties. By assuming that this is true, and there is very strong evidence that it is, the relative distances of many clusters of galaxies are obtained, and these contain a very large number of individual galaxies. These relative distances can be converted into absolute distances only if the luminosities of these brightest galaxies are known from some independent means, such as a calibration by one of the methods mentioned above. Unfortunately, the nearest of the large clusters, the Virgo cluster, is at about the maximum limit of the more direct distance determinations, as it has a distance of about 10^7 pc = 10 Mpc (megaparsecs). Thus the relative distances of the large clusters of galaxies are known more accurately than their absolute distances.

The Nearer Galaxies. The nearest galaxies are two irregular systems known as the Large Magellanic Cloud and the Small Magellanic Cloud. The fact

Figure 26.6 The Small (top) and Large Magellanic Clouds, nearby irregular galaxies. (Mount Stromlo Observatory)

that they are very close, plus the fact that they contain most if not all of the different types of objects that other galaxies contain, combine to make the Magellanic Clouds extremely valuable to astronomers. Nowhere else can such a rich selection of stars, star clusters, and gaseous nebulae be examined in detail side by side. The clouds have a distance of about 50,000 pc, and this is a distance modulus of about $(m - M) = 18.6$. Thus reasonably faint objects can be studied. It is unfortunate that the clouds cannot be observed from the Northern Hemisphere, where most of the large telescopes are located.

The two nearest spirals are M31 (the Andromeda galaxy) and M33. They are about 7×10^5 pc away, and this distance has a modulus of $(m - M) = 24.2$. If accurate work with the 200-in. telescope can be carried to apparent magnitude $m = 22$, then only objects brighter than absolute magnitude $M = -2.2$ can be studied, and interstellar absorption makes this worse by a few tenths of a magnitude. Thus, while M31 is a spiral that is nearly identical to the Galaxy, it is less useful than the Magellanic Clouds for studying stellar populations.

The Velocity-Distance Relation. V. M. Slipher was the first astronomer to notice the very large radial velocities typical of galaxies. Later Hubble and M. L. Humason had collected enough data to indicate that the radial velocity of a galaxy is closely related to its distance. The observations indicate that the radial velocity V_r is given by

$$V_r = Hr + R \tag{26.7}$$

where r is the distance of a galaxy, H is a constant known as the Hubble constant, and R is a random term which may be 100 km/sec or even larger. The difficulties in measuring the distances of the galaxies cause a large uncertainty in the value of H. The nearer galaxies have the best known distances, but for them the unknown random term R is too large a contributor to the radial velocity. (Measured velocities must be corrected for the Sun's orbital velocity in the Galaxy.) The random term is not important for the more distant galaxies, but there is a much larger uncertainty in r. Published values of H are usually around 100 km sec^{-1} Mpc^{-1} $= 0.3 \times 10^{-17}$ sec^{-1}, with an uncertainty of perhaps 25%.

Equation (26.7) is often known as Hubble's law. Since the radial velocities increase (greater velocity away) with greater distance, the spectral lines are shifted more and more to longer wavelengths or toward the red end of the visible spectrum. For this reason, equation (26.7) is also known as the red shift relation. This red-shift relation is well-enough established that it is the main source of information on distances of galaxies for which the previously mentioned methods fail. For example, suppose that a single galaxy is observed to have a red shift corresponding to a radial velocity of 3000 km/sec. The above mentioned value of H then suggests that this galaxy has a distance of

Figure 26.7 Galaxy NGC 5194, an Sc spiral with its companion NGC 5195. (Mount Wilson and Palomar Observatories)

about 30 Mpc. It should be emphasized that this is not an independent method of measuring distance, for it has to be calibrated by the methods discussed previously. It obviously does not work for the very nearest galaxies (in the Local Group), and there are problems in applying it to the extremely distant galaxies because of ambiguities in the definition of distance. The latter point is discussed in Section 27.

Masses of Galaxies. The mass of a galaxy can be found if the gravitational attraction of it on any object can be measured. The most accurate mass determinations are for those galaxies for which a rotation curve can be measured. This is a plot of the rotation velocity (usually assumed to be for circular orbits) as a function of the distance from the center of the galaxy. The rotation curve depends on the distribution of mass in the galaxy, and the data are analyzed in the same way as is described for the Galaxy in Section 25.

The rotation curve of a galaxy is obtained by measuring the radial velocity of many different parts of the galaxy. The measured velocity is the average over all emitting regions in the line of sight, and only spirals are thin enough for this to refer very effectively to one localized region. Also, interstellar emission lines are the best for this work, since the lines are quite narrow and easy to measure, and ellipticals do not usually show strong interstellar lines over large regions. Irregular galaxies often do not show strong evidence of rotation, so most of good rotation curves are for spirals.

In order to obtain the rotation velocity, the radial velocity must be corrected for the velocity of the galaxy as a whole, and the result must then be adjusted for the appropriate projection factor. The latter is obtained from the observed shape of the spiral on the assumption that it would be reasonably circular if seen from straight above.

There are problems in analyzing the rotation curves to obta masses. For example, the distance of the galaxy must be known before the correct linear scale can be found. There are cases in which large deviations from circular motion must occur. Even if the orbits are circular, the rotation velocity of a point is most sensitive to the amount of mass within that point, and there may be a considerable amount of mass outside the last measured point which goes essentially undetected. In this case the mass found from the rotation curve would only be a lower limit.

It is instructive to consider a simple example. Suppose that the matter in a galaxy is distributed in a spherically symmetric fashion. This is a very poor approximation to a spiral galaxy, but it can be handled easily and it illustrates the main points. Let $\mathcal{M}(r)$ be the mass within the sphere of radius r and $\bar{\rho}(r)$ be the average density of the matter in this sphere. Then if r_o is the radius of the galaxy, one has

$$\begin{aligned} \mathcal{M}(r) &= \tfrac{4}{3}\pi r^3 \bar{\rho}(r) && r \leqslant r_o \\ \mathcal{M}(r) &= \tfrac{4}{3}\pi r_0^3 \bar{\rho}(r_0) = \mathcal{M} && r > r_o \end{aligned} \tag{26.8}$$

where $\mathcal{M}$ is the total mass of the galaxy. If a star of mass m is at a distance r from the center of the galaxy, the gravitational force on it due to the entire galaxy is

$$F = G\frac{m\mathcal{M}(r)}{r^2}$$

Using (26.8), one finds

$$\begin{aligned} F &= \tfrac{4}{3}\pi G m \bar{\rho}(r) r && r \leqslant r_o \\ F &= G\frac{m\mathcal{M}}{r^2} && r > r_o \end{aligned} \tag{26.9}$$

The force on a mass increases with distance from the center until the edge of the galaxy is reached, and then it falls off as r^{-2}. Note that $\bar{\rho}(r)$ is not the density at r, but it is the average density of all points within radius r.

If it is now assumed that all orbits are circular, then the centrifugal force on mass m is (mV^2/r). Equating this to the expressions in (26.9), one finds

$$\begin{aligned} V^2 &= \tfrac{4}{3}\pi G \bar{\rho}(r) r^2 = \frac{G\mathcal{M}(r)}{r} && r \leqslant r_o \\ V^2 &= \frac{G\mathcal{M}}{r} && r > r_o \end{aligned} \tag{26.10}$$

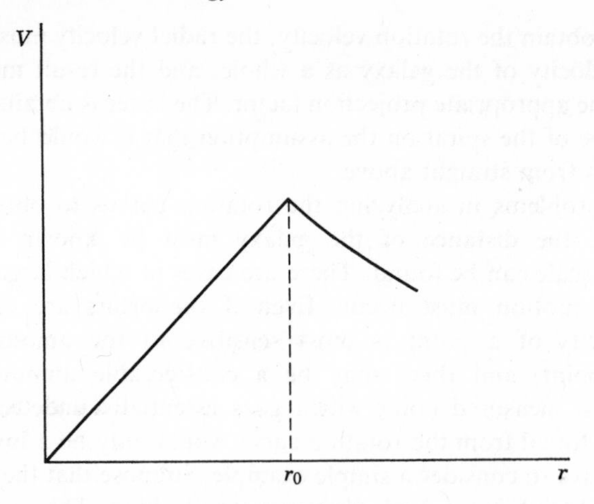

Figure 26.8 A schematic rotation curve.

Figure (26.8) shows the rotation curve of equation (26.10) for $\bar{\rho}(r)$ = constant. The velocity increases linearly up to the edge of the galaxy, and it falls off as $r^{-1/2}$ beyond. If the observed curve does not agree with this, then one can adjust the value of $\bar{\rho}(r)$ until there is agreement. In practice, one generally assumes that the galaxies have ellipsoidal rather than spherical distributions of mass, and the corresponding expressions (26.9) and (26.10) are more complicated; however, the procedure is much like that outlined above, and the observed rotation curves do appear rather similar to Figure 26.8. The velocity must eventually fall off as $r^{-1/2}$ for large enough values of r; when the observed curve does not do this, the indication is that there is considerable mass out beyond the farthest measured point, and the mass determination is only a lower limit.

Many galaxies are observed in pairs, and this occurs much too frequently for them to be separate objects seen by chance in the same line of sight. They must be gravitational pairs. Although orbital motions cannot be observed, radial velocity measurements do give statistical information on the masses.

Suppose that two galaxies of combined mass $\mathcal{M}$ are moving in circular orbits around each other with relative velocity V and with a separation of a. If they can be treated as mass points, then equation (26.10) indicates that

$$V^2 = \frac{G\mathcal{M}}{a} \tag{26.11}$$

V cannot be directly measured, but the difference in radial velocity ΔV_r of the two galaxies is the component of V in the line of sight. The observed separation r is the component of a normal to the line of sight. In Figure 26.9 one

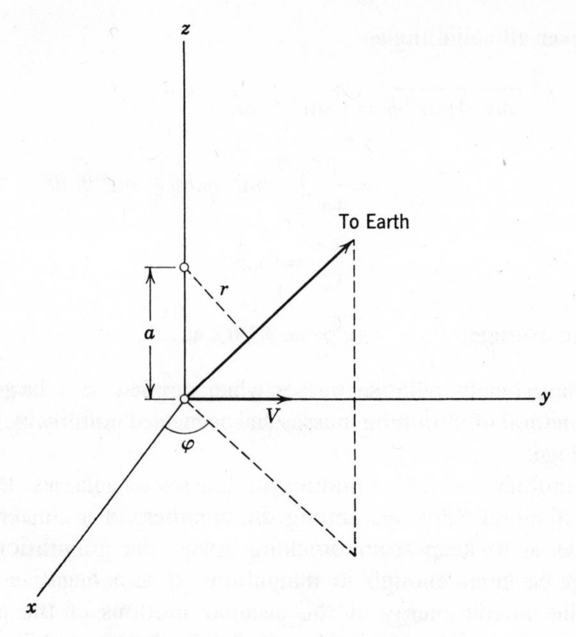

Figure 26.9 Geometry of a galaxy pair.

galaxy is at the origin and the other one is distance a above it on the z axis. The orbital plane is the yz plane, and the relative velocity V is taken along the y axis as shown. If (θ, ϕ) are the spherical angles defining the direction toward the Earth, then it is apparent that

$$\begin{aligned} r &= a \sin \theta \\ \Delta V_r &= V \sin \theta \sin \phi \end{aligned} \tag{26.12}$$

From equations (26.11) and (26.12), one has

$$\Delta V_r^2 \times r = G\mathcal{M} \sin^3 \theta \sin^2 \phi \tag{26.13}$$

where ΔV_r is observed and r can be found from the angular separation and the distance, if the latter is known. In most cases the distance must be found from the red-shift relation, (26.7). The orientation of the orbit with respect to the line of sight is not known, so the angles θ and ϕ are unknown. One can determine the average value of the $(\sin^3 \theta \sin^2 \theta)$ factor under the assumption that the orbits are random, and this can give galactic masses of statistical significance.

As usual, the probability that these angles fall in a given range is proportional to the solid angle of that range, so the average value is the function integrated over all solid angles:

$$\overline{\sin^3\theta \sin^2\phi} = \int \sin^3\theta \sin^2\phi \frac{d\omega}{4\pi}$$

$$= \frac{1}{4\pi}\int_0^{2\pi} \sin^2\phi \, d\phi \int_0^{\pi} \sin^4\theta \, d\theta$$

$$= \frac{3\pi}{32} = 0.29$$

Thus, on the average,

$$\overline{\Delta V_r^2 r} = 0.29G\mathcal{M} \tag{26.14}$$

This gives statistically reliable masses when applied to a large number of pairs. This method of obtaining masses has been used notably by E. Holmberg and by T. Page.

There is another way of estimating the masses of galaxies. It is based on the spread of radial velocities among the members of a cluster of galaxies. If the cluster is to keep from breaking apart, the gravitational potential energy must be great enough in magnitude (it is a negative quantity) to overcome the kinetic energy of the peculiar motions of the galaxies. The kinetic energy depends on the masses and the velocities, while the potential energy depends on the masses and the relative positions of the galaxies. An analysis of the radial velocities and positions of the members of a cluster can give a minimum value of the cluster mass if the cluster is assumed to be stable. More specifically, if the size of the cluster is not changing too rapidly, i.e., if it is not expanding or contracting, then the virial theorem holds for it. This theorem states that the kinetic energy is one-half of the magnitude of the potential energy if these quantities are suitably averaged with time. This is analogous to equation (21.22), which relates the thermal (or kinetic) energy of the gas in a star to the potential energy of gravity. If the virial theorem is satisfied for a cluster, then the observed radial velocities and positions of the members can be used to find the cluster mass and the average mass per visible galaxy.

The method of using rotation curves and the method of analyzing pairs of galaxies are in very good agreement with each other concerning the general run of galaxy masses. With few exceptions, the third method disagrees from the first two by very large amounts. If the virial theorem is assumed to be valid for a given cluster, or even if the cluster is only assumed to be stable, a less restrictive assumption, then the cluster mass usually is found to be so great that the average mass per galaxy is an order of magnitude or more greater than that found for similar types of galaxies by the other methods. This result is found for all types of clusters, regular and irregular, large and small. Even the Local Group has this characteristic.

There are many possible explanations for the above discrepancy, but none is completely satisfactory. Perhaps cluster galaxies are more massive than nonmembers of clusters. It is possible that some clusters may contain enough of the giant elliptical galaxies, systems far larger and more massive than the ordinary ellipticals or spirals nearby, to account for the extra mass, but this cannot be the answer in most cases. The most popular explanation is that clusters contain large amounts of unseen material in the form of intergalactic material and dwarf galaxies too faint to be seen from very far away. This would make the clusters more massive without increasing the masses of the galaxies actually observed. In most cases this unseen material would have to be much more abundant than the material that is visible, and there is no evidence to support this. J. A. Koehler and B. J. Robinson have detected the 21-cm line in absorption in sources which are in and beyond the Virgo cluster. They conclude that the amount of intergalactic hydrogen indicated by this is too small to resolve this discrepancy.

It has been suggested by V. A. Ambartsumian and others that some of the clusters actually are breaking up; if they are not stable, the virial theorem does not apply. The problem here is that the time needed for this breakup is often quite short, 10^9 years or less, while the galaxies involved are almost certainly older than this. The time needed for breakup is defined here as the size of the cluster divided by a typical velocity of a member. With breakup times this short, most galaxies would have by now escaped completely from any cluster in which they had been associated in the past; however, the observations clearly indicate that most if not all galaxies are members of clusters.

It is pointed out later in this section that evidence exists that galaxies eject matter, and it has been suggested that masses of galactic order of size can be ejected with high energy from the nuclei of certain giant galaxies. If some members of a cluster have been so ejected from another member, the cluster could be unstable even though the stars are all older than the cluster breakup time. This is a highly controversial idea, but it is considered by some persons to be in better accord with observations than any alternative which has been proposed.

Magnitude and Color Measurements. It has been pointed out that measures of distances and masses of galaxies involve large uncertainties. In contrast to the case for individual stars, the same is true of measures of apparent magnitudes and colors of galaxies. Galaxies show extended surface areas, and it is difficult to calibrate accurately the apparent magnitude of an extended source. Galaxies do not have sharp edges, but they gradually fade away; thus there may be an appreciable amount of luminous matter far beyond the easily seen inner regions, and the outer parts may have too low a surface brightness to be measured. This makes it very difficult to compare results obtained under different conditions, such as with different exposure times or

Figure 26.10 Galaxy M87, a giant elliptical galaxy in the Virgo cluster and a strong radio emitter. (Mount Wilson and Palomar Observatories)

with different equipment. Different parts of a galaxy have different surface brightnesses and colors, so the orientation of the galaxy in the line of sight has a significant effect on the measures. The distribution of obscuring dust in the galaxy can compound this effect.

In addition to the above effects, the magnitudes and colors of galaxies have to be corrected for the red shift. The entire radiation field of a distant galaxy is measured at longer wavelengths than those in which it is emitted. This is a particularly treacherous effect because it is directly related to distance through equation (26.7). Errors in correcting for this can mask other distance effects one is looking for, and they can produce apparent effects that do not actually exist.

Stellar Content. Baade's original suggestion of the two stellar populations was based on his observations of the differences between the brightest stars in the arms and in the nuclei of spiral galaxies. The arms contain massive supergiants and very hot, luminous stars associated with regions of dense interstellar matter. The brightest stars in the nuclear regions are the much fainter red giants, and little interstellar matter is evident. The former were called Population I and the latter Population II. Within our Galaxy, young galactic clusters and the large O associations were noticed to have the main characteristics of Population I, and the globular clusters seemed typical of Population II. Since it is easier to study the nearby objects in the Galaxy, these objects became the prototypes of the stellar populations. It was expected that ellipticals and the nuclei of spirals would be essentially the same as the globular clusters in the types of stars they contain, and the spiral arms and the irregular galaxies appeared identical to the very young galactic clusters.

It is now believed that all galaxies, including ellipticals and irregulars, contain a mixture of stellar populations. Most if not all galaxies show spectral lines indicating a much higher heavy-element abundance than is found in some of the extreme Population II globular clusters. Elliptical galaxies probably contain an appreciable amount of the Disc Population, and the Magellanic Clouds, which are irregular galaxies, contain RR Lyrae variables globular clusters, and other evidences of having a Population II component.

An accurate picture of the types of stars in galaxies cannot be found directly, for only the brightest stars, if any, can be studied individually. Instead, this must be inferred from the integrated properties of the galaxies. Colors, mass-luminosity ratios, and integrated spectral types are the sources of most of this information.

The colors of galaxies change rather smoothly with the Hubble type. The irregular galaxies are the most blue, and they become progressively redder toward Sc, Sb, Sa, S0, and the ellipticals. The spread is quite large, however, and there is a large overlapping for the different types. The three magnitudes (U,B,V) provide only a crude measure of the light properties of a galaxy. The measurements at a large number of wavelengths provide the basis for a much more complete study of the stellar content.

Stellar atmosphere theory can predict the flux distribution for the different types of stars. The problem is to find what kinds of stars can be combined in what proportions in order to synthesize the observed flux distribution of a galaxy. Stars of different temperatures make different contributions to the various spectral regions, so the observations should cover as wide a wavelength range as possible. The solution of this problem is not unique, but it can be quite informative. As an example, the near-infrared region of a galaxy may have a wavelength distribution similar to that of a K star of effective temperature around 4000°K, while the blue region may be similar to that of an F star of about 7000°K. The relative prominence of the blue and near-infrared regions can then indicate something about the relative proportions of these types of stars.

The same type of analysis is possible with the line spectra of galaxies, and it is probably more accurate. Certain lines are much more sensitive to temperature and luminosity than the continua, and the appearance of these lines in the various spectral regions of a galaxy can give a good indication of the main types of stars that are contributing to the light.

As expected, the main contributors to the long-wavelength regions are cooler than the stars which are responsible for the short-wavelength radiation of galaxies. While O and B star characteristics are sometimes observed in the

Figure 26.11 Galaxy NGC 4565, an Sb spiral seen edge-on. (Mount Wilson and Palomar Observatories)

blue region of galaxies which have a strong Population I component, K and M star characteristics dominate the red region. In the ellipticals and the nuclei of spirals, G and K giants dominate the shorter wavelengths and K and M dwarfs are important at longer wavelengths. The more massive galaxies seem to have a large proportion of cool dwarfs than the less massive ones. The stars do not show the strong deficiency in the heavy elements that is characteristic of Extreme Population II. This does not mean that this population is absent, only that it is not a major contributor to the light output.

The above types of analyses measure only those stars which contribute appreciably to the total radiation output of the galaxies. The intrinsically faint stars are not counted unless they have an extremely high abundance, and the interstellar matter may go largely unnoticed too. The mass-luminosity ratio, however, depends on the properties of all of the matter, both luminous and nonluminous, and so in some respects it is a better measure of the overall properties of galaxies.

Rather than the total luminosity, which is often difficult to obtain, the measures are usually given in terms of the luminosity in the blue region of the spectrum L_B. It is found that the ratio $(\mathcal{M}/L_B)$ varies quite smoothly along the Hubble sequence of galaxies. If these quantities are measured in solar units (so that $\mathcal{M} = L_B = 1$ for the Sun), then $(\mathcal{M}/L_B)$ is around 2 to 10 for irregulars and spirals, while its range is approximately between 20 and 100 for

TABLE 26.1. THE MASS-LUMINOSITY RATIO FOR DIFFERENT STARS

Star	$\mathcal{M}/L_B$	Star	$\mathcal{M}/L_B$
B0 V	0.002	M0 V	50
A0 V	0.05	M5 V	500
F0 V	0.2	K0 III	0.2
G0 V	0.8	White dwarf	100
K0 V	3		

ellipticals. These values are based on the mass determinations by the double-galaxy and by the rotation-curve methods. While the mass-luminosity ratio is rather uncertain for any individual galaxy, the trend of its increasing from irregulars through ellipticals is real.

Table 26.1 illustrates the approximate mass-luminosity ratios for different types of stars. These numbers show that all galaxies must contain a large number of faint stars with large ($\mathcal{M}/L_B$) values. To illustrate these data, suppose that all stars in a hypothetical galaxy were M0 V stars having $\mathcal{M} = 0.5$, $L_B = 0.01$ in solar units. This gives the mass-luminosity ratio of 50 as indicated in the table. (This value is typical for ellipticals, but one should not conclude from this that ellipticals are composed of entirely M0 dwarfs.)

Now suppose that a few B0 V stars are added to the galaxy. How many of these, each having $\mathcal{M} = 17$, $L_B = 6 \times 10^3$, are needed to bring the mass-luminosity ratio down to 5, which is typical for spirals and irregulars? Let N be the number of M stars needed for each B star. Then

$$\frac{\mathcal{M}}{L_B} = \frac{17 + 0.5N}{6 \times 10^3 + 10^{-2}N} = 5$$

The solution of this equation is $N = 6 \times 10^4$, which means that only one B star per 60,000 M stars is enough to cut the mass-luminosity ratio by an order of magnitude, from 50 to 5. This illustrates that the late-type dwarfs must be extremely numerous in order to make their presence known from the radiation. In this example the M stars supply essentially all of the mass of the galaxy, while the B stars provide about 90% of the blue light of the galaxy.

Suppose the same galaxy were observed in the visual region of the spectrum. From the definition of magnitude, one has

$$B - V = \text{constant} + 2.5 \log \frac{L_V}{L_B}$$

It must be remembered that these luminosities are defined in terms of the solar values (L_V is the visual luminosity of the star divided by the visual luminosity of the Sun, and so on), and the value of the constant must be determined accordingly. Equation (17.8) indicates that the Sun has $(B - V) = +0.62$. Since the Sun also has $L_V = L_B = 1$ by definition, the above constant is equal to 0.62. It follows that

$$L_V = 0.56 L_B \times 10^{0.4(B-V)} \tag{26.15}$$

The B and the M stars have $(B - V)$ colors of about -0.3 and $+1.4$, respectively. For the B stars, then, $L_V = 0.43 L_B = 2.6 \times 10^3$, and for the M stars $L_V = 2.0 L_B = 0.02$. In the visual light 6×10^4 M stars produce 1.2×10^3 solar units, and this a little under one-half that given off by one B star. The M stars produce only 10% of the blue light of the galaxy, but they provide about 30% of the visual light. This illustrates the statement that cooler stars make a greater contribution at the longer wavelengths.

The mass-luminosity ratio is affected by the interstellar matter, which has mass but which has very little luminosity. Thus $(\mathcal{M}/L_B)$ is very large for the interstellar matter, and the value for the stars alone must be somewhat smaller than the values found for entire galaxies. The abundance of interstellar matter is strongly correlated with the Hubble type in the sense that irregulars have the most, ellipticals the least. It follows that the mass-luminosity ratio has a greater variation among the stellar components of galaxies than among the galaxies as a whole.

Figure 26.12 Galaxy NGC 5128, a peculiar galaxy with strong radio emission. (Mount Wilson and Palomar Observatories)

One would like to have accurate data on the continua, the lines, and the mass-luminosity ratios for galaxies in order to find the types of stars they contain. As an example, one could start by assuming that the stars have the same relative frequencies as those observed in the solar neighborhood. One would then add or subtract special groups until the observations were well satisfied. An alternative procedure is to use the relative-luminosity functions as observed in star clusters for the initial trial. Globular clusters could be used for the Population II component and galactic clusters for the Population I component. These types of investigations have indicated that many conclusions of a specific nature can be made about the stellar content of galaxies, but there are still many details left to be filled in.

The Interstellar Matter. Observations of interstellar matter in galaxies come primarily from the emission lines they exhibit. These lines, on the whole, are similar to the ones which are observed in the Galaxy and discussed in Section 23.

Since most of the interstellar matter is probably neutral atomic hydrogen (unless molecular hydrogen is even more abundant), the 21-cm line is a good indicator of the total gaseous content of a galaxy. The analysis is similar to that discussed in Section 25 for the Galaxy. Irregulars have some 20–25% of their total mass in interstellar matter, although these masses are quite uncertain. Spirals have only 1–2% of their mass in the form of this gas, and most of this is in the spiral arms. Ellipticals have much less gas than the spirals.

The dust is apparently distributed in very much the same way as the gas. It is quite prevalent in irregulars, and decreases in amount through the spirals and ellipticals. In the spirals it occurs mainly in the arms, and this helps to cause the orientation of a spiral to have a strong effect upon its color and apparent magnitude.

The behavior of ionized gas in some galaxies is rather surprising. Irregulars and most spirals show a degree of excitation in the interstellar lines that is normal, considering the large number of hot stars in such galaxies which can produce H II regions. The nuclei of some spirals and most ellipticals show a much higher degree of excitation, and this is unexpected, since these have few or no very hot stars capable of producing this excitation. It has been suggested by E. M. and G. R. Burbridge that this anomalous excitation is due to evolutionary mass loss from stars having large random velocities. Random velocities of stars of several hundred kilometers per second do exist in the nuclear regions of galaxies, and any material ejected by these stars would produce considerable excitation in the ambient interstellar gas.

Perhaps related to this on an extreme scale is the set of characteristics of the so-called Seyfert galaxies, a group of peculiar objects first pointed out by C. K. Seyfert. These have small nuclei with very high surface brightness, they show interstellar emission lines of unusually high excitation, and the lines are extremely broad. It is difficult to understand how interstellar lines could get so broad except by the doppler effect, and this interpretation leads to velocities of expansion of several thousand kilometers per second. Other galaxies also show some of these features. C. R. Lynds and A. R. Sandage find gaseous filaments streaming out of the irregular galaxy M82 with velocities of 1000 km/sec, and they believe that some sort of explosion took place in the center of this galaxy about 1.5×10^6 years ago. M87, a giant elliptical galaxy in the Virgo cluster (see Figure 26.10) has a jet of material which strongly suggests that material was forcefully ejected from the galaxy sometime in the past. There are a large number of peculiar galaxies whose appearances suggest that they are unstable. The apparent peculiarities observed in visible light are often an indication that a galaxy is an unusually strong emitter of radio frequency radiation.

Thermal Radio Emission. Probably all galaxies give off appreciable amounts of radio-frequency radiation. The use of 21-cm radiation to determine rotation curves and the abundance of neutral hydrogen have been mentioned. Most of the radio emission from galaxies is continuous radiation, and it has been estimated that about 10^{-6} of the total energy output of most galaxies is in radio wavelengths.

Hot objects give off radiation at all wavelengths, and the stellar part of galaxies can be suspected of being the source of the radio emission; however, it is easy to show that the thermal emission of stars is far too small to account for this. A black body has the fraction $0.15T^{-3}$ of its energy in wavelengths longer than 1 cm, on the assumption that the temperature is high enough that the Rayleigh-Jeans approximation of equation (2.9) is valid at a wavelength of 1 cm. (See Problem 3 of Introduction.) If a black body is to radiate at least 10^{-6} of its energy in wavelengths longer than 1 cm, it must have a temperature of no more than 50°K. If this radiation is thermal in origin, it obviously does not come from the stars.

As in the case of the Galaxy, some of the radiation could come from thermal emission in the interstellar medium. Equation (23.27) gives the intensity in the radio region for a uniform thermal gas of optical thickness τ_λ. If this is expressed in frequency instead of wavelength terms, it becomes

$$I_\nu = \frac{2\nu^2 kT}{c^2}(1 - e^{-\tau_\nu}) \tag{26.16}$$

If the absorption is primarily due to free-free transitions of electrons in H II regions, then the absorption coefficient and τ_ν are proportional to ν^{-2}. If the radiation is measured at sufficiently high frequencies, therefore, the absorption should be very small, with the result that $\tau_\nu \ll 1$, and the material is optically thin. Going to very small frequencies increases the absorption, so the source should become optically thick ($\tau_\nu \gg 1$) at very small frequencies.

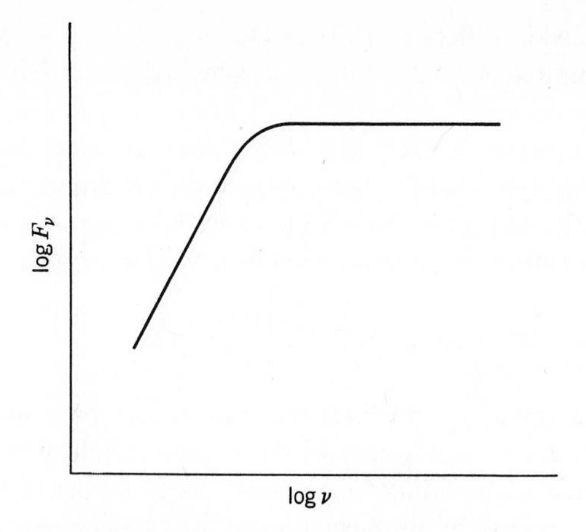

Figure 26.13 The thermal radio spectrum.

Equation (26.16) shows that if $\tau_\nu \gg 1$, the intensity is proportional to ν^2; if $\tau_\nu \ll 1$, then $(1 - e^{-\tau_\nu}) \cong \tau_\nu$, which is proportional to ν^{-2}, so the intensity is independent of frequency. Thus one has

$$
\begin{aligned}
I_\nu &\sim \nu^2 \qquad \text{(low frequency)} \\
I_\nu &\sim \text{constant} \qquad \text{(high frequency)}
\end{aligned}
\tag{26.17}
$$

The flux and the intensity should have the same frequency dependence, and the relations (26.17) are illustrated in Figure 26.13. Observations usually do not show a frequency dependence similar to the figure, so most of the radio radiation must be due to other sources. The thermal component is still present, however, and allowance must sometimes be made for it.

Synchrotron Radiation. If a particle having an electric charge has a component of velocity normal to a magnetic field, the field will accelerate it. Any charged particle emits radiation when it is accelerated, so the energetic interaction of such particles and magnetic fields produces radiation. Because the electrons have a much smaller mass than the positive ions, they move much more rapidly at the same energy. This causes them to be accelerated more by the magnetic field, so the radiation of the heavier ions is generally negligible compared with that of the electrons.

At low velocities the electron is accelerated around the magnetic field in a circle or a helix, and the radiation is emitted at the same frequency the electron has about the field. This radiation is emitted in all directions, with a slight preference for the direction of the field, and it is sometimes called cyclotron radiation. If the electron has a velocity nearly as great as that of light, then relativistic effects are important, and the character of the radiation is somewhat different. In the limit of extremely high electron energies the energy emitted is what is known as synchrotron radiation or magnetic bremsstrahlung.

Synchrotron radiation is extremely directional, as nearly all of the energy is emitted in a very small solid angle centered on the direction of the velocity of the electron. The energy is radiated in essentially a continuum of frequencies, but it has a rather sharp maximum at a frequency given by

$$\nu_{\max} = 0.07 \frac{eH_n}{mc}\left(\frac{E}{mc^2}\right)^2 \tag{26.18}$$

In the above equation e, m, and E are the charge, rest mass, and energy of the electron, and H_n is the component of the magnetic field which is normal to the electron velocity. The radiation is plane-polarized normal to the field. The total rate of energy loss by an electron due to synchrotron radiation in all frequencies is

$$\frac{dE}{dt} = -\frac{2e^4H_n^2}{3m^2c^3}\left(\frac{E}{mc^2}\right)^2 \tag{26.19}$$

The radiation from a group of electrons of different energies will now be considered. It will be assumed that all of the energy radiated by an electron is emitted at the single frequency given by equation (26.18). This assumption is not necessary, but it is a good approximation and it simplifies considerably the mathematics. Let $j(E)$ be the emission coefficient per unit energy interval for the synchrotron radiation; that is, $4\pi j(E)\, dE$ is the energy emitted within dE per second per unit volume into all directions by the relativistic electrons. Then if $N(E)\, dE$ is the number of relativistic electrons per unit volume having energy between E and $E + dE$, one has

$$4\pi j(E)\, dE = N(E)\left|\frac{dE}{dt}\right| dE \sim N(E)E^2\, dE \tag{26.20}$$

Equation (26.19) has been used in (26.20) to evaluate (dE/dt).

Equation (26.20) indicates that the emission coefficient per unit energy interval is proportional to the square of the electron energy times the distribution function of the electron energies. In order to compare this with observations, it is necessary to change from electron energy to the frequency of the radiation as the independent variable. Then if j_ν is the emission coefficient per unit frequency interval, $j_\nu\, d\nu = j(E)\, dE$, where ν is the frequency that corresponds to energy E. One has, from equation (26.18),

$$j_\nu\, d\nu = \tfrac{1}{2} j(E)\nu^{-1/2}\, d\nu$$

and so equation (26.20) gives

$$j_\nu\, d\nu \sim N(E)\nu^{1/2}\, d\nu \tag{26.21}$$

The observed radio fluxes of galaxies seem to have an exponential frequency dependence of the form

$$F_\nu \sim \nu^{-\alpha} \tag{26.22}$$

with α being approximately a constant and known as the spectral index. Spectral indexes lie in the range 0.2–1.2, with most galaxies preferring a value close to 0.7. As more and better observations become available, particularly over a wider frequency range, more examples of deviations from the form of equation (26.22) become apparent; nevertheless, the exponential form is a good approximation over a part of the radio spectrum of most galaxies.

Equation (19.6) shows that the intensity (as well as the flux) coming from an optically thin source has the same frequency dependence as the emission coefficient; therefore, one can obtain the form of equation (26.22) from (26.21) if the relativistic electrons have an exponential distribution in energy. If

$$N(E)\,dE = \text{constant} \times E^{-\gamma}\,dE \tag{26.23}$$

then $N(E) \sim E^{-\gamma} \sim \nu^{-\gamma/2}$, and the emission coefficient is

$$j_\nu \sim \nu^{-(\gamma-1)/2} \tag{26.24}$$

This is identical to (26.22) if

$$\alpha = \frac{\gamma - 1}{2} \tag{26.25}$$

The synchrotron mechanism is thus able to reproduce the main features observed in the radio spectra of most galaxies. No other plausible mechanism is known which can do this.

It should be emphasized that not all electrons produce the synchrotron radiation, only the relativistic ones, i.e., only those which have such large energies that they are moving with speeds that are very nearly as large as that of light. Most of the electrons have far smaller energies and do not produce synchrotron radiation. How the relativistic electrons received such high energies is a question that remains to be answered.

Of course, the spectrum (26.22) does not hold for all frequencies. There should be limits ν_1 and ν_2 for which (26.22) is valid only if $\nu_1 < \nu < \nu_2$. These limits are referred to as the cutoff frequencies. The low-frequency cutoff ν_1 can be caused by the emitting region becoming optically thick. The relativistic electrons can absorb energy by the synchrotron process, and if they follow the distribution given by equation (26.23), then the synchrotron absorption coefficient is proportional to

$$\sigma_\nu \sim \nu^{-(\gamma+4)/2} = \nu^{-(2\alpha+5)/2} \tag{26.26}$$

The absorption increases with decreasing frequency, and for sufficiently low frequencies the absorption will be large enough that the material will be no longer optically thin; i.e., self-absorption becomes important. Equation (19.6) shows that, for an optically thick region, the intensity is equal to the ratio of the emission and the absorption coefficients. The flux will have the same frequency dependence as the intensity, so equations (26.24) and (26.26) indicate that, for very low frequencies,

$$F_\nu \sim \frac{j_\nu}{\sigma_\nu} \sim \nu^{5/2} \tag{26.27}$$

This result is independent of the value of γ. Note that this is opposite to the type of frequency dependence shown by equation (26.22), so the flux should fall off on both sides of ν_1.

The high-frequency cutoff ν_2 can be due to any of a number of things. The flux in (26.22) is falling off with frequency, and eventually a frequency will be reached above which other mechanisms are more important than synchrotron radiation. For example, if a galaxy has a sufficient number of H II regions, the thermal emission illustrated in Figure 26.13 may dominate beyond some frequency ν_2.

Another possible cause of the high-frequency cutoff lies in the synchrotron mechanism itself. Suppose that at some instant of time t_0 relativistic electrons following the energy distribution (26.23) were injected into a region having a magnetic field, but that no new relativistic electrons were produced thereafter. According to equation (26.19), the energy loss is proportional to E^2, so the electrons with the highest energies will lose their energy most rapidly, and the distribution (26.23) will not be maintained. At a later time $t = t_0 + \Delta t$ all electrons of original energy greater than E_2 will have lost an appreciable fraction of their original energy if

$$\frac{\Delta t}{E_2}\left|\frac{dE}{dt}\right| \approx 1 \tag{26.28}$$

Substituting the energy loss rate from (26.19), one has

$$E_2 \approx \frac{3m^4c^7}{2e^4} H_n^{-2}\Delta t^{-1} \tag{26.29}$$

Equating electron energy with the frequency emitted through equation (26.18), one finds

$$\nu_2 \approx 0.16 \frac{m^5c^9}{e^7} H_n^{-3}\Delta t^{-2} \tag{26.30}$$

After the elapse of time Δt, electrons with energies greater than E_2 will be considerably less abundant than equation (26.23) predicts, and so frequencies greater than ν_2 will have less emission than equation (26.22) predicts. This is

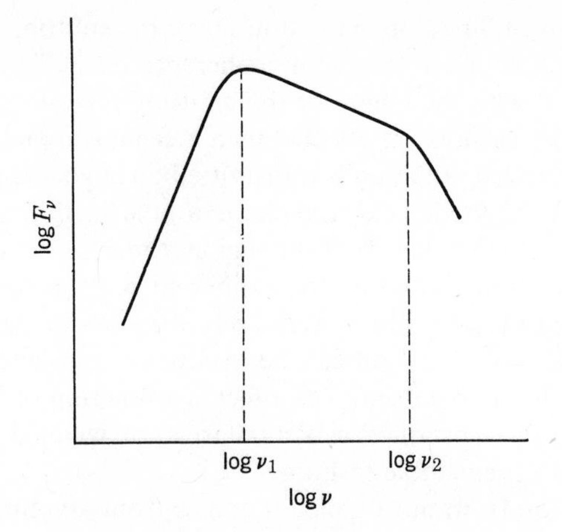

Figure 26.14 The synchrotron spectrum.

an aging effect, and the older the source, the smaller the frequency at which the aging effects are important. Observations of the high-frequency cutoff ν_2 plus some speculation about the size of the magnetic field are sometimes used to estimate Δt, the time scale of the synchrotron source. It may happen that the supply of relativistic electrons is continuously replenished, in which case the effect will be less pronounced; however, there will still be a deficit of high-energy electrons that grows with time.

Figure 26.14 shows a schematic plot of what the synchrotron spectrum should look like. For $\nu \ll \nu_1$, it can be seen that F_ν rises as $\nu^{5/2}$ in the optically thick region. (Of course, the low-frequency cutoff ν_1 could be due to a deficit of relativistic electrons with energy less than an amount E_1.) In the intermediate region the flux is proportional to $\nu^{-\alpha}$, and beyond ν_2 the flux falls even more steeply, because of the degradation of the high-energy region of the electron energy distribution. The latter region may be masked in actual galaxies by the rise of F_ν due to other mechanisms, such as thermal emission from H II regions or stellar radiation.

Radio Galaxies. Most galaxies, including our own, emit a modest amount of radio-frequency radiation. The spectra observed are often in fairly good approximation to Figure 26.14, but there are many differences of detail. Often one or both cutoff frequencies have not yet been observed, and sometimes the form of equation (26.22) does not apply at all. When this is the case, equation (26.23) is not valid for the electron-energy distribution, or possibly there are two or more separate sources being observed together.

The radio frequency radiation from galaxies is usually polarized, and this gives more evidence that synchrotron radiation is involved. Each electron emits polarized radiation, but the total radiation would not be polarized if the magnetic field changed direction many times over the emitting volume; therefore, the magnetic field must have some coherence over distances which are not small compared with the length of the emitting region.

The polarization measures are affected by a mechanism known as Faraday rotation. When polarized radiation is transmitted through a region containing free electrons and a magnetic field, the plane of polarization is rotated by an amount that depends on the length of the region, the density of electrons, and the size of the field. When radiation from different parts of the source suffers different amounts of Faraday rotation, the result is a partial depolarization of the radiation. This depolarization can be produced anywhere in the line of sight, even within the source itself. The effect is a function of frequency, and it can be recognized by its production of a polarization which depends strongly upon frequency in a predictable fashion.

The radio emission from most galaxies comes from a volume which is far larger than the source of most of the visible radiation from the same galaxies. It appears that a large, tenuous halo containing magnetic fields and high-energy electrons is a common feature of galaxies.

A small percentage of galaxies emit a much greater amount of radio energy than usual, and they are known as radio galaxies. In optical frequencies, radio galaxies are comparable in luminosity to the brightest of the normal galaxies, having some 10^{44} erg/sec emitted in the visible region. The emission in the radio region is in the range 10^{40}–10^{45} erg/sec, and so some of the radio galaxies emit most of their energy in the radio region. Most of the discrete radio sources which have been known for a number of years are radio galaxies, although only relatively few have been identified with objects in visible light.

In the cases in which radio galaxies have been identified with an object in visible light, the optical images usually show peculiarities of some kind, such as jets or long filaments. When the resolution is sufficiently great, it is usually found that there are two radio sources symmetrically placed on either side of the optical or visible light image. This strongly suggests that the sources of the strongest radio emission are masses which have been ejected from the parent galaxy in some violent explosion.

These double radio sources are at distances of up to about 10^6 light years from the central galaxies from which they were presumably ejected. Also, in those radio galaxies for which expansion velocities of the interstellar gas are available, the indication is that the expansion started in the center about 10^6 years ago. Of course, the numbers vary from one galaxy to another, but it is suggested by the data that the general time scale for strong radio emission is of the order of 10^6 years. Galaxies are certainly much older than this, so strong radio emission may be a short phase in the life of some or many galaxies, and it might be a recurrent phenomenon. Perhaps the gas observed expanding away from the center of our Galaxy is the result of a milder form of the same type of explosion. It is interesting that spirals rarely are among the strong radio emitters.

The time scale found above can be used in connection with equation (26.30) to estimate the strength of the magnetic field. With $\Delta t \sim 10^6$ yr $= 3 \times 10^{13}$ sec, one finds

$$H_n \sim 0.07\nu_2^{-1/3}$$

The high-frequency cutoff ν_2 is not always observed, but H is not very sensitive to its value. In many cases ν_2 appears to be around 10^{10}–10^{11} sec^{-1}, and this puts the magnetic field in the neighborhood of 10^{-5} gauss.

One can also estimate some of the properties of radio galaxies from energy considerations. The brightest radio galaxies have a radio luminosity of about 10^{45} erg/sec. If the time scale of 10^6 yr is assumed, then a total of 3×10^{58} erg is emitted over the radio lifetime. At least this much energy must have been stored in relativistic electrons, if the injection was a one-shot affair. If the energy in the protons is comparable to this, and if the energy was equally distributed to the particles and to the magnetic field, then the magnetic-field energy is about 6×10^{58} erg. (If the energy in the protons is two orders of magnitude greater than that in the electrons, as in the case of the cosmic rays, then the magnetic energy would be correspondingly greater.) The magnetic energy per unit volume is $H^2/8\pi$, as indicated by equation (1.15), so if the emitting volume is a sphere of radius r, one has

$$\tfrac{1}{6}r^3H^2 \approx 6 \times 10^{58}$$

With H of the order of 10^{-5} gauss, one finds that $r \approx 5 \times 10^4$ pc. It is encouraging that this does turn out to be of galactic dimensions. Often r can be checked. One can also estimate r from the observed frequency at which the region becomes optically thick.

The above calculations are not intended to be taken very seriously; they only illustrate some of the numbers which one can speculate with. The assumptions made about the interpretation of observed cutoff frequencies, about the time scale, and about equipartition of energy between particles and the magnetic field are all subject to major revision.

Quasistellar Objects. In 1963 a new type of object was recognized which is similar in some respects to the radio galaxies. This type of object has been called at various times a quasistellar object (QSO), quasistellar source, quasistellar radio source, and quasar. The name comes from the fact that these objects in visible light show a stellar image, i.e., they do not show an extended disc, although there are sometimes faint filaments seen associated with them. In radio frequencies QSO's look similar to the strong radio galaxies, and some of them have been known as strong radio sources for many years; it has only been in recent years that the increased resolution possible in radio frequencies has allowed the optical identifications to be more readily made.

The uniquess of the QSO's is based on their extremely large redshifts, which are obtained from optical emission and absorption lines. Some are known with a redshift $z = \Delta\lambda/\lambda$ of greater than 2. If these redshifts are of cosmological origin, i.e., if they are doppler shifts which are related to distance by Hubble's law (26.7) or a more accurate generalization of this equation, then the QSO's are the most distant and the most luminous objects known. Their radio output would be comparable to that of the strongest radio galaxies, but their optical luminosities would be some 100 times that of any previously known galaxy.

There are two other plausible explanations of the large redshifts of the QSO's besides the cosmological expansion: large doppler shifts of nearby objects, and gravitational redshifts. A photon leaving an object which has a large gravitational field will lose energy, and this produces a change in wavelength which is indistinguishable from the doppler shift of a receding object. The shift is given by

$$\frac{\Delta\lambda}{\lambda} = \frac{G\mathcal{M}}{c^2R} \tag{26.31}$$

where $\mathcal{M}$ and R are the mass and radius of the object. A relatively nearby object with sufficiently large $(\mathcal{M}/R)$ could produce the large redshifts observed; however, J. L. Greenstein and M. Schmidt showed that the gravitational interpretation is probably incompatible with the properties of the observed emission lines.

J. Terrell believes that the QSO's are nearby objects which were somehow accelerated to very large velocities. Since none of the QSO's has blueshifts or measurable proper motions, Terrell suggests that they originated in the nucleus of our Galaxy some 10^6 to 10^7 years ago. F. Hoyle and G. R. Burbidge also suggest that the QSO's may be nearby, but originating in the nuclei of nearby strong radio galaxies, in particular NGC 5128.

The QSO's have been found to be variable in both radio and optical frequencies. The variations are only semi-regular, with a long period of several years and shorter periods of as little as a week or so. These variations are up to half a magnitude, and they place upper limits on the possible sizes of the QSO's. If an object shows appreciable light variations over a certain time interval, then it is difficult to see how the object can be large compared to the distance light travels in that interval, at least if the variations have any regularity to them. This places difficulties on the cosmological interpretation of QSO's.

There are also difficulties with the local origin of QSO's. J. A. Koehler and B. J. Robinson observed 21-cm absorption in the radio spectrum of the

QSO 3C 273, and this absorption appears to be due to intergalactic hydrogen in the Virgo cluster of galaxies. If this interpretation is correct, then 3C 273 is beyond the Virgo cluster and is certainly not local.

The nature of the QSO's is a subject upon which opinion can change drastically in a very short time, as there are relatively few pertinent observations and new ones are being obtained at a very rapid pace. Many theories on the subject are obsolete long before they are published, and this is not the appropriate place to carry the speculations further. Whatever their nature turns out to be, they undoubtedly represent conditions which are unique to the experience of both astronomers and physicists.

Evolution of Galaxies. The very small galaxies may not have a strong enough gravitational field to retain the material that is shed by old stars. They would be similar to the galactic and globular clusters in that only one generation of stars could form. Most galaxies can retain this material, however, and they would be expected to evolve according to the general principles mentioned at the end of Section 25.

Stars of all luminosities are formed at all times in a galaxy, but those of higher luminosity have shorter burnout times; therefore, the light of a galaxy should become more strongly dominated by the cool, late-type stars of low luminosity as time goes on. Also, the interstellar matter should be slowly used up, and the abundance of the heavy elements should increase as a galaxy ages.

The colors and the gas and dust content of galaxies suggest on the above basis that irregulars, spirals, and ellipticals are galaxies of increasing age. The possibility has been mentioned by many persons that this is an evolutionary sequence, all galaxies starting as irregulars and later developing into spirals and then ellipticals. More recent and accurate data, however, do not agree with this. The data are not inconsistent with irregulars being able to keep large amounts of interstellar matter for periods of 10^{10} years or more, so the large amounts which they have do not necessarily mean that they are young. All galaxies which have been studied in detail, including some irregulars, show evidence of containing some very old stars. There is no direct evidence at present that any galaxy has an age which is significantly different from that of our Galaxy, i.e., about 10^{10} years. It should be emphasized, however, that the amount of direct evidence on this question is rather small.

Whatever the ages of the galaxies, there is strong evidence that the different Hubble types do not evolve to or from each other. Their differences, therefore, must be due to the conditions at their origin. Mass does not appear to be important in this regard, since there is a large overlapping in the masses of the different galaxy types. Composition is less certain. It is often assumed that the original material was pure hydrogen or hydrogen plus helium, the heavier elements being built up inside the stars. But a nonuniform distribution of original heavy elements could be of some importance, and the rate of production of heavy elements at very early times could affect the subsequent evolution of a galaxy. Angular momentum, turbulence, and magnetic fields are the most obvious agencies that could cause the observed differences between galaxies.

A galaxy is often thought of as starting as a large, very-low-density gas of nearly pure hydrogen. If the material is highly turbulent, it can resist gravitational contraction for a while; however, the collisions in the gas will fairly quickly damp out the chaotic motions, leaving only the orderly mass motions that represent the total angular momentum that the galaxy started with. This will lower the resistance to contraction perpendicular to the plane of rotation, so the mass will collapse to a flattened shape. The amount of flattening will depend on the amount of the total angular momentum. If the net angular momentum is zero, the collapse will be equal in all directions, and the object will maintain (or get, if it did not start with) a spherical shape. If nothing else happened, the collapse would continue until internal pressures built up or until the angular momentum stopped further collapse.

If part of the gas is constantly condensing to form stars, the stars will have the motions of the gas from which they were made. The future of the stars, however, may be quite different from that of the gas. The point is that collisions among the gas atoms will rather quickly damp out any large differential motions in the gas, but the corresponding effect on the stars is very inefficient. Thus the stars will keep the same velocity distribution they were born with, while the gas should settle down to a state having only very small differential motions, regardless of the motions it started with.

The above qualitative picture has many attractive features. Stars formed at very early times are expected to show little concentration toward the rotation plane. These stars should have a large velocity in the radial direction because of the collapse velocity of the gas from which they were formed, and this should result in orbits with large eccentricities. Stars formed after the collapse, however, should have much more nearly circular orbits, they should be confined to a smaller volume, and if the angular momentum of the galaxy is sufficiently great, should be confined to a much flatter volume. The more obvious features of the different stellar populations are thus explained. O. J. Eggen, D. Lynden-Bell, and A. R. Sandage have found observational evidence that the Galaxy had an early development along these lines.

It is tempting to suggest that objects with very little angular momentum per unit mass became ellipticals, while those with large rotation became spirals. This does not offer much of an explanation for the irregulars, nor does it explain why the gas in the rotation plane should take on the spiral formation. If a very large percentage of the gas condensed to form stars at very early times in ellipticals, then they would show less flattening than the spirals even if they had the same angular momentum. There is evidence that appreciable nonradial stellar motions do exist in elliptical galaxies.

Perhaps the variations in the star-gas ratio could be understood in terms of the star-formation function. If a large part of the material goes into high mass stars, then a large part of the star material will be returned to the interstellar medium, and heavy-element enrichment will take place at a rapid rate. If most of the material goes into low-mass stars, however, very little will be returned to the interstellar medium, and the medium will be used up more quickly.

The above discussion illustrates that little is known about the formation and evolution of galaxies other than some rather general ideas, and even these are not universally accepted. D. Layzer (see references at the end of the chapter) gives a review of some of the ideas in this field and presents some rather novel ideas of his own.

27. Cosmology

Cosmology is the study of the large-scale properties of the Universe. How large a scale does something have to have in order to be of cosmological significance? This depends on the nature of the Universe, which is unknown; therefore, the above definition allows a wide range of opinions on what is and what is not cosmologically significant.

There is probably no other field of science which is as free of the restraints of observational data as cosmology. New observations are very difficult to obtain and are not overly convincing, since they are necessarily made near the limit of present methods and equipment. This is a convenient argument of those whose theories are not substantiated by the new data. It is true that recent data have caused some theories to lose support, but evidence against a theory is not necessarily evidence in favor of its rivals.

The dearth of definitive observations results in a "Lets-wait-and-see" attitude among some, but it also encourages many persons to supplement the data with philosophical arguments. For example, one person might consider it "unthinkable" that the Universe can have a beginning or an end, while a second person might find it equally distasteful that the Universe can be self-sustaining for an infinite length of time. These two persons will find considerable disagreement over the relative merits of different cosmological theories regardless of the data. A statement to the effect that astronomers are on the threshold of discovering the nature of the Universe has been uttered many times in recent decades. I do not believe that this threshold will ever be crossed. A theory may receive the consensus support of the experts (even this is difficult to imagine at present), but cosmology leans too heavily on the philosophical attitude of the investigator for any one theory to be accepted with the same confidence that is given to other branches of physical theory.

There is another point in which cosmology differs from other fields of science. If the Sun is observed to have a certain property, one can ask whether this property is essential to all self-luminous gaseous masses. The answer to this question can at least be suggested by the observation of other stars; however, there is only one Universe, and so a similar study of its properties is not available to observation. As an example, it is generally agreed that the Universe is expanding. Some persons may consider this to be an essential rather than a random property of the Universe. Now, there are some theories which have the Universe expanding sometimes and contracting sometimes, and these theories must be considered inapplicable to the Universe by one who believes that the expansion is essential. This question and others could be answered if many universes could be observed.

Many differences of opinion which exist in cosmology arise because of the differences in emphasis placed upon the methods of investigation. The gravitational constant G does not vary with time, to the limit of observational accuracy. One point of view maintains that this gives an advantage to a theory which predicts G to be constant over a theory which has G varying slowly with time. The argument would be that the assumption that G is constant is firmly entrenched in physics, and it should not be given up without definite evidence for having to do so. The opposite point of view claims that any time variation of G which one wishes to assume, consistent with the accuracy of the observations, is as valid as any other and does not require special justification. The same arguments, of course, can be used on any conservation law suggested by physics.

Local Observations of Cosmological Significance. It is not true that observations which are of importance to cosmology are necessarily limited to extremely remote objects. Olbers' paradox provides an example.

In 1826 H. Olbers published a paper on the light received from distant stars. He assumed that infinite space is uniformly filled with stars and that the properties of the stars, averaged over a sufficiently large volume, do not change with position or time. Then if the stars do not have any large systematic motions and if the usually assumed laws of physics hold, one can easily show that the night sky should not be dark. Under these conditions a straight line drawn in any direction from an observer will eventually intersect a star, and so light of uniform intensity will be received from all directions. (As is noted in Section 1, it is the flux, not the intensity, which falls off as the square of the distance of the source.) This will cause the sky to have a uniform surface brightness, the same brightness as the disk of the Sun (if the Sun is an average star). Olbers considered his assumptions to be self-evident and attributed the dark sky to a thin absorbing medium which dims the light of distant stars; however, the medium would eventually be heated up until it radiates as much energy as it absorbs, so this does not clear up the paradox.

There are a number of possible answers to Olber's paradox. A finite limit on the size or the lifetime of the Universe would satisfy the observations. Light received from very distant stars originated very long ago, so if the Universe has a finite age, light from only a finite volume would be received at any one time. Also, the Universe could be of infinite size and age but with the density of stars falling off in an appropriate manner with distance. The expansion of the Universe could also be the main error in Olbers' assumptions, since the doppler effect causes the observer to receive energy from a receding source at a lower rate than that at which it is emitted. Of course, there is no limit to the possible ways in which one could modify the usual laws of physics in order to explain Olbers' paradox; however, the important point here is not that the paradox can be explained but that significant information is contained in the simple observation that it gets dark at night.

One is used to the idea that stating the position or the velocity of an object is useless unless it is clearly understood what standard these are measured with respect to. There do not appear to be any absolute standards which have a privileged nature with respect to position or velocity, and modern physics claims that there can be none. This is essentially what is meant by the statement that position and velocity are relative, not absolute.

When the driver of a vehicle suddenly slams on the brakes, the relativity-of-motion concept seems to suggest that it is equally valid to say that the vehicle is accelerated in a certain way or that the rest of the Earth is accelerated in the opposite way; however, experimental evidence shows that it is the occupants of the vehicle, not the rest of the people on Earth, who become shaken up by this experience. Any other experiments involving accelerations seem to point to the same conclusion: if only inertial accelerations are considered (accelerations due to gravitational and electromagnetic forces being excluded), then there does appear to be an absolute frame of reference for accelerations. If an object is accelerated with respect to this frame, it is subjected to measurable forces, the inertial forces; otherwise, it is not.

According to E. Mach this absolute frame of reference for the inertial accelerations is determined by the distribution of mass over the entire Universe, and this statement is often called Mach's principle. Whether or not Mach's principle is correct is a matter for debate. There are some experimental checks which can be made on it. For example, the rotation rate of the Earth can be measured by means of the inertial forces it produces, such as the centrifugal and coriolis forces which objects on the Earth are subjected to. Its rotation rate can also be measured with respect to the distant matter in the observable part of the Universe. These two rotation rates agree to within the accuracy of the experiments, and this is a necessary consequence of Mach's principle. Those who do not accept Mach's principle must consider this experimental result to be a coincidence.

According to Mach's principle, if there were no matter in the Universe except for one test particle, this particle would be subjected to no inertial forces. It would be meaningless to ask whether it is rotating or otherwise being accelerated, for these accelerations can arise only if there is other matter in the Universe. If a complete mathematical theory of Mach's principle were formulated, it would be possible to draw conclusions about the distribution of matter in the Universe from the observed properties of the inertial forces. Lacking this, one can still say with some conviction that the dizzyness one feels after spinning around too rapidly is somehow a manifestation of the large-scale properties of the Universe.

A local observation whose significance is not as obscure as that of inertia is the fact that the observed part of the Universe is mainly hydrogen. It was pointed out in Sections 21 and 22 that hydrogen is converted into heavier elements in stellar interiors, and there is no place known where the reverse takes place. One then expects the hydrogen content of the Universe to be diminishing with time. Estimates of the present hydrogen content and of the depletion rate indicate that the matter must have been nearly pure hydrogen (or hydrogen plus helium) at most a few times 10^{10} years ago. Thus matter must have gone through some significant changes in this time interval. If material is very much older than this, it must have been subjected to some kind of rejuvenation process within this time interval, a process which replenishes the supply of hydrogen (and possibly helium).

The Cosmological Principle. It is impossible to accurately assess the significance of observations unless something is known about the relation between the observable part of the Universe and the Universe as a whole. One can make any assumption about this relation without fear of being proved wrong, yet scientists are almost unanimously agreed on the importance of an assumption to the effect that the observed part is typical of the Universe as a whole. It is this assumption that makes observations relevant and keeps cosmology within the realm of science; without it cosmology becomes pure speculation.

Most cosmologists are not satisfied with the statement that the visible regions are typical, but insist on the much stronger statement which is known as the cosmological principle. If a fundamental observer is defined as an observer who is at rest with respect to the matter in his neighborhood, then the cosmological principle states that the Universe must appear the same to all fundamental observers, except for local irregularities. Also, the view to each fundamental observer must be the same in all directions, i.e., the Universe must be isotropic. Clusters of galaxies must be considered as local irregularities in this context. Because of the doppler effect, two observers at the same position but having a large relative velocity will not have the same view of the Universe, so the cosmological principle cannot apply to all observers. The obvious interest in the distribution of matter gives special significance to the fundamental observers. Any mass which is at rest with respect to its surroundings is a fundamental particle.

The cosmological principle is the basis of virtually all theories which are seriously proposed and considered today. As with any other point which cannot be proved, belief in its validity ranges from the opinion that it is somehow logically inevitable to the opinion that it is a convenient starting point. The main reason for its support lies in its fruitfulness: a large number of extremely important consequences result from it. One can ask whether this is any reason for supposing that it applies to the Universe, and the answer to this depends on one's philosophy. Certainly one has more confidence in a theory that explains some data on the basis of few special assumptions than in a theory which needs many special assumptions to explain the same data. The observation that the distribution of galaxies appears isotropic on a large-enough scale also lends some credance to the cosmological principle. This is a necessary but not sufficient condition for the principle to be valid.

The consequences of the cosmological principle are quite far-reaching. For example, the system of all fundamental observers (FO) must be without a most distant member, else his view would be different from the others. In this and in all other respects, no FO can have any property which distinguishes it from the other FO's. It follows that the FO's must uniformly fill an infinite space, or else the geometry of space must somehow allow each FO to view the Universe as if he were at its center. Since the FO's are moving with the average velocity of the mass in their vicinity, the distribution of the FO's also fixes the distribution of mass in the Universe.

The relative velocities of the FO's are not arbitrary; otherwise, two of them could come very close together while having quite different velocities, so the cosmological principle would be violated. The only motions allowed are a radial expansion or contraction of the system of FO's. It is thus seen that a strict application of the cosmological principle places very severe limitations on the possible spatial and velocity distributions of the matter in the Universe.

H. Bondi and T. Gold have carried the cosmological principle one step further to formulate what they like to call the perfect cosmological principle. This is the statement that the Universe appears the same to all FO's at any time as well as at any position. The statement requires the overall properties of the Universe to remain constant with time, although individual parts such as stars and galaxies can evolve with time. The cosmology which is based on this is known as the steady-state theory. It is apparent that the "perfect" cosmological principle is much more restricting than the ordinary cosmological principle mentioned first. Other properties of the steady-state theory are discussed later.

Relativity. Relativity is a branch of physics which deals with the measurement of position and time intervals. Many of the properties of these measurements seem almost intuitively obvious, yet, according to relativity, this intuition can be quite wrong if conditions are very different from those of everyday experience. In particular, for very large velocities or for very large accelerations, strange effects peculiar to relativity and contrary to intuition become important. This is why relativity is so closely connected with cosmology.

Relativity is divided into two parts, special and general relativity. Special relativity is concerned with the effects of velocities on measurements, while general relativity also includes the effects of accelerations. Special relativity was developed by A. Einstein in the early years of the twentieth century, and it was based on the results of an important experiment carried out by A. A. Michelson and E. W. Morley in 1887. The Michelson-Morley experiment indicated that the speed of propagation of light in a vacuum is independent of the velocity of the observer. Einstein used this result, plus the assumption that the equations decribing the laws of physics are not affected by the velocity of the observer, to derive equations which show how measurements do depend upon the velocity of the observer. These equations are known as the Lorentz transformations, after H. A. Lorentz who had derived them earlier from a different point of view.

Special relativity has a number of consequences which are in disagreement with classical or Newtonian physics and which have been confirmed experimentally many times. The rate of a clock, the length of a rigid body, the mass of an object, all depend on the relative velocity between the object and the observer. As noted in Section 6, special relativity also predicted the equivalence of mass and energy as expressed by equation (6.4). The equations of special relativity deviate appreciably from those of classical physics only when the relative velocities approach the speed of light. Thus the classical doppler effect, equation (11.2), is completely adequate for most astronomical applications; however, the velocities of recession of extremely distant galaxies are not a negligible fraction of the speed of light, so the relativistic doppler effect, equation (11.1), must be used. The experimental confirmation of special relativity is complete enough that there is no serious doubt of its validity among physicists today.

General relativity introduces accelerations to the theory. Its basic postulate is that it must be possible to formulate the laws of physics so that they are independent of the motion, accelerated or not, of the observer. A comparison with special relativity is helpful in explaining the meaning of this statement.

In the special theory, it is assumed that the gravitational force on a test particle can be determined from the distribution of mass around the particle. If this gravitational force is zero (electromagnetic and nuclear forces are excluded from consideration) the particle can be said to be unaccelerated in an absolute sense. Any reference system which is unaccelerated with respect to such a particle is called an inertial system. This gives a special significance to the inertial systems in that the laws of motion are simpler in form in them than in noninertial systems.

General relativity denies that inertial systems can have any special significance over other systems, so it must also deny that gravitational forces have any special significance over inertial forces. This placing of gravitational and inertial forces on an equal footing makes general relativity something of a unified theory, although electromagnetic and nuclear forces are still excluded. It may appear that the special significance of the system of fundamental observers violates the principle of equivalence of all reference systems, but this is not so. All systems must be equivalent for expressing the general laws of physics, specifically the equations of motion; however, the solutions of the equations need not be identical for all systems. The fundamental particles are significant in that they move with the matter in their vicinity, so the solution for the motion of matter takes its simplest form in the system of fundamental particles.

In the relativistic description of events it sometimes becomes necessary to drop the distinction between position and time. The two are combined into a four-dimensional space-time continuum, instead of the separate one-dimensional time plus three-dimensional space of classical physics. A "point" is specified by its four coordinates in a suitably chosen coordinate system.

Classical physics also assumes that the axioms of Euclid are valid, i.e., that space is euclidean. This is certainly true to a good approximation in the nearby regions of space, but it is a special assumption which is not necessarily true for the Universe as a whole. General relativity takes account of the possibility that the geometry of space may be noneuclidean. The surface of a sphere is an example often given of a two-dimensional noneuclidean space, but it is not possible to visualize one of three or more dimensions; nevertheless, the properties of such spaces can still be mathematically determined, and there is no assurance that they are any less physically real than the three-dimensional euclidean space that is more familiar.

Einstein made a plausible generalization of classical mechanics, consistent with the principles of relativity, and derived the basic equations of general relativity. These are called the field equations, and they are equivalent to the equations of motion plus the law of gravitation of classical physics. They relate the geometry of space-time to the properties of the matter in the Universe, and they reduce to the equations of classical physics in the first approximation. Since there is no distinction between gravitational and inertial effects, both being related to the distribution of matter, the field equations do incorporate Mach's principle to some extent; there is debate, however, over whether this incorporation is as complete as some cosmologists believe it should be.

General relativity is much more difficult to subject to observational test than special relativity. This is partly because most of the effects peculiar to general relativity are important only on an extremely large scale, where observations are difficult. Another reason is that, like most equations of physics, the field equations are very difficult to solve, and solutions have been found only under some especially simple conditions. A check of these solutions would be a test of both the assumed conditions and the theory, and it could prove difficult to disentagle the two effects.

Shortly after Einstein published the general relativity theory, K. Schwarzschild solved the field equations for a mass in an otherwise empty universe. This should be a good approximation to the gravitational field near a star, and the analysis came up with three effects which differ from Newton's theory: (1) planetary orbits should be slightly different from those predicted by Newton's theory; (2) light rays coming from a distant star and passing near the Sun should be deflected by a predicted amount from a straight line; and (3) the spectral lines of a star should be observed to be shifted toward longer wavelengths by the gravitational field of the star.

The first effect is strongest for the innermost planet, Mercury. The major axis of Mercury's orbit around the Sun should rotate by an amount that differs from that predicted by Newton's theory by 43″ per century. This is in very good agreement with observations. The second effect should cause the relative positions of stars as observed from the Earth to depend on their angular distance from the Sun. Stars seen very close to the Sun's disc should have their apparent positions displaced outward by about 1″.76. Stars can be observed at small angular distances from the Sun only during total solar eclipses, and many eclipse expeditions have given a good confirmation of this effect. The third effect is very difficult to measure in stars. Normal stars like the Sun have a very small effect, while white-dwarf stars, which have very large gravitational fields, also have such broad spectral lines that their wavelengths cannot be accurately determined. This is the effect illustrated by equation (26.31), and very accurate measurements by means of the so-called Mössbauer effect have confirmed the wavelength change as produced by the Earth's gravitational field. The total evidence is thus very strong in favor of general relativity.

Relativistic Models. The properties of the Universe are investigated by the same methods that are used to study any other physical system: a theoretical model of the Universe is calculated by whatever means is needed to determine a unique model, and its properties are compared with the observations. The comparison with observations provides the only evidence which can pertain to the question of the relevance of a particular model to the Universe.

Models based on general relativity must be consistent with the relativistic field equations. The field equations are not sufficient to determine a model by themselves, as they can only determine the properties of a system whose initial configuration is sufficiently well specified. The configuration of the Universe is not well enough known for this to be of much value; however, when the field equations are combined with the cosmological principle and all irregularities ignored, the problem simplifies considerably. Whether the clustering tendency of matter into stars, galaxies, clusters of galaxies, and the like is relevant to any of the fundamental properties of the Universe is not known, but neglecting the tendency by assuming that the matter is uniformly spread throughout the Universe is a simplification that suggests itself as an obvious starting point.

One can imagine space filled with fundamental observers, each of which is at rest with respect to the matter in his vicinity. According to the cosmological principle, all of the FO's should have identical views of the Universe. One FO could measure the various properties of the Universe and tabulate them as a function of the time. Other FO's must be able to obtain identical tabulations. If the Universe is evolving or changing with time, then it can be used as a clock for the measurement of a universal or cosmic time. Any observer can refer unambiguously to the time for which the observations gave certain tabulated values.

An important property of the geometry of space is what is known as its curvature. This is easily visualized only for spaces of less than three dimensions, but it can be defined mathematically for spaces of any number of dimensions. The cosmological principle requires that the curvature be the same for all FO's, and this greatly simplifies the geometry. In two dimensions a space of constant curvature must be a sphere or a plane. (The latter is a sphere of infinite radius.) The sphere is closed and has a finite size or extent, while the plane, which has zero curvature, is of infinite extent. In three dimensions a space of constant curvature can be curved and of finite extent, it can have zero curvature and be of infinite extent, or it can be curved and of infinite extent. The first possibility is analogous to the spherical surface, and it is generally known as a spherical space. The second geometry is analogous to the plane surface and is known as flat or euclidean space. The third geometry has no analogy in two dimensions and is called a hyperbolic space.

The properties of space are quite confusing when one tries to visualize them, and analogies to two dimensions are often not very accurate. For example, the surface of a sphere may have only two dimensions, but it is impossible to visualize such a surface except as imbedded in a three-dimensional space. The curvature of such a surface seems to be a measure of its extension into the third dimension. Likewise the curvature of four-dimension space-time (which may be different from that of three-dimensional space) seems to imply the existence of spaces of yet higher dimensions. Mathematicians like to say that one should not try to visualize the properties of space, one should only calculate them. It is important to realize that in the present context the properties of space are determined by the results of measurements made with physical instruments. This is quite different from the concept of space as having existence and properties independent of any matter which it may or may not contain.

Consider the figure formed by connecting the positions of a number of fundamental particles at a fixed value of the cosmic time. As cosmic time progresses, this figure can expand or contract without changing shape, but any distortion of the shape would violate the cosmological principle. Thus it is possible and convenient to define a three-dimensional coordinate system which is fixed to the fundamental particles. This system must expand or contract along with the system of the fundamental particles, and so it is called a co-moving coordinate system. Three coordinates then specify the position of a fundamental particle with respect to the origin, and these coordinates do not change with time. If u is the radial coordinate of a given particle, a measure of its distance from the origin is the product $uR(t)$, where $R(t)$ is a scale factor which increases or decreases with time as the system expands or contracts.

The coordinate u is simply a number or label which indicates the position of an object, and it is not a distance. The term distance must be used with care, for it is not a meaningful term unless it is defined in terms of a physical measuring process. There are several ways to do this, and in noneuclidean geometries they are not equivalent. For this reason any measure of distance must be uniquely defined from possible measurements before it is used, and the quantity $uR(t)$ is not a good distance in this respect. It is usual to take u as without dimensions, so R has the dimensions of length.

Consider a fundamental particle of distance r, defined as $r = uR$. Then

$$\frac{dr}{dt} = u\frac{dR}{dt} = uR \times \frac{1}{R}\frac{dR}{dt} = r \times \frac{1}{R}\frac{dR}{dt}$$

In local regions of space the geometry is known to be euclidean to a very good approximation, so the above ambiguities in the definition of distance do not occur. If r is small enough for this to be the case, then dr/dt is the velocity of the particle with respect to the origin. Hubble's law of equation (26.7), taken without the random term since a fundamental particle is being considered, then indicates that

$$H = \frac{1}{R}\frac{dR}{dt} \tag{27.1}$$

Hubble's constant is the ratio of the time derivative of the scale factor to the factor itself. If the ambiguities in the definition of distance are important, then equation (26.7) is no longer valid, although equation (27.1) remains valid.

When the cosmological principle is applied to the field equations of general relativity, they reduce to the following:

$$\frac{2}{R}\frac{d^2R}{dt^2} + \left(\frac{1}{R}\frac{dR}{dt}\right)^2 + \frac{kc^2}{R^2} = -\frac{8\pi G}{c^2}p + c^2\Lambda \tag{27.2}$$

$$\left(\frac{1}{R}\frac{dR}{dt}\right)^2 + \frac{kc^2}{R^2} = \frac{8\pi G}{3}\rho + \tfrac{1}{3}c^2\Lambda \tag{27.3}$$

In these equations p is the pressure and ρ the density of the matter and radiation, and G is the gravitational constant. The quantity Λ is a constant which known as the cosmological constant, and relativity theory does not specify its value. The constant k is fixed by the geometry of space. It is positive, zero, or negative according to whether the geometry is spherical, flat, or hyperbolic, respectively. It is apparent from the above equations that changing the absolute value of k is the same as changing the size of the scale factor R. The ratio R^2/k is the square of the radius of curvature of the space, and it is standard to set R equal numerically to this radius (for $k \neq 0$) so that the permissible values of k are $+1$, 0, and -1.

A physical interpretation for Λ can be obtained from the equation obtained by subtracting (27.3) from (27.2). This is

$$\frac{2}{R}\frac{d^2R}{dt^2} = -\frac{8\pi G}{3}\left(\rho + \frac{3p}{c^2}\right) + \tfrac{2}{3}c^2\Lambda \tag{27.4}$$

The left side of this equation is $(2/R)$ times essentially the radial acceleration of the Universe. The first term on the right side is due to the gravitational force of the matter and radiation content. Gravitation is a force of attraction, so it always tends to make a negative acceleration, i.e., to slow down the expansion or speed up the contraction, as the negative sign indicates. The second term on the right side tends to increase the acceleration if $\Lambda > 0$, so positive values of the cosmological constant represent a repulsive force acting on all parts of the Universe. Unlike the gravitational force, the force represented by Λ does not vanish as the density and pressure of matter and radiation go to zero. Negative values for Λ represent an attractive force which adds to gravitation. The fact that gravitational effects are observed to be dominant in the nearby regions of space indicates that, if the theory leading to the above equations is correct, the cosmological constant must be small enough that it becomes important only over very large distances.

A relativistic model is given by any unique solution of the above equations. It is seen that Λ and k must be given, and some information about p and ρ must also be known before a model can be determined. It is usually assumed that the pressure term in these equations can be ignored, and it is true that the pressure term in equation (27.4) is very much smaller than the density term in the easily observed part of the Universe.

If one multiplies equation (27.3) by R^3, differentiates each term with respect to the time, and compares the result with equation (27.2), one easily finds that the assumption $p = 0$ leads to the result that $\rho R^3 =$ constant. This is to be expected from the conservation of mass and energy, and it is nice to see that this conservation law is contained in the field equations. (If the pressure is not negligible, the work done by the pressure forces in the expansion or contraction must be included.)

The importance of the assumption $p = 0$ in the determination of models can be seen from equation (27.3). With the definition

$$K = \frac{8\pi G}{3}\,\rho R^3 = \text{constant} \tag{27.5}$$

the equation can be written

$$\left(\frac{dR}{dt}\right)^2 = \frac{K}{R} + \frac{\Lambda c^2 R^2}{3} - kc^2 \tag{27.6}$$

In other words, the assumption $p = 0$ is all the information about pressure and density that is needed to solve the field equations. Solutions of equation (27.6) give the scale factor as a function of the time. What type of function this is depends on the values of the three parameters K, k, and Λ.

The different types of solution of equation (27.6) are nicely tabulated in Chapter IX of Bondi's book. (See the references at the end of the chapter.) For some of the solutions, R increases indefinitely with time, for some it increases to a maximum and then decreases, and for some it approaches some asymptotic value. The simplest case occurs for $\Lambda = k = 0$. Then

$$\left(\frac{dR}{dt}\right)^2 = \frac{K}{R} \tag{27.7}$$

If the time is set to zero when $R = 0$, this solution is

$$R(t) = (\tfrac{9}{4}K)^{1/3}t^{2/3} \tag{27.8}$$

This models expands indefinitely as the time to the power $\frac{2}{3}$. One must not extrapolate these models to values of R which are very small, for the assumption that the pressure is zero will break down.

When the function $R(t)$ is known, then it is possible to calculate all of the properties of the model, in particular those properties which can be checked by observation. This is not always easy to carry out, for the results of a measurement will depend on the geometry of space, which is fixed by k and $R(t)$. A description of the comparison of theory and observation is given at the end of the section.

The Steady-State Theory. There are models of the Universe which are based on theories other than general relativity. The most important of these is the steady-state theory which was first proposed by H. Bondi and T. Gold and which was considerably extended by F. Hoyle. Steady-state theory is based on the so-called perfect cosmological principle, which is an extension of the ordinary cosmological principle to include time as well as space. The large-scale properties of the Universe are to remain constant in time. This does not mean that motions are ruled out, for a steady state need not be a static one. A river can be in a steady state even though water is flowing. The velocity at any fixed point must not change with time, although the velocity can change from one point to another. Steady-state theory requires that the Universe as a whole not evolve with time, so it has no equivalent to cosmic time.

The observations indicate that the Universe is expanding. The conservation of mass then leads to the conclusion that the matter is thinning out, and the average density is decreasing with time. Since this is contrary to the basic tenets of steady-state theory, the theory requires that new matter be created at such a rate that the average density remains constant. This continual creation of matter is the most interesting property of steady-state theory.

Many objections have been raised against continual creation, and they are generally based on philosophical prejudice. All theories must contend either with the problem of the creation of matter or with the problem of how the Universe can be self-sustaining for an infinite period of time. Continual creation certainly does not put steady state at any particular scientific disadvantage. Of course, it must be compatible with the observations, and the creation of matter out of nothing has not been observed.

If a sphere of radius r were considered with an observer at the center, then the expansion of the Universe would carry matter out of this sphere (r is not a co-moving coordinate). A particle with an expansion velocity V will leave the sphere in the time interval Δt if it is in the shell of thickness $V\,\Delta t$ and outer radius r. This shell has a volume $4\pi r^2 V\,\Delta t$, so if the density of matter is ρ, then the rate at which mass leaves is

$$\frac{\Delta \mathcal{M}}{\Delta t} = 4\pi r^2 V \rho$$

If Hubble's law of equation (26.7) is valid for the expansion velocity with the random term left out, then

$$\frac{\Delta \mathcal{M}}{\Delta t} = 4\pi r^3 H \rho$$

where H is Hubble's constant. The volume of the sphere is $4\pi r^3/3$, so the average decrease of mass per unit volume per second is

$$\frac{\Delta \rho}{\Delta t} = 3H\rho \tag{27.9}$$

Steady-state theory then requires that mass be created per unit volume at the rate given by equation (27.9) in order that the mean density can remain constant with time. Hubble's constant H has a value of about $10^{-10}\,\mathrm{yr}^{-1}$. The mean density of matter in the nearby region of the Universe is rather uncertain, but a value of $10^{-30}\,\mathrm{g\,cm}^{-3}$ does not do violence to the observations.

New matter should then be created at a rate of about 3×10^{-40} g cm^3 yr^{-1}, which is one hydrogen atom per cubic centimeter every 0.5×10^{16} years. Even if this number can be as much as two orders of magnitude larger, it is far below anything that could possibly be detected. Steady state does not violate the observed basis for the conservation of mass and energy.

Steady-state theory denies the validity of general relativity, so equations (27.2) (27.6) do not hold; however, many properties of the steady-state model can be readily determined. For example, the Hubble constant must not change its value with time. From equation (27.1), therefore, one has

$$\frac{1}{R}\frac{dR}{dt} = H = \text{constant}$$

The solution of this equation is

$$R = R_o e^{H(t-t_0)} \tag{27.10}$$

The quantity R cannot here be identified with any observable aspect of the Universe, but it does determine how the observable properties of the fundamental particles change with time. The geometry of the steady-state model must be flat, corresponding to $k = 0$; otherwise, the square of the radius of curvature, R^2/k, would change with time according to equation (27.10).

Observational evidence today is rather strongly against the steady-state theory. Still, the theory is of interest because of the great influence it has had on astronomical thought in recent years, much of it of a rather controversial nature.

There are other nonrelativistic theories which have been proposed, including the kinematic relativity of E. A. Milne and the so-called fundamental theory of A. S. Eddington, but they do not play a major role in present cosmological thought.

The Observations. There are three observable relationships which can check the cosmological models: (1) the red-shift/apparent-magnitude relation for galaxies; (2) the angular-size/apparent-magnitude relation for galaxies and clusters of galaxies; and (3) counts of galaxies as a function of apparent magnitude.

The redshift/apparent-magnitude relation is an obvious generalization of the velocity-distance relation implied by Hubble's law; it differs from the latter in that only quantities which are observable are involved, at least in principle. In practice the observed magnitudes are subject to a considerable correction as mentioned in Section 26, although observers feel that the correction can be well determined. An additional correction is needed for the aging effects of the galaxies. Galaxies are observed as they were a light-travel time in the past, so the more remote objects are systematically younger as observed than the nearer ones. (This is not true for the steady-state theory.) Any uncertainty in the aging effect will be transmitted to the apparent magnitudes, since the redshift/magnitude relation must be corrected to objects of uniform luminosity if it is to be meaningful. In spite of these uncertainties, the observed redshift/apparent-magnitude relation is probably the best present check on cosmological models.

The angular-size/apparent-magnitude relation illustrates an interesting cosmological effect. Both quantities are measures of distance, but they are not identical measures of it. The relation between the two is a function of the geometry of space. For example, for some models the apparent size of an object has a minimum value, and it increases for both larger and smaller distances as measured by the apparent magnitude. Accurate angular sizes of galaxies are very difficult to obtain, but the angular separations of members of clusters of galaxies should show much more promise.

Galaxy counts are similar to the star counts described in Section 24. The number of objects seen at different apparent magnitudes is a function of the density of them at positions that correspond to these magnitudes, and the relation to be expected is obviously a function of the cosmological model. Unfortunately, the magnitude difficulties are apparently too great for reliable values to be obtained in the large numbers needed for galaxy counts. The clustering tendency of galaxies also makes it difficult to obtain meaningful counts.

Counts of radio galaxies can also be made, and they present a different set of advantages and problems. Most radio galaxies are point sources as seen by the receivers, so the apparent brightnesses are subject to less error than in the case of extended sources. The radio receivers can measure nearly monochromatic fluxes, and the radio spectra of galaxies are generally simpler than the optical spectra, so the correction of measured fluxes for the redshift are easier at radio frequencies than at optical frequencies. Most radio galaxies do not have measured redshifts, so they must be estimated from the redshift/apparent-brightness relation established by those that do. This requires the assumption that radio galaxies which have been identified with optical images be typical of all radio galaxies.

Perhaps the greatest uncertainty in the radio counts is the effects of aging. It is pointed out in Section 26 that there are reasons for believing that strong radio emission lasts only for a short time in a given galaxy, and it may be recurrent. If this occurrence or the strength of it were correlated with the age of a galaxy, then the observed counts would obviously be affected. Since very little is known about radio galaxies, there is enough leeway in the the aging effects to make almost any cosmological model in reasonable agreement with the observations—with the exception of steady state.

To illustrate some of these points, equations (8.7) and (24.1) indicate that the number of sources with measured fluxes greater than F is

$$N(F) = \frac{DL^{3/2}}{6\sqrt{\pi}} \frac{1}{F^{3/2}} \tag{27.11}$$

where D is the average number of sources per unit volume, and L is their luminosity. Equation (27.11) is valid only for Euclidean geometry, and it ignores the redshift of distant sources. With constant values for D and L, the number of sources is proportional to $F^{-1.5}$. If one postulates that strong radio emission is more likely to occur in very young galaxies than in old ones, then evolutionary cosmologies will predict that D will appear to increase with distance, since the more distant sources are observed at earlier cosmic times. As a result, fainter sources will be more abundant than (27.11) predicts, and the measured exponent of F will be less than -1.5. The steady-state model does not have this flexibility, since it requires the proportion of galaxies of different ages to be the same for all distances. The observations of B. Y. Mills, M. Ryle, and others indicate that the flux exponent is about -1.8. Cosmological models which have been investigated are not able to reproduce this value without assuming that radio galaxies were more abundant at earlier cosmic times, so this is evidence against the steady-state theory.

The most obvious procedure is to compare the predictions of specific models with observations. An alternate procedure is to make a general Taylor expansion of the observable quantities in terms of each other and the other relevant parameters. One can in this way place limits on the properties of acceptable models.

The scale factor $R(t)$ plays a central role in this, so the expansions of interest will involve R and its time derivatives, all evaluated at the time of observation, i.e., now. The subscript zero will be used to designate a quantity evaluated for the present epoch; however, instead of the the derivatives of R directly, the following quantities are more convenient:

$$H_0 = \frac{1}{R_0} \frac{dR_0}{dt} \qquad q_0 = -\frac{1}{R_0 H_0^2} \frac{d^2 R_0}{dt^2} \tag{27.12}$$

Strictly speaking, H_0 is what is known as the Hubble constant, not the time-dependent quantity given by equation (27.1). Since q_0 is proportional to the second time derivative of R, it is known as the deceleration parameter.

The redshift of an object is defined by the quantity

$$z = \frac{\Delta\lambda}{\lambda} \tag{27.13}$$

The apparent magnitude can be expanded in a Taylor series which involves z. If z is small enough so that terms in z^2 and higher powers can be neglected, then the following equation can be derived:

$$m = M - 5 - 5 \log \frac{cz}{H_0} + 1.086(1 - q_0)z + \cdots \qquad (27.14)$$

where M is the absolute magnitude of the galaxy. The quantity M must be corrected for the evolution of the galaxy during the light-travel time, and the apparent magnitude m must be corrected for the redshift of the spectrum through the wavelengths of the given magnitude system. The first three terms on the right side follow from the definition of magnitude and Hubble's law.

A plot of the corrected magnitudes vs. redshift could determine H_0 and q_0. If the absolute magnitudes were not known, this would fix q_0 only, since an error in M cannot be distinguished from an error in H_0; however, H_0 can be found from the nearer galaxies for which this is less important. It is important to realize that equation (27.14) is quite general for relativistic models, as it does not depend on the values of the cosmological constant or on the curvature constant k.

In view of the many uncertainties in the corrected data, it is not surprising that a definitive value of q_0 is not yet available. The evidence seems to be rather strong that q_0 is not negative, and values in the range $0 \leqslant q_0 < 3$ are usual. As stated in Section 26, the quantity H_0 has a value of about 100 km sec^{-1} Mpc^{-1} = 0.3×10^{-17} sec^{-1}, and it seems unlikely that this is in error by more than 50% at most.

These data allow some speculations on the properties of the Universe. For example, equation (27.12) shows that a positive value for q_0 means that (d^2R_0/dt^2) is negative, so the expansion is slowing down. If the cosmological constant is positive, it is not large enough to overcome the attraction due to gravity. This still does not decide whether the expansion will continue indefinitely or whether it will eventually stop and be followed by contraction.

If equation (27.4) is written for the present epoch $t = t_0$ and the definitions in (27.12) are used, one finds for $p = 0$

$$H_0^2 q_0 = \frac{4\pi G}{3} \rho_0 - \tfrac{1}{3}c^2\Lambda \qquad (27.15)$$

with ρ_0 being the present average density of matter and radiation. If the observed density of matter is essentially equal to this total density, then ρ_0 is of the order of 10^{-30} g cm^{-3}. The above value of H_0 then leads to the relation

$$\Lambda \simeq 10^{-55}(10^{-2} - \tfrac{1}{3}q_0) \qquad \text{cm}^{-2}$$

For any appreciably positive value of q_0 the cosmological constant is negative, i.e., it represents an attractive force. If the total density were 10^{-28} g cm^{-3}, two orders of magnitude greater than assumed above, then the above relation would be

$$\Lambda \cong 10^{-55}(1 - \tfrac{1}{3}q_0) \qquad \text{cm}^{-2}$$

and the favored values for q_0 would indicate a near-zero value for the cosmological constant. The present observations probably do not rule out the possibility of a mean density this large, and theoretical reasons related to the stability of clusters of galaxies are given in Section 26 as favoring large amounts of unseen matter; however, the arbitrary assumption of some persons that Λ cannot be negative would appear to unduly restrict the possible values of ρ_0 or q_0, or both.

The physical arguments relating q_0, ρ_0, and Λ are not difficult to understand. Moderately large positive values of q_0 indicate that the expansion is slowing down by an appreciable amount. If gravitation is the only agency causing this, as is the case when negative Λ is ruled out, then large material densities are needed. If a negative Λ is permitted, then it works with gravity slowing down the expansion, and smaller material densities are sufficient.

Another critical role of the density can be seen as follows: if the cosmological constant is eliminated between equations (27.2) and (27.3), and if the result is evaluated at $t = t_0$, one finds

$$\frac{kc^2}{R_0^2} = -H_0^2(1 + q_0) + 4\pi G\rho_0 \qquad (27.16)$$

Once again the pressure has been set to zero. Putting in the numerical value for H_0 and $4\pi G$, one has

$$\frac{kc^2}{R_0^2} \cong -10^{-35}(1 + q_0) + 10^{-6}\rho_0$$

The important property of this relation is its sign, for k is restricted to the values $+1$, 0, and -1 according to the geometry of space. The critical value of ρ_0 is seen to be about 10^{-29} g cm^{-3}: for a much larger mean density k is positive and the Universe would be spherical, while a much smaller value would indicate $k = -1$. The mean density is not known well enough for any of the three possible k values to be strongly indicated. It should be emphasized that the value of k does not determine whether the expansion will continue indefinitely or whether it will be followed by a later contraction. Both possibilities exist for all three values of k.

Another quantity of interest is t_0, the time since the expansion began. This if often referred to as the age of the Universe, but the term does not imply that possible events which occurred before $t = 0$ are "outside the realm of science," as is sometimes suggested. The present understanding of physics may or may not be sufficient to allow reasonable extrapolations to negative values of the cosmic time, but attempts to make them will certainly be made.

The quantity t_0 is the cosmic time interval in which R increased from zero to its present value of R_0. Models with $\Lambda > 0$ may not have $R = 0$ at any finite time in the past, so they could be of infinite age by the present definition. If $\Lambda \leqslant 0$, then there is a finite time in the past for which $R = 0$. In this case one has

$$t_0 = \int_0^{R_0} \frac{dR}{dR/dt} \tag{27.17}$$

For any given model $R(t)$ is known, and so dR/dt can be found, and the above integral can be evaluated. The very high densities and pressures which occur near $t = 0$ would seem to rule out the possible existence of large structures like galaxies, so t_0 must be at least as great as the age of the Galaxy. This would make any model unacceptable if it required t_0 to be significantly less than about 10^{10} yr.

Suppose that the expansion takes place at a uniform rate, and thus that $dR/dt = C =$ constant. Then equation (27.17) gives

$$t_0 = \frac{R_0}{C} = \frac{R_0}{dR_0/dt} = H_0^{-1} \tag{27.18}$$

In this case the age is simply the reciprocal of the Hubble constant, so $t_0 \simeq 10^{10}$ yr.

For models with $q_0 > 0$, it can be seen that d^2R_0/dt^2 is negative, and the expansion was more rapid in the past than it is now. Thus t_0 is somewhat less that H_0^{-1}. For example, consider the simple model for which $\Lambda = k = 0$. This model is represented by equations (27.7) and (27.8). The latter equation shows that

$$t_0 = \frac{2}{3}\frac{R_0^{3/2}}{K^{1/2}}$$

But, from (27.7), one has

$$H_0 = \frac{1}{R_0}\frac{dR_0}{dt} = \frac{1}{R_0}\left(\frac{K}{R_0}\right)^{1/2} = \frac{K^{1/2}}{R_0^{3/2}}$$

so one finds

$$t_0 = \tfrac{2}{3}H_0^{-1} \tag{27.19}$$

This is under 10^{10} years according to the value of H_0 which is being used here. A value this small is dangerously close to being ruled out by the calculated ages of objects in the Galaxy, although there is possibly enough uncertainty in both H_0 and the age of the Galaxy that these can be made compatible. A negative value for the cosmological constant will lower t_0, and this makes the fit even more difficult; however, going to a model with $k = -1$ again increases t_0 to some extent.

The presently accepted value of the Hubble constant is considerably smaller than the value which was in vogue some years ago. At that time a very serious discrepancy existed between the time scales of the relativistic models and the age of the Galaxy, and this was a major impetus for the steady-state theory. The discrepancy does not exist today, but the time-scale arguments do place important limitations on acceptable models of the Universe.

The 3°K Radiation and the Brans-Dicke Theory. A. A. Penzias and R. W. Wilson and others have recently detected a weak, isotropic radiation field that corresponds to black-body radiation at a temperature of about 3°K. This radiation field is swamped by much stronger sources at all wavelengths except around 1–10 cm. A reasonable explanation of the 3°K radiation is that it is the present-day residue of the high-temperature radiation field which existed in the very early phases of the expanding Universe.

In the early high-temperature and high-density phase of the Universe, there must have been a period in which helium nuclei were formed by the fusion of simpler particles. The amount of this primordial helium which was produced depends on the density and temperature of the Universe during this critical period, and it should be consistent with the helium abundance found in the oldest Population II stars. For any assumed model of the Universe, the primordial helium abundance is a unique function of the density and temperature at the present epoch. P. J. E. Peebles has carried out calculations for the present radiation temperature of 3°K, and he concluded that relativistic models with present matter densities consistent with observations produce too much helium.

Some cosmologists believe that the above arguments carry very strong evidence against general relativity. A modification of relativity due to C. Brans and R. H. Dicke avoids the above difficulty. The Brans-Dicke theory predicts that gravitational interactions were stronger in the past than they now are, and one consequence is that the Universe would evolve more rapidly through the early helium-producing stage than predicted by general relativity. The overproduction of helium would thus be avoided.

The Brans-Dicke theory predicts the same gravitational redshift as general relativity, and present observations of the gravitational deflection of light are not accurate enough to distinguish between the two theories; however, the rotation of the orbit of Mercury provides some interesting if not definitive differences. As stated previously, the rotation of Mercury's orbit is consistent with general relativity if the Sun is spherically symmetric. If the matter in the Sun were appreciably distorted by rapid rotation, however, then the evidence would be against relativity and consistent with the Brans-Dicke theory.

Although the outer layers of the Sun rotate very slowly, Dicke and associates believe that it is possible for the interior regions to be rotating much more rapidly. The interaction of the solar wind and the solar magnetic field tends to slow down the rotation of the outer layers. If this braking action were great enough, then the outer layers would have stopped rotating by now unless they receive additional angular momentum from a rapidly rotating solar interior. Even a very small flattening of the visible disc would be consistent with this picture, and Dicke and H. M. Goldenberg have obtained observational evidence of such a flattening; however, a number of persons have questioned whether the Sun would be able to maintain the very large gradient in angular velocity which is required by the new theory. The question is unsettled at the present, and it may remain so for a long time to come.

> "There is something fascinating about science. One gets such wholesale returns of conjecture out of such a trifling investment of fact."—Mark Twain

PROBLEMS

1. There are 16 stars brighter than apparent visual magnitude 1, and there are about 5000 stars brighter than $V = 6$. If interstellar absorption is ignored, what does this indicate about star densities?

2. A classical Cepheid variable has a period that corresponds to $M_V = -4.0$. Its apparent magnitude is $V = 10.6$, and it has a radial velocity corrected for the solar motion of -21 km/sec. If it has a galactic longitude of $l = 160°$, determine its interstellar extinction on the assumption that it has a circular orbit around the center of the Galaxy.

3. Derive equation (26.2).

4. Compare the rates of evolution of the different types of galaxies.

5. Suppose that all galaxies were shot out of a small volume at time $t = 0$, and that all accelerations since then have been zero. Determine the form of the observed velocity-distance relation.

REFERENCES

Section 24 Star counts are considered in the following:

1. Bok, B. J. *The Distribution of the Stars in Space*, Univ. of Chicago Press, Chicago, 1937.
2. Trumpler, R. J., and H. F. Weaver. *Statistical Astronomy*, Dover, New York, 1962.

Section 25 Articles on all phases of galactic structure are contained in the following:

3. Blaauw, A., and M. Schmidt (Eds.). *Galactic Structure*, Univ. of Chicago Press, Chicago, 1965.

Other references include the following:

*4. Bok, B. J., and P. F. Bok. *The Milky Way*, Third Edition, Harvard Univ. Press, Cambridge, Mass., 1957.
*5. Petrie, R. M. *Sky and Telescope*, **26,** 330, 1963.
*6. Berge, G. L., and G. A. Seielstad. *Scientific American*, **212,** 6, 46, 1965.

Reference 6 discusses the magnetic field of the Galaxy. For other references, see any general astronomy textbook plus Reference 3 of Chapter II and Reference 2 above.

Section 26 General references to galaxies in addition to elementary textbooks include the following:

*7. Hodge, P. W. *Galaxies and Cosmology*, McGraw-Hill, New York, 1966.
8. McVittie, G. C. (Ed.). *Problems of Extra-Galactic Research*, Macmillan, New York, 1962.
*9. Shapley, H. *Galaxies*, Second Edition, Harvard Univ. Press, Cambridge, Mass., 1961.

Reference 8 consists of the proceedings of a conference held in 1961. Clusters of galaxies are discussed in the following:

10. Abell, G. O. *Ann. Rev. Astron. Astrophys.*, **3,** 1, 1965.

The population content of galaxies is discussed in the following:

11. Baade, W. *Evolution of Stars and Galaxies*, Harvard Univ. Press, Cambridge, Mass., 1963.
12. Roberts, M. S. *Ann. Rev. Astron. Astrophys.*, **1,** 149, 1963.
*13. Roberts, M. S. *Scientific American*, **208,** 6, 94, 1963.

A pictoral atlas and a description of the Hubble types is found in the following:

14. Sandage, A. R. *The Hubble Atlas of Galaxies*, Carnegie Institution, Washington, D.C., 1961.

Synchrotron radiation is the subject of the following:

15. Ginzburg, V. L., and S. I. Syrovatskii. *Ann. Rev. Astron. Astrophys.*, **3,** 297, 1965.

Exploding galaxies and cosmic rays are discussed in the following:

*16. Sandage, A. R. *Scientific American*, **211,** 5, 38, 1964.
*17. Burbidge, G. R. *Scientific American*, **215,** 2, 32, 1966.

The QSO's are discussed in the following:

18. Burbidge, E. M. *Ann. Rev. Astron. Astrophys.*, **5,** 399, 1967.
*19. Greenstein, J. L. *Scientific American*, **209,** 6, 54, 1963.
*20. Burbidge, G. R., and F. Hoyle. *Scientific American*, **215,** 6, 40, 1966.
21. Terrell, J. *Science*, **154,** 1281, 1966.

Galaxy formation and evolution is considered in the following:

22. Eggen, O. J., D. Lynden-Bell, and A. R. Sandage. *Astrophys. J.*, **136,** 748, 1962.
23. Layzer, D. *Ann. Rev. Astron. Astrophys.*, **2,** 341, 1964.
*24. Arp, H. C. *Scientific American*, **208,** 1, 70, 1963.
*25. Page, T. L. *Sky and Telescope*, **29,** 4 and 81, 1965.

Section 27 The following are general accounts of cosmology:

26. Bondi, H. *Cosmology*, Second Edition, University Press, Cambridge, England, 1961.

*27. McVittie, G. C. *Fact and Theory in Cosmology*, Macmillan, New York, 1961.

28. Zeldovich, Ya. B. *Adv. Astron. Astrophys.*, **3**, 241, 1965.

The observational aspects of cosmology are covered in the following:

29. Sandage, A. R. *Astrophys. J.*, **133**, 355, 1961.

The effects of galactic evolution on observed magnitudes is considered in the following:

30. Sandage, A. R. *Astrophys. J.*, **134**, 916, 1961.

The Brans-Dicke theory is discussed in the following:

31. Dicke, R. H. *Physics Today*, **20**, 1, 55, 1967.

Finally, the article by O. Heckmann on p. 429 of Reference 8 above is pointed out for its interesting account of different cosmological theories.

* These are on the popular or semi-popular level.

Appendix **A**

Values of Constants

Physical Constants*

Speed of light	$c = 2.997925 \times 10^{10}$ cm sec^{-1}
Planck's constant	$h = 6.6256 \times 10^{-27}$ erg sec
Boltzmann's constant	$k = 1.38054 \times 10^{-16}$ erg °K^{-1}
Electron charge	$e = 4.80298 \times 10^{-10}$ g$^{1/2}$ cm$^{3/2}$ sec^{-1}
Electron mass	$m_e = 9.1091 \times 10^{-28}$ g
Proton mass	$m_p = 1.67252 \times 10^{-24}$ g
Mass of hydrogen atom	$m_H = 1.67343 \times 10^{-24}$ g
Mass of unit atomic weight	$m_u = 1.66044 \times 10^{-24}$ g
Avogadro's number	$N_A = 6.02252 \times 10^{23}$ atoms mol^{-1}
Gas constant	$R = 8.3143 \times 10^{7}$ erg °K^{-1} mol^{-1}
Bohr radius	$a_0 = 5.29167 \times 10^{-9}$ cm
Stefan-Boltzmann constant	$\sigma = 5.6697 \times 10^{-5}$ erg cm^{-2} sec^{-1} °K^{-4}
Gravitational constant	$G = 6.670 \times 10^{-8}$ dyn cm^2 g^{-2}

1 eV = 1.60210×10^{-12} erg
= energy of photon of wavelength 1.23981×10^{-4} cm
= energy of photon of frequency 2.41804×10^{14} sec^{-1}

Astronomical Constants

Solar mass	$\mathcal{M}_o = 1.989 \times 10^{33}$ g
Solar radius	$R_o = 6.960 \times 10^{10}$ cm
Solar luminosity	$L_o = 3.90 \times 10^{33}$ erg sec^{-1}
Solar effective temperature	$T_e = 5800$°K
Mass of the Galaxy	$\mathcal{M} \cong 1.5 \times 10^{11} \mathcal{M}_o$
Hubble constant	$H_o \cong 100$ km sec^{-1} Mpc^{-1}
	$\cong 0.3 \times 10^{-17}$ sec^{-1}
	$\cong 10^{-10}$ yr^{-1}
Mean density of the Universe	$\rho \cong 10^{-30}$ g cm^{-3}

Colors and magnitudes of the Sun:

$M_{bol} = +4.77$	B.C. = 0.07
$M_V = +4.84$	$m_V = -26.73$
$(B - V) = +0.62$	$(U - B) = +0.10$

1 sidereal year = 3.1558×10^{7} sec

1 pc = 3.086×10^{18} cm = 206,265 AU = 3.262 lt yr

* After NAS-NRC recommended values, *Physics Today*, **17,** 48, 1964.

Appendix **B**

Properties of Normal Stars

MAIN-SEQUENCE STARS (LUMINOSITY CLASS V)

Spectrum	$B-V$	BC	M_V	M_{bol}	$\log L$	$\log T_e$	$\log R$	$\log \mathcal{M}$
O9	−0.31	3.34	−4.8	−8.1	5.15	4.50	1.10	1.27
B0	−0.30	3.17	−4.4	−7.6	4.95	4.48	1.04	1.22
B2	−0.24	2.23	−2.5	−4.7	3.79	4.34	0.74	0.93
B5	−0.16	1.39	−1.0	−2.4	2.87	4.21	0.54	0.70
B6	−0.14	1.21	−0.7	−1.9	2.67	4.19	0.48	0.65
B8	−0.09	0.85	+0.2	−0.6	2.15	4.13	0.34	0.52
A0	0.00	0.40	+0.8	+0.4	1.75	4.03	0.34	0.42
A2	+0.06	0.25	1.2	1.0	1.51	3.99	0.30	0.36
A5	0.15	0.15	1.8	1.6	1.27	3.94	0.28	0.30
A7	0.20	0.12	2.0	1.9	1.15	3.91	0.28	0.27
F0	0.33	0.08	2.4	2.3	0.99	3.86	0.30	0.23
F2	0.38	0.06	2.8	2.7	0.83	3.84	0.26	0.19
F5	0.45	0.04	3.2	3.2	0.63	3.82	0.20	0.14
F7	0.50	0.04	3.8	3.8	0.39	3.80	0.12	0.08
G0	0.60	0.06	4.4	4.3	0.19	3.77	0.08	0.03
G2	0.64	0.07	4.7	4.6	0.07	3.76	0.04	0.00
G5	0.68	0.10	5.1	5.0	−0.09	3.75	−0.02	−0.04
G8	0.72	0.15	5.5	5.4	−0.25	3.74	−0.08	−0.08
K0	0.81	0.19	5.9	5.7	−0.37	3.72	−0.10	−0.11
K2	0.92	0.25	6.3	6.0	−0.49	3.68	−0.08	−0.14
K3	0.98	0.35	6.5	6.2	−0.57	3.66	−0.08	−0.16
K5	1.18	0.71	7.2	6.5	−0.69	3.60	−0.02	−0.19
K7	1.38	1.02	8.1	7.1	−0.93	3.55	−0.04	−0.25
M0			8.7					
M2			10.1					
M4			11.2					

NORMAL GIANTS (LUMINOSITY CLASS III)

Spectrum	$B - V$	BC	M_V	M_{bol}	log L	log T_e	log R
B0			−5.0				
B5			−2.2				
A0			−0.6				
F0			+0.6				
G0			+0.6				
G5	+0.86	0.22	+0.3	+0.1	1.87	3.72	1.02
G8	0.93	0.27	+0.4	+0.1	1.87	3.70	1.06
K0	1.01	0.37	+0.8	+0.4	1.75	3.67	1.06
K2	1.16	0.54	+0.8	+0.3	1.79	3.63	1.16
K3	1.29	0.76	+0.1	−0.7	2.19	3.60	1.42
K4	1.40	1.00	−0.1	−1.1	2.35	3.57	1.56
K5	1.52	1.18	−0.3	−1.5	2.51	3.54	1.70
M0			−0.4				
M2			−0.4				
M4			−0.5				

In the above tables, the units of T_e are °K, while L, R, and $\mathcal{M}$ are in solar units. Data for M_V are from A. Blaauw and from P. C. Keenan; ($B - V$), BC, and T_e are from D. L. Harris III; and $\mathcal{M}$ is from D. L. Harris III, K. Aa. Strand, and C. E. Worley, all appearing in the volume *Basic Astronomical Data* (K. Aa. Strand, Ed.), Univ. of Chicago Press, Chicago, 1963.

Appendix **C**

Element Abundances and Masses

ABUNDANCES OF ELEMENTS IN THE SUN*

Atomic No.	Element	$\log \frac{10^{12}N_z}{N_H}$	Atomic No.	Element	$\log \frac{10^{12}N_z}{N_H}$
1	H	12.00	28	Ni	5.91
2	He	10.98	29	Cu	5.04
3	Li	0.96	30	Zn	4.40
4	Be	2.36	31	Ga	2.36
6	C	8.72	32	Ge	3.29
7	N	7.98	37	Rb	2.48
8	O	8.96	38	Sr	2.60
10	Ne	8.72	39	Y	2.25
11	Na	6.30	40	Zr	2.23
12	Mg	7.40	41	Nb	1.95
13	Al	6.20	42	Mo	1.90
14	Si	7.50	44	Ru	1.43
15	P	5.34	45	Rh	0.78
16	S	7.30	46	Pd	1.21
19	K	4.70	47	Ag	0.14
20	Ca	6.15	48	Cd	1.46
21	Sc	2.82	49	In	1.16
22	Ti	4.68	50	Sn	1.54
23	V	3.70	51	Sb	1.94
24	Cr	5.36	56	Ba	2.10
25	Mn	4.90	70	Yb	1.53
26	Fe	6.57	82	Pb	1.33
27	Co	4.64			

* The table gives the logarithm of 10^{12} times the abundance of the given elements relative to hydrogen, by numbers of atoms. The helium abundance is from R. L. Sears, *Astrophys. J.*, **140,** 477, 1964. The neon abundance is from several investigations of very hot stars as reported by L. H. Aller, *The Abundance of the Elements*, Interscience, New York, p. 115, 1961. All other elements are from the analysis of the Sun by L. Goldberg, E. A. Müller, and L. H. Aller, *Astrophys. J. Suppl.*, V, 45, 1, 1960.

MASSES OF SOME STABLE ISOTOPES*

Atomic No.	Isotope	Mass	Atomic No.	Isotope	Mass
0	n	1.0087	7	N^{14}	14.0031
1	H^1	1.0078	7	N^{15}	15.0001
1	H^2	2.0141	8	O^{16}	15.9948
2	He^3	3.0160	8	O^{17}	16.9991
2	He^4	4.0026	8	O^{18}	17.9992
3	Li^6	6.0150	9	F^{19}	18.9984
3	Li^7	7.0159	10	Ne^{20}	19.9923
4	Be^9	9.0122	11	Na^{23}	22.9888
5	B^{10}	10.0130	12	Mg^{24}	23.9847
5	B^{11}	11.0093	14	Si^{28}	27.9776
6	C^{12}	12.0000	26	Fe^{56}	55.9353
6	C^{13}	13.0034			

* These are on the C^{12} system for which unit atomic mass is 1.66044×10^{-24} g. To obtain the atomic masses in the O^{16} system, the above figures should be multiplied by 1.00032. The data are from *Handbook of Chemistry and Physics*, 41st Edition, Chemical Rubber Publishing Co., Cleveland, 1959.

Appendix **D**

Ionization Properties of the Elements

IONIZATION POTENTIALS*

Element	I	II	III	Element	I	II	III
1 H	13.60			11 Na	5.14	47.29	71.65
2 He	24.58	54.40		12 Mg	7.64	15.03	80.12
3 Li	5.39	75.62	122.42	13 Al	5.98	18.82	28.44
4 Be	9.32	18.21	153.85	14 Si	8.15	16.34	33.46
5 B	8.30	25.15	37.92	16 S	10.36	23.4	35.0
6 C	11.26	24.38	47.87	19 K	4.34	31.81	46
7 N	14.53	29.59	47.43	20 Ca	6.11	11.87	51.21
8 O	13.61	35.11	54.89	22 Ti	6.82	13.57	27.47
10 Ne	21.56	41.07	63.5	26 Fe	7.87	16.18	30.64

* The ionization potentials are given in electron volts. The source of the data is C. W. Allen, *Astrophysical Quantities*, Second Edition, Athlone Press, London, 1963.

The logarithm of the partition function of a completely stripped nucleus is 0.00. The following ions have partition functions which are nearly independent of temperature and electron pressure as long as the abundance of the ion is not negligibly small:

PARTITION FUNCTIONS*

Ion	$\log B_i$	Ion	$\log B_i$	Ion	$\log B_i$
H I	0.30	O II	0.60	Mg III	0.00
He I	0.00	O III	0.95	Al II	0.00
He II	0.30	Ne I	0.00	Al III	0.30
Li II	0.00	Ne II	0.78	Si II	0.78
Li III	0.30	Ne III	0.95	Si III	0.00
C II	0.78	Na II	0.00	K II	0.00
C III	0.00	Na III	0.78	K III	0.78
N II	0.96	Mg II	0.30	Ca III	0.00
N III	0.78				

The following table gives log B_i for other ions as a function of temperature and electron pressure:

		Temperature (°K)					Temperature (°K)		
Ion	P_e	5040	7200	10,080	Ion	P_e	5040	7200	10,080
Li I	1	0.33	0.75	1.68	Si I	1	0.97	1.00	1.21
	10^2	0.32	0.44	0.91		10^2	0.96	0.99	1.06
	10^4	0.32	0.37	0.51		10^4	0.96	0.98	1.04
C I	1	0.96	0.98	1.01	K I	1	0.43	1.16	2.02
	10^2	0.95	0.98	1.01		10^2	0.35	0.62	1.19
	10^4	0.95	0.98	0.99		10^4	0.33	0.45	0.69
N I	1	0.60	0.62	0.66	Ca I	1	0.08	0.46	1.36
	10^2	0.60	0.62	0.66		10^2	0.07	0.30	0.79
	10^4	0.60	0.62	0.66		10^4	0.04	0.27	0.62
O I	1	0.95	0.95	0.97	Ca II	1	0.34	0.43	0.57
	10^2	0.95	0.96	0.97		10^2	0.33	0.42	0.56
	10^4	0.95	0.96	0.97		10^4	0.31	0.38	0.52
Na I	1	0.32	0.68	1.53	Fe I	1	1.47	1.85	2.83
	10^2	0.31	0.41	0.81		10^2	1.47	1.78	2.36
	10^4	0.30	0.36	0.51		10^4	1.47	1.74	2.26
Mg I	1	0.00	0.02	0.30	Fe II	1	1.38	1.55	1.79
	10^2	0.00	0.00	0.09		10^2	1.38	1.54	1.79
	10^4	0.00	0.00	0.06		10^4	1.38	1.52	1.75
Al I	1	0.78	0.84	1.33	Fe III	1	1.40	1.42	1.49
	10^2	0.78	0.79	0.91		10^2	1.40	1.42	1.49
	10^4	0.78	0.73	0.82		10^4	1.40	1.42	1.49

* The electron pressure in the above table is in dynes per square centimeter. Approximate partition functions for some of the ions not given can be obtained by extrapolating along isoelectronic sequences. Thus, C I, N II, and O III all have 6 bound electrons, and their partition functions are similar. The data are based on unpublished calculations of A. N. Cox, Los Alamos Scientific Laboratory, Los Alamos, New Mexico.

Appendix E

Solutions of Problems

Introduction

1. The mean intensity and the flux are given by

$$J(r) = \frac{1}{4\pi}\int I\, d\omega$$

$$F(r) = \int I \cos\theta\, d\omega$$

where the integrals are over the solid angle extended by the star. Since the star radiates uniformly in all directions, I is a constant for all directions at r which come from the star, and it is zero for all other directions. Then if θ_0 is the angular radius of the star, an application of equation (1.2) shows that

$$J(r) = \frac{I}{4\pi}\int_0^{2\pi} d\phi \int_0^{\theta_0} \sin\theta\, d\theta = \tfrac{1}{2}I(1 - \cos\theta_0)$$

$$F(r) = I\int_0^{2\pi} d\phi \int_0^{\theta_0} \cos\theta \sin\theta\, d\theta = \pi I \sin^2\theta_0$$

But the angular radius of the star obviously satisfies

$$\sin\theta_0 = \frac{R}{r}$$

so the above expressions reduce to

$$J(r) = \frac{I}{2r}[r - (r^2 - R^2)^{1/2}]$$

$$F(r) = \frac{\pi I R^2}{r^2}$$

If $r \gg R$, then $(r^2 - R^2)^{1/2} \cong r - R^2/2r$, and the mean intensity becomes

$$J(r) \cong \frac{IR^2}{4r^2}$$

2. The average energy $\bar{\varepsilon}$ is given by

$$\bar{\varepsilon} = \frac{\int_0^\infty \varepsilon_\nu N_\nu \, d\nu}{\int_0^\infty N_\nu \, d\nu}$$

where ε_ν is the energy of a photon of frequency ν and $N_\nu \, d\nu$ is the number of photons per unit volume with frequency between ν and $\nu + d\nu$. The energy density $u_\nu \, d\nu$ gives the energy per unit volume within $d\nu$, so the number of photons is given by

$$N_\nu \, d\nu = \frac{u_\nu \, d\nu}{\varepsilon_\nu} = \frac{u_\nu \, d\nu}{h\nu} = \frac{8\pi\nu^2/c^3}{e^{(h\nu/kT)} - 1}$$

Then one has

$$\bar{\varepsilon} = \frac{kT \int_0^\infty \frac{x^3 \, dx}{e^x - 1}}{\int_0^\infty \frac{x^2 \, dx}{e^x - 1}}$$

The integral in the numerator has the value $\pi^4/15 = 6.4939$, and the denominator is equal to $2(1 + \frac{1}{8} + \frac{1}{27} + \frac{1}{64} + \cdots) = 2.4041$. Thus the mean energy of a photon is

$$\bar{\varepsilon} = 2.7012kT$$

3. Using equations (2.9) and (2.10), one finds that the ratio is given by

$$\frac{\int_\lambda^\infty B_\lambda \, d\lambda}{\int_0^\infty B_\lambda \, d\lambda} = \frac{2ckT \int_\lambda^\infty \lambda^{-4} \, d\lambda}{\frac{\sigma}{\pi} T^4} = \frac{2\pi ck}{3\sigma} \lambda^{-3} T^{-3} = 0.153\lambda^{-3} T^{-3}$$

4. For He II, one has $Z = 2$ and the frequencies of the Pickering series members are found from equation (3.13) to be

$$\nu = 3.290 \times 4 \times 10^{15} \left(\frac{1}{16} - \frac{1}{n^2} \right)$$

with $n > 4$. When n is even, say $n = 2n'$, the above reduces to

$$\nu = 3.290 \times 10^{15} \left(\frac{1}{4} - \frac{1}{n'^2} \right)$$

which is identical to the expression for the members of the Balmer series of hydrogen.

5. For the Balmer series of hydrogen, the wavelengths in Angstroms are found from equation (3.13) to be

$$\lambda = 3645 \frac{n^2}{n^2 - 4}$$

Overlapping of the lines begins for the n for which

$$\lambda(n) - \lambda(n + 1) = 1 \text{ Å}$$

One could find this by trial and error, but the following approximate method is quicker. Considering λ to be a function of n as given above, one has

$$3645\lambda^{-1} = 1 - 4n^{-2}$$
$$-3645\lambda^{-2}d\lambda = 8n^{-3}\, dn$$
$$n^3 = \frac{8\lambda^2}{3645}\left(-\frac{dn}{d\lambda}\right)$$

What is desired is the value of n for which $dn = 1$, $d\lambda = -1$. It is apparent that this n is considerably greater than 1, so λ must be rather close to 3645 Å. Then $n^3 \simeq 8 \times 3645$, $n \simeq 31$, or about 30 Balmer lines are visible.

6. If the partition functions can be taken as constants, then equation (4.10) indicates that the ionization degrees will remain unchanged if

$$2.5 \log T - \frac{5040}{T} I - \log P_e$$
$$= 2.5 \log (T + \Delta T) - \frac{5040}{T + \Delta T} I - \log (P_e + \Delta P_e)$$

This reduces to

$$\log\left(1 + \frac{\Delta P_e}{P_e}\right) = 2.5 \log\left(1 + \frac{\Delta T}{T}\right) + \frac{5040I}{T}\,\frac{\Delta T}{T + \Delta T}$$

For small values of x, one has

$$\log (1 + x) = 0.4343\left(x - \frac{x^2}{2} + \cdots\right)$$

If $\Delta P_e/P_e$ and $\Delta T/T$ are very small, the above becomes

$$\frac{\Delta P_e}{P_e} = \frac{\Delta T}{T}\left(2.5 + \frac{11600\, I}{T}\right)$$

An increase in temperature thus requires an increase in the electron pressure in order to cancel ionization changes. It is not possible to satisfy this condition for two or more elements because of the dependence upon the ionization potential I, although it can be approximately satisfied by elements having nearly the same values of I.

7. Half hydrogen and half helium by mass means that there are four H nuclei for each He nucleus. The average mass per nucleus is then $m_n = 2.67 \times 10^{-24}$ g. The number of nuclei per unit volume is $N_n = \rho/m_n = 3.75 \times 10^{16}$ cm^{-3}. The partial pressure of the nuclei (both atoms and ions) is found from equation (5.11) to be $P_n = N_n kT = 5.18 \times 10^4$ dyn cm^{-2}. The pressure due to the free electrons must be found and added to this in order to obtain the total pressure.

The number of free electrons must be found from the ionization equation applied to hydrogen and helium; however, it can be used only if N_e or P_e is already known, so an iterative procedure must be used. From Tables 4.1 and 4.2, one might guess that hydrogen is completely ionized for the present conditions, and Figure 4.1 suggests that helium might be completely neutral. This guess requires N_e to equal the number of hydrogen nuclei per unit volume, which is $\frac{4}{5}$ of the total number of nuclei per unit volume:

$$N_e = N_H = 0.8N_n = 3.00 \times 10^{16} \qquad \text{ptcls cm}^{-3}$$

This guess can be checked by the ionization equation (4.9). For the present conditions one has

$$\log \frac{N_{\text{H II}}}{N_{\text{H I}}} = 14.53 - \log N_e = -1.95$$

$$\log \frac{N_{\text{He II}}}{N_{\text{He I}}} = 9.59 - \log N_e = -6.89$$

The data for the ionization potentials and the partition functions are taken from Appendix D. The helium is all neutral, but the hydrogen is mostly neutral also, so the above guess on N_e was a poor one.

Since helium is certainly all neutral, all of the free electrons must come from hydrogen, and therefore $N_e = N_{\text{H II}}$. The total hydrogen abundance is still 3.00×10^{16} nuclei cm^{-3}; so further guesses on N_e can be tested as follows:

Guess on $\log N_e$	$\log \frac{N_{\text{H II}}}{N_{\text{H I}}}$	$\log N_{\text{H II}}$
16.00	−1.47	14.99
15.50	−0.97	15.46
15.48	−0.95	15.48

The iteration converges on the value $\log N_e = 15.48$. Also,

$$\frac{N_{\text{H II}}}{N_{\text{H I}} + N_{\text{H II}}} = \frac{N_{\text{H II}}}{N_{\text{H}}} = 0.101$$

so the hydrogen is about 10% ionized. The electron pressure is

$$P_e = N_e kT = 4.17 \times 10^3 \qquad \text{dyn cm}^{-2}$$

and the total gas pressure is

$$P = P_n + P_e = 5.60 \times 10^4 \qquad \text{dyn cm}^{-2}$$

The total number of free particles per unit volume is

$$N = N_n + N_e = 4.05 \times 10^{16} \qquad \text{ptcls cm}^{-3}$$

The average mass per particle is

$$m = \frac{\rho}{N} = 2.47 \times 10^{-24} \qquad \text{g/ptcl}$$

Finally, the mean molecular weight of the mixture is

$$\mu = \frac{m}{m_o} = 1.49$$

If the total pressure had been given instead of the density, the procedure would have been much the same, as an iteration on the electron pressure would still have been necessary. If the electron pressure had been given, however, the ionization of hydrogen and helium could have been found directly, and the pressure, density, and mean molecular weight could have been determined without an iteration.

8. If the four charges are numbered 1–4, then the force between the two dipoles is

$$F = F_{13} + F_{14} + F_{23} + F_{24}$$
$$= e_2\left[\frac{1}{r^2} - \frac{1}{(r+d)^2} - \frac{1}{(r-d)^2} + \frac{1}{r^2}\right]$$

It is assumed that the dipoles are oriented so that charges 2 and 3 are opposites; otherwise, all four terms above would have their signs changed. Since $r \gg d$, one can use the binomial expansion

$$(r \pm d)^{-2} = r^{-2} - 2r^{-3}(\pm d) + 3r^{-4}(d^2) + \cdots$$
$$= \frac{1}{r^2} \mp \frac{2d}{r^3} + \frac{3d^2}{r^4} + \cdots$$

Dropping the terms of higher order, one finds that the force between the dipoles reduces to

$$F = -\frac{6d^2e^2}{r^4}$$

The potential energy is then found by applying equation (3.5):

$$PE = +6d^2e^2 \int_{\infty}^{r} r^{-4}\, dr$$

$$= -\frac{2d^2e^2}{r^3}$$

If one of the dipoles were turned around so that charges 2 and 3 were the same, the net force would be repulsion and the PE would be positive.
9. The man's potential energy in the Earth's gravitational field increases by mgh, where g is the acceleration of gravity and h is the height climbed. Two hundred pounds equals 9×10^4 g, 10^4 ft $= 3 \times 10^5$ cm, and $g = 980$ cm sec^{-2}, so

$$\Delta PE = 2.7 \times 10^{13} \quad \text{erg}$$

The mass-energy equivalence then indicates that his mass increases by

$$\Delta m = \frac{\Delta PE}{c^2} = 3 \times 10^{-8} \quad \text{g}$$

Chapter I

1. According to equations (2.4) and (2.10), each unit area radiates $\pi B = \sigma T^4$ erg/sec. The total area of the disc is $2\pi R^2$ (the radiation from both sides of the disc must be included), so

$$L = 2\pi\sigma R^2 T^4$$

which differs by a factor of 2 from the corresponding equation for a sphere (8.3). The flux at distance r is

$$F(r) = \int I \cos\theta\, d\omega = \frac{\sigma}{\pi} T^4 \int \cos\theta\, d\omega$$

where the integral is over the solid angle subtended by the disc. At very large distances, θ is a very small angle and $\cos\theta \simeq 1$. Then

$$F(r) \simeq \frac{\sigma}{\pi} T^4 \omega$$

where ω is the solid angle of the disc. This depends on the orientation of the disc. If α is the angle between the direction to the disc and the normal of the disc, then $\pi R^2 \cos\alpha$ is the area of the disc projected normal to its direction. By equation (1.1) $\omega = \pi R^2 \cos\alpha / r^2$, and

$$F(r) = \frac{\sigma T^4 R^2 \cos\alpha}{r^2} = \frac{L\cos\alpha}{2\pi r^2}$$

This is to be contrasted with equation (8.7) for the sphere.

2. Equation (9.8) readily indicates that $r = 132$ pc. Equation (9.7) shows that

$$\log \frac{L}{L\,(\text{Sun})} = 0.4[M\,(\text{Sun}) - M]$$

In Appendix A it is indicated that $L\,(\text{Sun}) = 3.9 \times 10^{33}$ erg/sec, $M_{\text{bol}}\,(\text{Sun}) = +4.8$. It follows that

$$L = 19L\,(\text{Sun}) = 7.4 \times 10^{34} \quad \text{erg/sec}$$

Equation (8.5) then gives $R = 1.32 \times 10^{11}$ cm $= 1.9R\,(\text{Sun})$.

3. Equation (9.8) indicates that

$$M = m + 5 + 5 \log p$$

where p is the parallax in seconds of arc. If ΔM is the uncertainty in M due to the uncertainty Δp in the parallax,

$$M + \Delta M = m + 5 + 5 \log (p + \Delta p)$$

$$\Delta M = 5 \log \left(1 + \frac{\Delta p}{p}\right)$$

According to Section 7, the quantity Δp is about $\pm 0\overset{''}{.}005$. At 20 pc, $\Delta p/p \cong \pm 0.1$, from which one finds $\Delta M \cong \pm 0.2$ mag.

4. The cluster has a luminosity of 10^5 times that of the Sun, from which one finds that its absolute bolometric magnitude is $-12.5 + 4.8 = -7.7$. The apparent and absolute magnitudes together indicate a distance of about 3.5×10^3 pc.

5. The angular radius of a star is $\alpha = R/r$, so the flux measured at the Earth is

$$F(r) = \sigma \alpha^2 T_e^4$$

Applying (9.1) to two stars, one has

$$m_{\text{bol}}(2) - m_{\text{bol}}(1) = 2.5 \log \frac{F_1}{F_2}$$

$$= 5 \log \frac{\alpha_1}{\alpha_2} + 10 \log \frac{T_e(1)}{T_e(2)}$$

The known apparent magnitudes and angular sizes show that

$$T_e = 0.27T_e\,(\text{Sun}) = 1600^\circ\text{K}$$

6. In a given volume there are 10 times as many $M = 5$ stars as $M = 0$ stars; therefore,

$$\bar{M} = \frac{10 \times 5 + 1 \times 0}{11} = 4.5$$

Suppose that the apparent magnitude is chosen so that all of the stars of the faint type within the distance r are brighter than it. Then, since $M = 0$ stars are 100 times more luminous, the stars of the luminous type are brighter than the given apparent magnitude if they are within a distance of $10r$. If there are D of the stars of the faint type per unit volume, the sample brighter than the given apparent magnitude will contain $4\pi r^3 D/3$ of the faint stars and $4\pi(10r)^3 \times 0.1 D/3 = 100 \times 4\pi r^3 D/3$ of the luminous stars. The mean absolute magnitude of this sample is then

$$\bar{M}(m) = \frac{100 \times 0 + 1 \times 5}{101} = 0.05$$

Luminous stars can be seen over a far greater volume than the intrinsically faint stars, so any sample chosen because of the apparent magnitude will be heavily loaded with those of high luminosity.

7. The electric field for $x > x_0$ can be considered to be a simple sine wave, but with an amplitude $E_0 e^{-a(x-x_0)}$ which falls off with increasing x. According to equation (1.18) the flux carried by the wave is $(c/8\pi)$ times the square of the amplitude:

$$F(x) = \frac{cE_0^2}{8\pi} e^{-2a(x-x_0)}$$

This decrease in flux is brought about by the absorption of energy from the wave by the material. By equation (10.1), the absorption coefficient is

$$\sigma = -\frac{1}{F}\frac{dF}{dx}$$

From the above relations one finds $\sigma = 2a$.

8.
$$V = C - 2.5 \log F_V$$

where V is the apparent magnitude of the combined light of the three stars, C is the constant of the visual system, and $F_V = F_1 + F_2 + F_3$ is the total visual flux from the three stars. One has

$$2.0 = C - 2.5 \log F_1$$
$$2.5 = C - 2.5 \log F_2$$
$$3.0 = C - 2.5 \log F_3$$

or

$$F_1 = 0.158 \times 10^{0.4C}$$
$$F_2 = 0.100 \times 10^{0.4C}$$
$$F_3 = 0.063 \times 10^{0.4C}$$

Then $F_V = 0.321 \times 10^{0.4C}$, and $V = C - 2.5 \log F_V = 1.22$.

9. A straightforward substitution into equation (11.20) gives $M_V = -2.55$. The individual parallaxes are then found by comparing the apparent and absolute magnitudes. The results are $p_1 = 0''.0123$, $p_2 = 0''.0098$, and $p_3 = 0''.0078$.
10. Large values of H imply faint apparent magnitudes and large proper motions; however, the former suggest large distances, while the latter suggest stars that are relatively nearby, so the correlation of H with distance is not obvious from this. If apparent magnitude and proper motion are eliminated by means of equations (9.8) and (11.4), then

$$\mathrm{H} = \text{constant} + M + 5 \log V_t$$

Stars selected because of large H values then are intrinsically faint stars with large tangential velocities; H is not correlated directly with distance at all.
11. Applying equation (11.2) to the four lines, one finds: Hα, $V_r = -1500$ km/sec; Hβ, $V_r = -1600$ km/sec; Hγ, $V_r = -690$ km/sec; Hδ, $V_r = -1600$ km/sec. The star has a velocity of approach of about 1600 km/sec, which is an extremely large velocity. The large discrepancy shown by the Hγ result makes its measurement highly suspect.

Chapter II

1. For the solar type star $V = +8.0$, $M_V = +4.8$; therefore $r = 44$ pc. The semi-major axis of the orbit is then

$$A = r \times a'' = 44 \text{ AU}$$

Kepler's third law then gives, for the masses,

$$\mathcal{M}_1 + \mathcal{M}_2 = \frac{A^3}{P^2} = 8.5$$

in units of the solar mass. Since the solar type star must have $\mathcal{M}_1 = 1$, the other has $\mathcal{M}_2 = 7.5$.
2. The orbital plane is in the line of sight, so the radial velocities can be converted directly into orbital velocities. The center of mass has a radial velocity of $\frac{1}{2}(V_{max} + V_{min}) = 30$ km/sec. The orbital velocities of the two stars are then 27.1 km/sec and 38.3 km/sec. The radii of the absolute orbits are found, from the velocities and the period, to be

$$a_1 = 3.73 \times 10^{12} \text{ cm} = 0.249 \text{ AU}$$

$$a_2 = 5.27 \times 10^{12} \text{ cm} = 0.352 \text{ AU}$$

$$a = a_1 + a_2 = 9.00 \times 10^{12} \text{ cm} = 0.601 \text{ AU}$$

The subscript 1 refers to the more luminous and the more massive star. From Kepler's third law,

$$\mathcal{M}_1 + \mathcal{M}_2 = \frac{a^3}{P^2} = 2.89$$

Also,
$$\frac{\mathcal{M}_1}{\mathcal{M}_2} = \frac{a_2}{a_1} = 1.41$$

The masses are then $\mathcal{M}_1 = 1.69$, $\mathcal{M}_2 = 1.20$.

During one entire eclipse the relative orbital motion covers the distance $2(R_L + R_S)$, where the subscripts L and S refer to the large and the small star, respectively. Likewise, during the interval of constant light in the middle of each eclipse the relative orbital velocity covers the distance $2(R_L - R_S)$. The relative orbital velocity is $27.1 + 38.3 = 65.4$ km/sec, so

$$2(R_L + R_S) = 4.59 \times 10^{11} \quad \text{cm}$$

$$2(R_L - R_S) = 0.97 \times 10^{11} \quad \text{cm}$$

Then
$$R_L = 1.39 \times 10^{11} \quad \text{cm}$$

$$R_S = 0.90 \times 10^{11} \quad \text{cm}$$

It is not yet known whether the small star is the hot one or the cool one.

Let F_c, F_h, $F = F_c + F_h$ be the fluxes received at the Earth from the cool star, the hot star, and from the two combined. Also, F_p and F_s are the fluxes received during the primary eclipse and during the secondary eclipse. Then it is apparent from equation (15.3) that

$$F_p = F - \left(\frac{R_S}{R_h}\right)^2 F_h = F_c + F_h\left(1 - \frac{R_S^2}{R_h^2}\right)$$

$$F_s = F - \left(\frac{R_S}{R_c}\right)^2 F_c = F_h + F_c\left(1 - \frac{R_S^2}{R_c^2}\right)$$

These relations follow from the fact that the area eclipsed in both cases is πR_S^2, the projected area of the small star. Then if $b = F_h/F_c$, one has

$$\frac{F}{F_p} = \frac{1 + b}{1 + b(1 - R_S^2/R_h^2)} \qquad \frac{F}{F_s} = \frac{b + 1}{b + (1 - R_S^2/R_c^2)}$$

The light curve dips by 0.445 mag during primary minimum and by 0.245 mag during secondary minimum, so

$$0.445 = 2.5 \log \frac{F}{F_p} \qquad 0.245 = 2.5 \log \frac{F}{F_s}$$

Suppose that the small star is also the hot one. Then $R_S = R_h$, and the above relations show that $b = 0.51$, $R_S/R_c = 0.548$. But this cannot be, for it has already been determined that $R_S/R_L = 0.65$; therefore, the small star must also be the cool one, and $R_S/R_h = 0.65$. This gives

$$b = \frac{F_h}{F_c} = 4.0$$

Since the hot star is also the large one, it must be the more luminous of the two stars and can be identified with star 1. The luminosities of the stars are proportional to their fluxes as seen from the Earth, so

$$\frac{L_1}{L_2} = 4.0$$

From the apparent bolometric magnitude of the system and the distance, one finds for the combined light of the two stars $M_{\text{bol}} = +2.05$. This gives a total luminosity of

$$L = L_1 + L_2 = 12.3L_o = 4.8 \times 10^{34} \quad \text{erg/sec}$$

$$L_1 = 9.8L_o = 3.8 \times 10^{34} \quad \text{erg/sec}$$

$$L_2 = 2.5L_o = 9.8 \times 10^{34} \quad \text{erg/sec}$$

where L_o is the solar luminosity. Finally, the effective temperatures follow from the luminosities and radii. The results are

$$T_e(1) = 7250°\text{K} \qquad T_e(2) = 6400°\text{K}$$

One can ask whether or not the properties of the stars have been appreciably influenced by their being in a binary system. There are two ways in which the stars can influence each other: through the gravitational field and through the radiation field. The former at any point is proportional to the mass of the source divided by the square of its distance, and the latter is proportional to the luminosity of the source divided by the square of its distance. Star 1 will modify the effective surface gravity and the surface flux of star 2 by the following amounts:

$$\text{relative change of gravity} = \frac{\mathcal{M}_1}{a^2} \times \left(\frac{\mathcal{M}_2}{R_2^2}\right)^{-1} = 1.4 \times 10^{-4}$$

$$\text{relative change of flux} = \frac{L_1}{a^2} \times \left(\frac{L_2}{R_2^2}\right)^{-1} = 4 \times 10^{-4}$$

These amounts are so small that one can be confident that the stars do not appreciably influence each other's physical properties.

3. With $T_e R$ = constant, one has, for any two times,

$$\frac{L_1}{L_2} = \frac{R_1^2 T_e^4(1)}{R_2^2 T_e^4(2)} = \left(\frac{R_2}{R_1}\right)^2$$

Let the subscripts 1 and 2 refer to maximum light and minimum light, respectively. The total range in the light curve is 0.5 mag, so

$$0.5 = 2.5 \log \frac{L_1}{L_2} = 5.0 \log \frac{R_2}{R_1}$$

or $R_2 = 1.26 R_1$. Also,

$$R_2 - R_1 = 10^{12} \text{ cm}$$

so

$$R_2 = 4.84 \times 10^{12} \quad \text{cm} = 70 R_o$$

$$R_1 = 3.84 \times 10^{12} \quad \text{cm} = 55 R_o$$

The period-density relation, (16.8), indicates that

$$\frac{\rho}{\rho_0} = \frac{\mathcal{M}}{\mathcal{M}_0}\left(\frac{R_o}{R}\right)^3 = \frac{Q^2}{P^2} = 2.5 \times 10^{-5}$$

With the mean radius of the star being about $62 R_o$, one finds the mass to be about $6 \mathcal{M}_o$. This value of the mass is quite uncertain.

Note that the assumption about the method of pulsation requires maximum light to occur at minimum radius. This is contrary to observations, so the above analysis does not apply to observed pulsating stars.

Chapter III

1. The behavior of the iron lines suggests that the relative abundances of Fe I and Fe II is the same in the two stars. Problem 6 of the Introduction then shows that

$$\frac{P_e(2) - P_e(1)}{P_e(1)} \cong \frac{100}{6000}\left(2.5 + \frac{11{,}600 \times 7.87}{6000}\right) = 0.29$$

If calcium were to have the same degree of ionization in the two stars, it would require

$$\frac{P_e(2) - P_e(1)}{P_e(1)} \cong 0.24$$

as calcium has a smaller ionization potential than iron. Thus the greater electron pressure in star 2 is more than enough to compensate for its greater temperature in the ionization of calcium. The ratio Ca I/Ca II therefore might be expected to be greater in star 2 than in star 1.

2. The intensity coming radially out from the star is

$$I(\theta = 0) = \int_0^\infty B(T)e^{-\tau}\,d\tau = \frac{\sigma}{\pi}\int_0^\infty T^4 e^{-\tau}\,d\tau$$

$$= \frac{3\sigma}{4\pi}T_e^4\int_0^\infty (\tau + \tfrac{2}{3})e^{-\tau}\,d\tau = \frac{5\sigma}{4\pi}T_e^4$$

The part of this which originates above the layer τ_0 is

$$I(\tau < \tau_0) = \frac{3\sigma}{4\pi}T_e^4\int_0^{\tau_0}(\tau + \tfrac{2}{3})e^{-\tau}\,d\tau$$

$$= \frac{3\sigma}{4\pi}T_e^4[\tfrac{5}{3} - (\tau_0 + \tfrac{5}{3})e^{-\tau_0}]$$

The fraction originating above any optical depth is then

$$\frac{I(\tau < \tau_0)}{I} = 1 - (0.6\tau_0 + 1)e^{-\tau_0}$$

τ_0	$I(\tau < \tau_0)/I$	τ_0	$I(\tau < \tau_0)/I$
0.0	0.000	1.5	0.576
0.2	0.083	2.0	0.702
0.4	0.169	3.0	0.861
0.6	0.254	5.0	0.973
0.8	0.335	10.0	1.000
1.0	0.411	∞	1.000

It is seen that 70% of the emitted intensity originates above optical depth 2, while only 3% originates below optical depth 5. It is more significant to determine the layers which give rise to the flux, rather than the intensity. The results are similar to the above, but the deeper layers are somewhat less important than indicated above.

3. Equations (19.5) and (19.9) show that

$$I_\nu = B(T)(1 - e^{-\sigma_\nu L})$$

Then

$$\frac{I_{\nu 0}}{I_\nu} = \frac{1 - e^{-\sigma_{\nu 0}L}}{1 - e^{-\sigma_\nu L}}$$

if ν_0 and ν do not differ by very much.

There are three limiting cases of interest. If the emitting region is optically thin for all frequencies, then $\sigma_\nu L \ll 1$ for all ν, and

$$\frac{I_{\nu 0}}{I_\nu} = \frac{\sigma_{\nu 0}}{\sigma_\nu} \gg 1$$

The intensity is proportional to the absorption coefficient, and the emission line at ν_0 stands out strongly. An intermediate case occurs if the slab is optically thick in the line at frequency ν_0 but optically thin elsewhere. In this case $\sigma_{\nu_0}L \gg 1$, $\sigma_\nu L \ll 1$, and

$$\frac{I_{\nu_0}}{I_\nu} = \frac{1}{\sigma_\nu L} \gg 1$$

The line is still in emission, but less strongly. The other extreme occurs when the slab is optically thick in all frequencies, and

$$\frac{I_{\nu_0}}{I_\nu} = 1$$

The emission line disappears in this case. Note that the strength of the emission line is here being measured in terms of its contrast with the surrounding continuum, not in terms of the amount of energy being emitted at the frequency ν_0.

4. Since both lines have the same lower levels and essentially the same wavelength, their only difference is in the f-values. When the lines are very weak and on the linear part of the curve of growth, then the equivalent widths are proportional to the f-values, and D_2 will have twice the equivalent width as D_1. This is indicated by equations (20.10) and (20.13).

If the number of absorbing atoms is increased, saturation will eventually become important, first for D_2 and then for D_1. Then the ratio of the equivalent widths will decrease until in the limit of complete saturation it becomes essentially one.

Finally, if the lines are extremely strong they develop damping wings, and the strength becomes proportional to the square root of the f-value; therefore, D_2 will have an equivalent width $\sqrt{2}$ times that of D_1. The latter ratio will be modified somewhat by the fact that the 2 lines will not have identical damping constants, as γ depends on the upper level of the transition. It should be pointed out that, although equations (20.13), (20.15), and (20.16) are valid only for the simple model of line formation assumed in Section 20, the general properties of the curve of growth are generally valid. Thus very weak lines have equivalent widths proportional to Nf regardless of the details of line formation, etc.

5. Their equivalent widths are proportional to $N_L f$, where N_L is the number of atoms in the lower level of the transition, and the corrections for induced emission are negligible. Thus

$$(N_L f)_{\mathrm{Na}} \cong (N_L f)_{\mathrm{Al}}$$

or

$$\frac{(N_L)_{\mathrm{Na}}}{(N_L)_{\mathrm{Al}}} \cong 2.2$$

The total abundance of sodium and aluminum must now be found. From the excitation equation and the partition functions given in Appendix D, one finds for the given conditions

$$\log \frac{(N_L)_{\mathrm{Na}}}{N_{\mathrm{Na\,I}}} = -2.13 \qquad \log \frac{(N_L)_{\mathrm{Al}}}{N_{\mathrm{Al\,I}}} = -3.63$$

Very few of the neutral atoms of either element are excited to the lower levels of the given transitions. The ionization equation gives

$$\log \frac{N_{\mathrm{Na\,II}}}{N_{\mathrm{Na\,I}}} = 3.13 \qquad \log \frac{N_{\mathrm{Al\,II}}}{N_{\mathrm{Al\,I}}} = 1.83$$

One finds from the above that

$$\log \frac{N_{\mathrm{Na}}}{(N_L)_{\mathrm{Na}}} = 5.26 \qquad \log \frac{N_{\mathrm{Al}}}{(N_L)_{\mathrm{Al}}} = 5.47$$

Both elements are almost completely ionized, and less than one atom in 10^5 of either type is able to absorb the line in question. The total abundances are then

$$\log \frac{N_{\mathrm{Na}}}{N_{\mathrm{Al}}} = 0.13$$

Sodium is some 1.3 times more abundant in the Sun than aluminum by numbers of atoms and ions. Because of the rather similar values of the excitation and ionization potentials of these 2 elements, this result is not very sensitive to the assumed values of temperature and electron pressure; however, the excellent agreement of this very crude example with the more detailed investigation whose result is given in Appendix C is largely accidental.

6. The equation of state (21.30) substituted into equation (21.5) yields the following temperature distribution for this model:

$$T(r) = \frac{Gm\mathcal{M}}{2kR}\left(1 - \frac{r^2}{R^2}\right)$$

Thus,

$$\frac{T(r)}{T_c} = 1 - \frac{r^2}{R^2}$$

if m is assumed to be constant. The energy generation is proportional to T^n, so

$$\frac{\varepsilon(r)}{\varepsilon_c} = \left(1 - \frac{r^2}{R^2}\right)^n$$

To find the value of r/R for which this is 0.1, one has

$$\log\left(1 - \frac{r^2}{R^2}\right) = -\frac{1}{n}$$

One finds $\frac{r}{R} = 0.66\ (n = 4) \qquad \frac{r}{R} = 0.33\ (n = 20)$

This illustrates that the energy sources are much more concentrated toward the center of a star using the CN cycle, for which $n = 20$ is typical, than one using the PP reaction, for which $n = 4$ is typical. In real stars the temperature falls off much more rapidly with increasing r than in the constant density model, so the energy generation is not important over nearly as large a volume as in the above example.

7. The star has a measured color of $(B - V) = +0.3$. According to the data in Appendix B, A0 V stars have intrinsic colors of 0.0. This star, therefore, has a color excess of $E_{BV} = 0.3$, and by equation (23.13), its visual magnitude is dimmed by about 0.9 mag by the interstellar absorption. A0 V stars also has $M_V = +0.8$, so equation (23.4) gives a distance of 180 pc, and a parallax of 0".0056. Trigonometric parallaxes have uncertainties of at least ±0".005, so it is apparent that the above spectroscopic parallax is far more accurate than any trig parallax which the star may have.

Chapter IV

1. From equation (24.3), one has

$$D = 2.4 \times 10^{-4} N(m) \times 10^{-0.6(m-M)}$$

The data for $m = 1$ and $m = 6$ give

$$D_1 = 10^{-3} \times 10^{0.6M} \qquad D_6 = 3 \times 10^{-4} \times 10^{0.6M}$$

If it is further assumed that the Sun is average, so that $M = 5$, then

$$D_1 = 1 \qquad D_6 = 0.3$$

The two magnitudes would agree on a star density of the order of 1 star/pc^3; however, these numbers are grossly misleading.

As problem 6, Chapter I indicates, stars chosen by apparent magnitude are heavily overloaded with those of high luminosity, so one is not justified in assuming that the Sun is typical of the above groups. The average visual absolute magnitude of stars having $V < 1$ is about -1.5, while that of stars having $V < 6$ is probably about $+1.0$. With these numbers one finds

$$D_1 = 10^{-4} \qquad D_6 = 10^{-3}$$

These can be compared with the total star density near the Sun of about 0.05 star/pc^3. The large spread in absolute magnitude makes this kind of analysis nearly meaningless unless the spread is taken into account.

2. If the star has a circular orbit, the Oort equation, (25.9), gives a distance of 2.2 kpc. Then the apparent magnitude in the absence of interstellar absorption would be $V_0 = 7.7$. The observed magnitude $V = 10.6$ indicates that the starlight has been dimmed by 2.9 mag in the visual region.

3. Referring to Figure 26.3, it can be seen that the line which is tangent to the ellipse has the slope

$$\frac{dy}{dx} = \text{ctn}\,\theta$$

and it intersects the y axis at $y = b' \csc\theta$; thus the equation of the line is

$$y = x\,\text{ctn}\,\theta + b' \csc\theta$$

The equation of the ellipse is

$$b^2x^2 + a^2y^2 = a^2b^2$$

and so the slope at any point is

$$\frac{dy}{dx} = -\frac{b^2x}{a^2y}$$

Let (x_0, y_0) be the point at which the above tangent line intersects the ellipse. Then the slope at this point must be ctn θ, so

$$\frac{x_0}{y_0} = -\frac{a^2\,\text{ctn}\,\theta}{b^2}$$

If this is substituted into the equation for the ellipse, one finds

$$x_0 = -\frac{a^2\,\text{ctn}\,\theta}{(a^2\,\text{ctn}^2\,\theta + b^2)^{1/2}} \qquad y_0 = \frac{b^2}{(a^2\,\text{ctn}^2\,\theta + b^2)^{1/2}}$$

A straight line through this point and having the slope ctn θ is

$$y = \text{ctn}\,\theta\,(x - x_0) + y_0 = x\,\text{ctn}\,\theta + (a^2\,\text{ctn}^2\,\theta + b^2)^{1/2}$$

Comparing the two equations for the same straight line, one finds

$$b' \csc\theta = (a^2\,\text{ctn}^2\,\theta + b^2)^{1/2}$$

which reduces to equation (26.2).

4. As with single stars, the nuclear fuel available to a galaxy is proportional to its mass, while it is using that fuel at a rate which is proportional to its luminosity. Thus the mass-luminosity ratio should be a measure of the time

it takes a galaxy to significantly change its characteristics. One then expects ellipticals, spirals, and irregulars to be in order of increasing rate of evolution. This does not mean that an irregular galaxy will burn out more quickly than an elliptical one; an M dwarf will last just as long in an irregular galaxy as in an elliptical one. One does not expect, however, a galaxy with $\mathcal{M}/L \simeq 1$ to be able to continue producing high-luminosity stars at its present rate for longer than a few times 10^{10} yr.

5. A galaxy having velocity V has a true position r at time t, where

$$V = \frac{r}{t} = Hr$$

This is in the form of the Hubble relation with $t = H^{-1}$ being the time since the expansion started. A galaxy is not observed at position r, however, but at the position r_0 which it had at a light travel time in the past. Then the galaxy travels the distance $(r - r_0)$ in the time that light travels the distance r_0, so

$$\frac{r - r_0}{V} = \frac{r_0}{c}$$

It follows that

$$V = H\left(r_0 + \frac{r_0 V}{c}\right)$$

or

$$V = \frac{Hr_0}{1 - (Hr_0/c)} = Hr_0\left(1 + \frac{Hr_0}{c} + \cdots\right)$$

The observed velocity-distance relation would curve above the straight line $V = Hr_0$ for very distant galaxies.

Subject Index